Rapid Intervention Company Operations (R.I.C.O.)

Rapid Intervention Company Operations (R.I.C.O.)

Michael R. Mason and Jeffrey S. Pindelski

THOMSON
DELMAR LEARNING

Australia Canada Mexico Singapore Spain United Kingdom United States

Rapid Intervention Company Operations (R.I.C.O.)

Michael R. Mason and Jeffrey S. Pindelski

Vice President, Technology and Trades ABU:
David Garza

Director of Learning Solutions:
Sandy Clark

Acquisitions Editor:
Alison Weintraub

Product Manager:
Jennifer A. Thompson

Channel Manager:
Bill Lawrensen

Marketing Coordinator:
Mark Pierro

Production Director:
Mary Ellen Black

Production Editor:
Toni Hansen

Editorial Assistant:
Maria Conto

Printed in the United States of America

1 2 3 4 5 XX 09 08 07 06 05

For more information contact Thomson Delmar Learning
Executive Woods
5 Maxwell Drive, PO Box 8007,
Clifton Park, NY 12065-8007
Or find us on the World Wide Web at
www.delmarlearning.com

Library of Congress Cataloging-in-Publication Data

Mason, Michael R. (Michael Ross), 1953-

Rapid intervention company operations, (R.I.C.O.) / Michael R. Mason and Jeffrey S. Pindelski.

p. cm.

Includes bibliographical references and index.

ISBN 1-4018-9503-4

1. Lifesaving at fires--Case studies. 2. Command and control at fires--Case studies. 3. Fire fighters--Protection--Case studies. 4. Fire extinction--Accidents--Case studies. I. Pindelski, Jeffrey S. II. Title.

TH9402.M37 2006

628.9'2--dc22

2005026725

NOTICE TO THE READER

Publisher does not warrant or guarantee any of the products described herein or perform any independent analysis in connection with any of the product information contained herein. Publisher does not assume, and expressly disclaims, any obligation to obtain and include information other than that provided to it by the manufacturer.

The reader is expressly warned to consider and adopt all safety precautions that might be indicated by the activities herein and to avoid all potential hazards. By following the instructions contained herein, the reader willingly assumes all risks in connection with such instructions.

The publisher makes no representation or warranties of any kind, including but not limited to, the warranties of fitness for particular purpose or merchantability, nor are any such representations implied with respect to the material set forth herein, and the publisher takes no responsibility with respect to such material. The publisher and the authors shall not be liable to the reader for any special, consequential, or exemplary damages resulting, in whole or part, from the readers' use of, or reliance upon, this material.

CONTENTS

FOREWORD

THE RAPID INTERVENTION REALITY

Michael O. McNamee
District Chief
Worcester, MA Fire Department

The rapid intervention team is described by many different acronyms—RIT, RIC, or FAST, for example. This crew is specifically trained and stands ready for immediate deployment at an incident. The team's sole function is to perform the rescue of one, or several, of our own, firefighters who find themselves in life-threatening situations. However, we do not work in a world that guarantees ideal outcomes. The incident scene is a dynamic, rapidly changing environment. Constant monitoring and effective adjustments in tactics and strategies are necessary to bring the situation to a safe conclusion. Experience, knowledge, skill, training, and effective communications coupled with a thorough understanding and implementation of defined and practiced procedures is necessary to achieve the desired outcome.

Rapid intervention is one of the hot-button topics in the fire service today. Many articles have been written and much training has been conducted. Skills essential to various modes of search, hot-bottle changes in zero visibility, and effective packaging and extraction methods are being taught at firefighter safety and survival seminars and fire academies across the country. This is long overdue. In the past, the fire service did not place a high enough priority on the safety of its own members. Instead, rescue training focused on occupants and citizens. The fire service culture treated line-of-duty deaths as "acceptable losses . . . part of the job." This is no longer a mode of thinking to which we subscribe. Attitudes are finally starting to change. Our objective is to be as well prepared as is humanly possible to accomplish a successful rapid intervention operation. Before we go further on this topic, let's back up a little bit.

The ultimate goal should be to *avoid* the need for the rescue of one of our own. Consistent use of an accepted Incident Management System along with a reliable, up-to-the-minute personnel accountability system will allow for tighter control of emergency operations. Establishment of practical rehab facilities and procedures will help to ensure that we don't overwork our firefighters and place them in a position where mistakes and injuries are more probable. The employment of risk management principles more clearly defines the level of peril that we will subject our members to when we are involved with fire or other emergencies. We can do a better job of setting risk limits for given situations. However, we work in a business that offers no absolutes. Even when operating at minimal calculated risks, things can go suddenly and drastically wrong.

Fortunately, most incident commanders (ICs) will go through their careers without experiencing the gut-wrenching reaction to the dreaded call of "MAYDAY – MAYDAY – MAYDAY". In an instant, all who hear this message undergo a series of powerful emotions. An adrenalin rush transports you from the normal intensity that accompanies emergency situations to disbelief, anger, intensity, apprehension, fear, hope, and desperation. In the midst of this internal commotion, the incident commander must process the information, analyze the situation, and direct appropriate actions. You are about to order one of the most critical and dangerous operations on the fireground, the deployment of the rapid intervention team. How well have you and your members been trained?

Too often, unrealistic expectations are placed on that two-, three-, or four-member team to successfully locate, supply air to, and extract the Mayday victim(s). Except in very limited, unusual, and lucky situations, that is just not going to happen. This is especially true in large-area, commercial occupancies. If the truth were told, an analysis of real-world RIT operations in these structures would probably not list many successful outcomes. The principle factor working against the team is unstoppable and limited—it is *time*. A firefighter that is lost or injured in this situation has very little time before the air supply is exhausted. Even with the best air management efforts, there could be precious few minutes of breathable air left . . . or less. Under the best conditions, the RITs reaction time to respond to the last known general location of the victim takes several minutes. Being confronted with the lack of time and a vast area presents nearly insurmountable odds for the searchers. So, are we to surrender all hope for the rescue of one of our own? That is not even in the realm of thinking in the fire service. What can be done to maximize efficiency to improve those odds.

We must put the *rapid* back into the RIT. Training must be frequent enough to perfect the RITs required skills. Teams must not be too large. They must be as mobile as possible, carrying a minimum of equipment that facilitates rapid, effective penetration. Their singular goal should be to locate and get air to the member in order to buy time. The training of team members is only part of the equation.

The preparation of command officers to manage the Mayday efficiently is essential. Without effective control and accountability of personnel, the situation rapidly deteriorates into an emotionally driven freelance fiasco that can lead to further losses. One should never assume that everything falls into place with conventional command tactics in such situations. Operations of this sort are extremely labor intensive. A prearranged and practiced plan must be in place before the incident. Whether striking additional alarms, summoning mutual aid, or diverting personnel from the firefight, it is up to the IC to ensure that enough resources are available to complete the effort, including fire control. The proper management of those resources thus rests with the IC. A designated RIT commander, focusing only on the RIT deployment, should be employed whenever possible. The importance of anticipating needs and thinking at least two steps ahead of where the operation is unfolding at the moment cannot be overstated. Plan B and C must be developed and prepared and ready to implement, in the event that plan A falters or fails. In some cases, more than one approach must be undertaken simultaneously. Meanwhile, time is ticking by and the window of opportunity for rescue continues to close. This forces the decision that can be referred to as the RIT dilemma: At what point is the decision made to halt the operation? What are the factors that would dictate that a RIT should not be deployed at all? The fire is gaining control of the structure. The building's stability is compromised and too much time has passed to allow for victim survival. Does the risk outweigh the benefit?

The fire service is a tough, dangerous business that makes physical, psychological, and emotional demands on its members on a regular basis. Nobody ever said that it would be easy. Are you ready for it?

Chief McNamee has served on the Worcester FD since 1972 and is currently assigned as District Chief in charge of Health and Safety. He has over ten years experience as a District Chief and served as the department training officer from 1982 to 1988. Chief McNamee was the initial Incident Commander of the Worcester Cold Storage fire that occurred on December 3, 1999, in which six firefighters unselfishly gave their lives initially searching for civilians and later in courageous and valiant efforts to save their own. We are sincerely grateful for his friendship and assistance in the completion of this text.

PREFACE

Rapid Intervention Company Operations (*R.I.C.O.*) was designed with firefighter safety in mind. It is intended to train firefighters at departments and academies to respond quickly to a fellow downed firefighter or firefighters when timing—and the difference between life and death—counts. During these critical moments of rapid intervention you will be forced to make vital decisions that will directly influence the outcome of the rescue. Be trained, and be ready.

HOW TO USE THIS BOOK

This book is structured to provide you with the necessary background knowledge, training, and guidance for improving the survival rate of your fellow firefighters, as well as for self-survival. Important survival techniques are integrated throughout, along with tactical advice and other valuable information to guide you through the rapid intervention situations you may encounter in your career.

Each chapter contains several features to facilitate your learning experience:

Learning Objectives

A set of ***Learning Objectives*** highlights chapter goals.

Introduction/Summary

An ***Introduction*** to the topics begins each chapter, and a ***Summary*** of important concepts provides essential review for training and the fireground.

Key Terms

Key Terms enable you to communicate effectively in rescue situations.

Review Questions

Practice and refresh your knowledge by answering the ***Review Questions.***

Additional Resources

Further your understanding of the material by researching the wide array of ***Additional Resources*** listed at the end of each chapter.

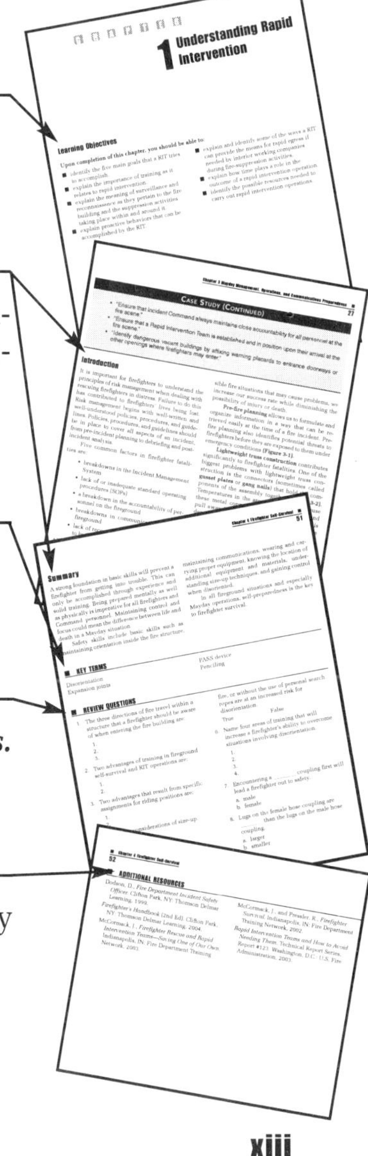

Each chapter provides comprehensive explanations of important rapid intervention techniques:

Chapter 1 Understanding Rapid Intervention—The concept of what a rapid intervention team is and what is expected from the team is outlined in Chapter 1, as well as the variables working against the team.

Chapter 2 Compliance and Standards—To gain a full understanding of rapid intervention it is important to know and understand the rules set forth by agencies with an interest in firefighter safety. Standards and regulations set forth by OSHA and the NFPA are discussed at length in this chapter.

Chapter 3 Mayday Management, Operations, and Communication Preparedness—Managing the fireground Mayday is discussed from the perspective of what is expected from incident Commanders and RIT officers. Model guidelines and procedures are discussed that can be adapted and utilized by any fire department.

Chapter 4 Firefighter Self-Survival—Before a firefighter can be concerned about rescuing fellow firefighters, that firefighter must understand the basics of why firefighters get into trouble and what can be done to help prevent that from happening. These basic areas are covered in this chapter.

Chapter 5 SCBA Emergencies and Survival Skills—The self-contained breathing apparatus is the most important tool utilized by firefighters but is unfortunately also one of the most misunderstood. This chapter outlines the basics of air consumption and emergency procedures for crisis situations such as entanglement, low air supply, and escaping through tight or constricted spaces. In keeping with the philosophy of the book, several methods are presented and discussed in detail to provide the reader with as much information as possible. Consumption course layouts, checklists, and forms are provided in this chapter for firefighters to utilize or adapt for their own departments.

Chapter 6 Ropes, Knots, Anchoring, and Basic Mechanical Advantage for Rapid Intervention—The use of ropes and basic knots is required knowledge for all rapid intervention teams. It is unfortunate that this subject area is a major weakness for many firefighters. The basic knots needed are broken down into their simplest form and presented. The photos in this chapter outline each knot step by step so that they can easily be understood, by the rookie and the seasoned veteran.

Chapter 7 Firefighter Emergency Escape Maneuvers—Reaching back to Chapter 4, specific skills for firefighter self-survival are now introduced in detail. In keeping with the philosophy of the book and our training course, several methods are presented and discussed in detail to provide the reader with as much information as possible. Areas approached include: interior wall breaching, emergency egress through windows, ladder bails, and emergency escape rappels.

Chapter 8 Removal Preparation and Emergency Air Supply—This chapter discusses the concepts necessary to be performed initially when the downed firefighter is located.

Chapter 9 Moving the Downed Firefighter: Carries and Drags—One of the biggest challenges of any rapid intervention operation is actually moving the downed firefighter. Skills and concepts covering basic and advanced drags and carries are discussed, including moving up and down stairs.

Chapter 10 Subfloor Rescues—Raising the Downed Firefighter—Numerous case studies have been presented involving firefighters falling through floors or staircases and not being able to be rescued. The handcuff knot has been the standard method instructed across the country for firefighter removal in these situations. This chapter discusses the handcuff knot but also introduces numerous variations and techniques that have proven to be successful and are very easy to perform with minimal equipment. Some techniques covered in detail include hoseline retrievals, double handcuff variation, "W" technique, single rope, and using 2-to-1 mechanical advantage systems.

Chapter 11 Removal and Lowering Firefighters from Windows—Techniques used to raise a firefighter up and over a windowsill are presented and discussed in detail. Confined or restricted spaces such as presented in the "Denver Drill" are also discussed. Techniques for this particular situation will be covered in great detail, including the use of high-point anchor systems and simple mechanical advantage.

Chapter 12 Removing Distressed Firefighters Using Ladders and Aerial Devices—Another difficult and challenging task for firefighters is to safely move victims down ladders. Portable ground ladders and techniques are discussed in detail. Included are options for raising ladders with limited manpower. Also covered are techniques and concepts utilizing aerial devices for firefighter rescue.

Chapter 13 Roof Operations—The area above the fire is a dangerous position on the fireground. Firefighters getting into trouble by falling through a weakened roof or suffering a cardiac-related event on the roof pose a unique challenge for the rapid intervention team. Several techniques are discussed in detail to bring the downed firefighter back to ground level to be safely treated by emergency medical personnel. Techniques that are covered range from simple firefighter level II techniques to advanced methods utilizing simple mechanical advantage and lowering systems.

Chapter 14 Enlarging Openings—Enlarging existing openings can sometimes prove to be the quickest way of removing a downed firefighter. Construction features and the challenges they present to firefighters are outlined and techniques to overcome them are presented in detail.

Chapter 15 Rapid Intervention and the Thermal Imaging Environment—This chapter explains the technology and concepts of thermal imaging in simple terms so that all firefighters have a better understanding of how to use it to their advantage.

Chapter 16 Search Techniques and Search Rope Systems—Large-area search is a task that is not understood by many firefighters but is a necessity as single family residences are being built larger and larger and fires in retail and industrial buildings are becoming more common. This seldom-approached topic is presented with several techniques outlined to safely perform large-area searches in an efficient manner.

Chapter 17 Rapid Intervention and the Collapse Environment—The basic concepts of structural collapse are discussed with the role and duties of the rapid intervention team taken into consideration.

Chapter 18 Rapid Intervention Training—Training is the most important aspect of rapid intervention, after safety. Without proper training, a RIT will never be able to carry out its task of providing for the safety and rescuing of a downed firefighter. Concepts and techniques to facilitate better adult learning are introduced as well as thoughts and ideas to challenge firefighters in a training environment.

FEATURES OF THIS BOOK

Safety

An emphasis on safety is standard throughout this book. ***Safety Tips*** throughout the text ensure the appropriate execution of various rescue techniques to prevent or minimize injury and loss of life, and Chapter 2, Compliance and Standards, stresses safety.

Technology

Up-to-date technology is covered thoroughly, including an entire chapter dedicated to the safe and efficient use of thermal imaging cameras.

Case Studies

Case Studies at the beginning of each chapter describe various rapid intervention situations, relating the chapter content to actual events that have occurred and highlighting the important lessons learned from the incident.

Skills

Skills integrated throughout the text illustrate the step-by-step sequences necessary for self-survival and rapid intervention tactics.

Rescue Tips

Rescue Tips offer important technical advice in the execution of various skills to improve efficiency of rescue operations, and ***Notes*** highlight important, need-to-know information.

SUPPLEMENT TO THIS BOOK

An *Instructor's Guide on CD* provides instructors with various training materials to enhance classroom presentation:

- An editable ***Instructor's Guide*** in Microsoft Word format includes *Lecture Outlines* integrating *Training Tips* and *PPT references* to the accompanying PowerPoint presentations, as well as *Answers to the Review Questions.* Instructors may print out the *Instructor's Guide* or add their own notes electronically.
- Editable ***PowerPoint*** presentations, including color photos and graphics, outline important concepts in each chapter and correlate to the lecture outlines contained within the *Instructor's Guide.*
- ***Quizzes and Tests*** are available to instructors to enable them to evaluate student knowledge of the content contained in each chapter.
- An ***Image Library*** containing all of the art from the book, in color, including the *Skill* sequences, allows instructors to highlight important techniques in classroom presentations.

ACKNOWLEDGMENTS

This textbook is the end result of nine years of collecting material, experimenting, modifying, instructing, and training with some of the best that the fire service has to offer. Many obstacles, attitudes, and mind-sets were encountered and overcome along this journey. Alone, we could never have overcome these potential setbacks to turn our goals into reality. This text is dedicated to the past, present, and future members of the most noble profession—the fire service. May the words *brotherhood, commitment, pride, honor,* and *integrity* continue to define it in every way.

A special thank you is to be extended to our families, who exhibited unbelievable patience, understanding, and support throughout this endeavor. Without their love and never-ending support, this project would not have become a reality.

The help, support, and contributions of many people made the development of this textbook possible. Their input is greatly appreciated and we are forever grateful. Please forgive us if we have failed to mention one of the numerous individuals or entities.

Kent Adams, Chief, Romeoville Fire Department

Larry Andersen, Hurst Rescue Tools

William Andersen III, Firefighter, Willow Springs Fire Department

Jeff Bailey, Firefighter, Willow Springs Fire Department

Gary Bark, Firefighter, Somonauk Fire Department

Bedford Park Fire Department

Dennis Bergman, Chief, Lena Fire Department

Kelly Browning, Firefighter, Willow Springs Fire Department

Tim Camp, Firefighter, Somonauk Fire Department

Peter Chong, Information Specialist

Robert "Butch" Cobb, Battalion Chief, Jersey City Fire Department

James Crawford, Firefighter, Pittsburgh Bureau of Fire

Tom Connely, IEP Fire

Mitch Crooker, Captain, Chicago Fire Department

Brad Cummings, Lieutenant, Somonauk Fire Department

Chris Cwik, Fire Cadet, Frankfort Fire District

Rich Czajkowski, Captain, Palos Heights Fire Protection District

Al Dagys, Lieutenant, Palos Heights Fire Protection District

Darien Woodridge Fire Protection District

Davis Fire District

De Kalb Fire Department

Ken Dempsey, Sling-Link, Inc.

Downers Grove Fire Department

Rich Edgeworth, Battalion Chief, Chicago Fire Department

Fire Chief Magazine

Fire Department Training Network

Firehouse Magazine

Firehouse.com

Fire Rescue Magazine—Jems Communications

William Friedrich, Battalion Chief, Downers Grove Fire Department

Dave Gallagher, Lieutenant, Huber Heights Fire Department

Robert Gasparas, Firefighter, Willow Springs Fire Department

Walter Giard, District Chief, Worcester, MA Fire Department

Dan Gilbert, Lieutenant, Downers Grove Fire Department

Gurnee Fire Department

Matt Hackett, Firefighter, Justice Fire Department

Lucas Hansen, Somonauk Fire Department

John Hardy, Battalion Chief, Downers Grove Fire Department

Don Harper, Captain, Consolidated Fire District 2, Kansas

Dave Hayward, Firefighter, Indianapolis Fire Department

Bob Hoff, Battalion Chief, Chicago Fire Department

Jon Ibrahim, Firefighter, Downers Grove Fire Department

James Jackson, Deputy Chief, Downers Grove Fire Department

Brian Jankowski, Firefighter, Somonauk Fire Department

Omar Jordan, R.I.T. Systems Inc.

Justice Fire Department

Patrick Kenny, Chief, Hinsdale Fire Department

Rick Kolomay, Lieutenant, Schaumburg Fire Department

Mark Kuzmicki, Firefighter, North Palos Fire Department

Rick Lasky, Chief, Lewisville Fire Department

Lena Fire Department

Joe Levey, Air One Equipment

Lockport Township Fire and Ambulance District

John Lovato, Engineer, Downers Grove Fire Department

Pat Lynch, Lieutenant, Chicago Fire Department

Ken Mains, S.C.B.A.'s Inc.

Clifford May, Firefighter, Justice Fire Department

Jim McCormack, Lieutenant, Indianapolis Fire Department

Jim McKinnon, 5 Alarm Emergency Equipment

Mark McLees, District Chief, Syracuse, NY Fire Department

Mike McNamee, District Chief, Worcester, MA Fire Department

Gayle Mikel, Assistant Professor, Lewis University

Dave Mooney, R.I.T. Systems Inc.

Robert Murphy, Firefighter, Willow Springs Fire Department

Sean Murphy, Firefighter, Roberts Park Fire Protection District

Jack Nagle, Chief, Palos Heights Fire Protection District

Dan Noonan, Firefighter, Fire Department City of New York (Ret.)

Ray Orozo, District Chief, Chicago Fire Department

Bret Pashnick, Lieutenant, Roberts Park Fire Protection District

Palos Heights Fire Protection District

James Pelliteri, Training Officer, Gurnee Fire Department

Bob Pressler, Lieutenant, Fire Department City of New York (Ret.)

Rapid Intervention.com

Gary Rauch, Firefighter, Downers Grove Fire Department

Tyler Riffell, Firefighter, Somonauk Fire Department

Roberts Park Fire Protection District

Phil Ruscetti, Chief, Downers Grove Fire Department

Lanson Russell, Chief, De Kalb Fire Department

Brother Pierre St. Raymond, Lewis University

Safe IR

Paul Segalla, University of Illinois Fire Service Institute

John Schultz, Firefighter, Downers Grove Fire Department

Darrin Shaw, RIT Bag Inc.

Tom Shervino, Div. Chief, Oak Lawn Fire Department (Ret.)

Mike Smith, Deputy Chief, District of Columbia Fire Department (Ret.)

Somonauk Fire Department

Brian Spini, Lieutenant, Consolidated Fire District 2, Kansas

Ronald Taft, Firefighter, Somonauk Fire Department

Dan Tasso, Firefighter, Downers Grove Fire Department

Steve Tellefson, Captain, Davis Fire District

Chris Thomas, Firefighter, Geneva Fire Department

Dennis Weidler, Firefighter, Oak Park Fire Department (Ret.)

Robb Weise, Division Chief, Downers Grove Fire Department (Ret.)

Will Trezek, Lieutenant, Chicago Fire Department

Jim Votteler, Firefighter, Willow Springs Fire Department

Lisa Zoellner, Firefighter, Willow Springs Fire Department

A special thank you to Chief Robert Kristie, Captain Mike Kuzmicki, and the Willow Springs Fire Department for going the "extra mile." The countless hours spent helping to illustrate the techniques and provide support to the program is a true testament to the dedication and professionalism of your department.

Unfortunately, Chief Robert Kristie passed away prior to completion of this project—it is hoped that his efforts and dedication to help other firefighters will live on through this text.

Alison Weintraub, Jennifer Thompson, Stacey Wiktorek, and Toni Hansen and the team from Thomson Delmar Learning—thank you for helping to make this text the best that it could be.

And to the reviewers who offered their content expertise, we are truly grateful:

Rich Edgeworth
Battalion Fire Chief
Chicago Fire Dept
Chicago, IL

Mike Fazio
Firefighter
Grants Pass Fire and Rescue
Grants Pass, OR

Tom Freeman
Fire Chief
Lisle-Woodridge Fire District
Lisle-Woodridge, IL

Mike Kuzmicki
Firefighter/Paramedic
Gurnee Fire Department
Gurnee, IL

Todd Livingston
Lieutenant
St. Petersburg Fire and Rescue
St. Petersburg, FL

Harry Rountree
Fire Chief (Retired)
Burton Fire District
Burton, SC

Paul Segalla
Fire Chief (Retired)
Lockport Township Fire District
Lockport, IL

A special thanks to Doug Cline, Level II Instructor and Battalion Chief of Chapel Hill Fire Department, Chapel Hill, North Carolina, for his technical review to ensure accuracy of the content, and for his contribution to the development of the instructor materials to accompany this book.

ABOUT THE AUTHORS

Jeffrey S. Pindelski

Jeffrey Pindelski has been a student of the fire service for over sixteen years. Jeff is currently a Battalion Chief with the Downers Grove Fire Department in Illinois. He previously served for twelve years as a Firefighter and Lieutenant on the Truck and Heavy Rescue Company. In addition to his background in a career position, he has also served on departments in a volunteer and part-time capacity.

Jeff is a staff instructor at the College of Du Page and also instructs courses at the Downers Grove Fire Academy. He is a Certified Instructor III and Fire Officer II through the International Fire Service Accreditation Congress, and is also certified as a Fire Suppression Incident Safety Officer by the National Board on Fire Service Professional Qualifications. Chief Pindelski holds a MS in Public Safety Administration from Lewis University, a Graduate Certificate in Managerial Leadership, and a BA from Western Illinois University.

He has been involved with the design of several training programs dedicated to firefighter safety and survival including R.I.C.O. (Rapid Intervention Company Operations), which is a 40-hour rapid intervention training program held on a national level. Chief Pindelski was a recipient of the State of Illinois Firefighting Medal of Valor in 1998 and has been published in several trade journals on various topics related to the fire service.

Michael R. Mason

Lieutenant Michael Mason is a twenty-three-year plus veteran of the fire service. As an officer he was initially assigned to Truck 1 / Squad 1 of the Downers Grove Fire Department in Illinois just outside of Chicago. Presently he is assigned to Engine Co. 3, which is located in a high-rise district. He is a Certified Instructor III / Fire Officer II and is currently working toward his Bachelor of Science degree in Fire and Emergency Services.

Lt. Mason is continually lecturing and instructing hands-on training through several avenues. He has helped his department and other area departments in developing training programs, policies, and procedures in the areas of rapid intervention, high-rise firefighting, self-survival, and advanced self-contained breathing apparatus.

He is a staff instructor for the Downers Grove Fire Academy and has independent instructor relationships with various fire departments throughout the Midwest. He has also authored articles in different publications in the fire service.

Lt. Mason is known for his spirited involvement and belief in the firefighters he instructs. In addition, Mike is currently involved in developing the Company Officer Development/Leading and Nurturing program for the Downers Grove Fire Academy and other affiliates.

SKILLS TABLE OF CONTENTS

INTRODUCTION

The road to completing this text has not been an easy one by any means. Nine years of researching, learning, modifying, and trying the various techniques introduced has led to the development of this text as well as friendships with some of the greatest people to be associated with the fire service. The obstacles that have come before us have caused us to continually raise our goals and expectations to the next level. They have helped us to grow both as students of the fire service and as human beings. Our goal has always been to do what is right and stand for what we believe: to train firefighters! The gender, race, department insignia on the sleeve, career status, rank, or experience level has never made a difference in that goal, for we all have been called to this profession for the same purpose.

The need for a text such as this is genuine. The job of today's firefighters is not getting any less challenging. Fire-prevention measures have accomplished outstanding milestones over the last thirty years as the number of fires each year decreases. This decrease in fires, in turn, has led to a decrease in the experience level of today's firefighter. As the number of fires decreases, the number of firefighter deaths remains constant at around 100 each year (with the exception of 2001). What this means is that a breakdown exists somewhere within our society. As long as the human element is present, we will respond to fires. Personal and political agendas do not always serve the best interests of the fire service. When it comes to rapid intervention on the fireground, there is no place for egos or self-serving agendas. The material presented in this text has been gathered from across the United States and has been proven to work. Most of it, unfortunately, has been developed as a result of firefighter fatalities in a particular area. Each member of the fire service is owed this knowledge. More important, each firefighter is owed the knowledge of how to stay out of trouble and how to recognize it. Even when experience is prevalent, continuous training is the only way to accomplish these objectives.

Training today's fire service will need to be a constant, ongoing process that is evaluated and kept on a pace with the ever-changing conditions and challenges that are being faced. Coupled with new construction techniques, use of chemical products for fabrication of goods, and the expansion of public expectations of the fire department, this challenge will only increase further. It should be realized that it is important to be creative and open-minded in order to move forward to generate new ideas and concepts to improve rapid intervention company operations.

There are many thoughts on the actual positioning and deployment of rapid intervention teams throughout the fire service. What works for one particular fire department may not be successful for another. Staffing, response times, and resources are just some of the factors that will play important roles in these decisions. The thought processes and techniques in this book are not meant to be the end to all problems faced by the fire service, but are intended as a beginning, to help catalyst changes and new thought processes that will improve the safety and well-being of our members.

The case studies presented prior to each chapter are not intended to insinuate blame on any party or dishonor any firefighter in any capacity. They are simply meant to present situations and challenges that can be presented in relation to the material covered in the chapter. It is recommended that the full case reports be researched where available and used to lead drills and discussions.

Safety should always be the number one priority when training on any of the skills mentioned in this text. A little common sense will go a long way in this area. Never perform any skills in training without the proper safety measures in place—no exceptions! Make certain that only qualified people that are fully trained and capable lead the training. Fancy titles or positions in fire service organizations do not qualify an instructor to teach rapid intervention. If not done properly, serious injuries and even death can occur in training.

The firefighter that opens this book for reading is to be commended. Thank you for taking a step to attempt to make yourself better in your profession and improve in the safety of your fellow workers. We look forward to learning further in our careers and to the success stories that readers can relate to the information shared. Please be vigilant and aggressive in your pursuit of knowledge. May God bless and watch over each and every firefighter.

Battalion Chief Jeffrey S. Pindelski

Around 1995 I became heavily involved in instructing firefighters through several different departments and fire academies as well as teaching independently in the areas of survival training. The predominant area involved self-contained breathing apparatus. At that time I was also attending training in the areas of self-survival brought forth by some very talented instructors in the state of Illinois. It was, without question, training that made us aware of our precarious positions on the fireground and the need for sound practical skills in self-survival. It was during this time that the idea of rapid intervention began to evolve over a wider spectrum throughout the fire service. It has been well documented that the idea of rapid intervention in various forms was introduced much earlier on different types of rescue squads located throughout urban cities in America and abroad. Their concepts and ideas were directly related to civilian rescue as well as the safety of firefighters on the fireground. It may not have been called "rapid intervention" at the time, but nonetheless its existence in its infancy should still be appreciated today for what it has brought forth.

During my teachings and studies involving the tactics and strategies of these early self-survival classes, I began to assemble and seek out information in the areas of self-rescue as a progression to a team concept. Between the years of 1996 and 1998, I began to train on techniques and maneuvers introduced by various departments and instructors from all over the country. I knew that I wanted to present training that would provide firefighters a team-based concept for rescuing one of their own. I realized that my endeavors and sincerity in developing this task was of interest to other fellow firefighters and instructors as well. If there was one push forward that finally put rapid intervention into the mainstream of the fire service, it was the deaths that were needlessly occurring up to this point—such death provided us the insight that we needed to do better.

The deaths of three Pittsburgh firefighters that occurred in 1995 at Bricelyn Street led me to communicate with those involved at the Pittsburgh Bureau of Fire who had since provided instruction and information in the areas of rapid intervention. My communications with James Crawford took place two years after the tragic incident. By this time I had collected and received information on many additional techniques and maneuvers along with strategic concepts in rapid intervention. I realized how fast the information was being communicated throughout the fire service. With the help of many fire departments and individuals across the country, I ended up with approximately 120 pages of strategies, techniques, and maneuvers. This grew quickly to 200 pages and became the beginnings of a training manual entitled *R.I.C.O.* (*Rapid Intervention Company Operations*) in 1999. I envisioned this information becoming a week-long training program. The program I envisioned would incorporate self-survival skills but would also concentrate on the areas of rapid intervention teams in a scenario-based environment. By this time my partner and I, who is also the co-author of this labor of love, were teaching and incorporating rapid intervention skills along with self-survival skills through our fire academy and other agencies on a continuous basis throughout the Midwest.

Between 1999 and 2000, rapid intervention was flourishing within the fire service. Information on various techniques and maneuvers as well as team concepts was being brought forth by many individuals, departments, and instructors. We found ourselves teaching, trying, and testing everything. Through hundreds of hours of training, drilling, and instructing, we were able to separate the techniques and maneuvers that worked from the ones that did not. What we came to realize through our instructing and teachings is that what may work for one department or individual may not necessarily work for another.

The most important aspect of teaching and training in rapid intervention maneuvers and techniques is the proper safety measures that must be taken into consideration at all times. All of the maneuvers and techniques involved in self-survival and rescuing the downed firefighter require sound and qualified instructors. Serious injury can result from many of these last-resort emergency maneuvers if they are not carried out properly. Responsible instructors that are safety conscious are required in any hands-on application.

All the techniques, maneuvers, and systems available under the term "rapid intervention" can be mastered as long as you train on the specifics of the choices you make. In other words, it is not a matter of one particular technique or maneuver carrying the opinion of being bad or good; it is rather that a particular maneuver, technique, or system possesses safety, simplicity in its application, and reveals repeated success in its outcome. This guarantees its application and success under extreme conditions.

By the year 2001, my partner and I had both been heavily involved in attending classes and sharing knowledge with instructors from the east coast to the west coast. The R.I.C.O. manual was becoming larger and larger by now. Our thoughts and directions began to give way to the origination of possibly a forty-hour or week-long course of instruction in rapid intervention. There were other courses being developed throughout the area and around the country, but they were still centered on the self-survival aspects and some limited rapid intervention techniques. Most of these programs were between one to three days long. By this time we had definitely realized that while both self-survival techniques and rapid intervention maneuvers coincided with each other, to deliver a quality program in both areas within one day or even three days would be impossible. It was at this point that we were invited to participate in developing an existing three-day program that was being presented by a local area fire academy into a five-day program. We enthusiastically and heartfully provided our services in helping generate the first five-day program in Illinois in 2001 and 2002 with many other local instructors.

At the end of the 2002 five-day program, we determined that a national-level program was needed in the Midwest. We had tried to introduce new concepts, techniques, and instructors from all over the country into the existing local programs in our area, but were met with resistance. Whether the resistance to change and new ideas was political or egotistical, one thing was clear: Rapid intervention, like all things in fire service, is subject to ever-changing ideas.

Progress is the saving grace to us as firefighters, as well as for our students. Whether that progress is provided by a probationary firefighter student or the thirty-year veteran, it should not matter if it moves us forward and provides clarity to our future. This is especially true when our lives depend on it. This progressive attitude and belief in training firefighters without pertinence to rank, gender, experience, race, or affiliation should always be prevalent in our minds. So many caring, dedicated, and unselfish individuals are responsible for what you read and hold in your hands—as well as the countless endearing lives that have been lost—to provide you the opportunity to learn. The first forty-hour national program entitled R.I.C.O. was presented in June of 2003 through our department and fire academy and continues to be held every year.

In 2003 this text was still being worked on by adding, deleting, and exchanging information on techniques and maneuvers as well as training. Many times we would analyze an article or receive communications from other instructors prompting us to test and train their ideas. There are so many avenues of discovery between the covers of this text that we are hoping to provide a resource that not only could be continually used by firefighters and their departments but also a text that would encourage and spawn new ideas as well.

We also realized early on the vastness of our subject matter. With the opinions and contributions of so many talented and knowledgeable associates it became apparent that the need for this text was quite clear. We could see that rapid intervention would be in a constant state of change over time. Even during the writing of this introduction my partner and I are still conversing and preparing last-minute entries on new techniques, maneuvers, products, and information. It is our hope that we have supplied the objectives of our endeavors to our fellow firefighters.

This text has been thoughtfully laid out in a specific progressive order, spanning eighteen chapters. Rapid intervention is a team concept and this text provides and explains that concept in depth for many areas. We also felt that the importance of depth in the areas of managing a Mayday along with fireground operations and communications were vital to success. Extensive information has also been provided in the areas of firefighter self-survival skills and SCBA emergencies as well as emergency escape. It is important that every firefighter recognize the need to be well versed and prepared in these techniques and maneuvers before the full scope of rapid intervention team concepts are employed. By thoroughly training firefighters in these areas, we can diminish the potential of the Mayday or least begin to control it where it starts—with the individual having the problem. Training also creates a known predictability—the rapid intervention team will know what a distressed firefighter will do/is capable of doing during the course of his or her Mayday. The text also provides techniques and maneuvers that will help increase the likelihood of success in reaching the downed firefighter. If we cannot find or locate each other, we cannot rescue each other.

As stated earlier, rapid intervention is a team concept. This text starts with the individual and brings you to the ultimate challenge in teamwork as it relates to rescuing one of our own. The motivation of teamwork is the basis of the fire service; without it there would be little evidence of success when providing our services. The teamwork required in rapid intervention takes talented, aggressive, dedicated, and envisioned individuals. Every individual within the team needs to contribute 100 percent back to the team when faced with the problems confronted on the fireground. Rescues are not usually made by one individual when it comes to the downed firefighter, but rather a team of firefighters and in many cases two or three teams. We hope that this text provides you with the tools to function at a very high level of performance when rescuing a fallen brother.

The components of this text are intended for all levels and individuals involved in fire and rescue activities. The ultimate objective of its content is to develop the abilities of not only firefighters but all ranks within the fire service to provide for the protection and rescue of a firefighter in distress—to bring everyone home, no matter what.

This text, as well as the hands-on information it provides, is the responsibility of everyone to pursue individually as well as together in order to provide the framework for teamwork. The success of any team or program is based on many factors—they are easily recognizable, but success requires everyone to participate. We, as well as all of our associates, hope this text reveals what can be accomplished by a set of dedicated individuals assigned as our guardians on the fireground. Getting involved in this material and its concepts is an investment in yourself, your career, your coworkers, and your loved ones. We hope our efforts in bringing to you the knowledge and sincerities of so many throughout the fire service will be shared by all firefighters everywhere. In so doing all of us can continue to grow and strengthen the common bond we have together in serving mankind.

Lieutenant Michael R. Mason

CHAPTER

1 Understanding Rapid Intervention

Learning Objectives

Upon completion of this chapter, you should be able to:

- identify the five main goals that a RIT tries to accomplish.
- explain the importance of training as it relates to rapid intervention.
- explain the meaning of surveillance and reconnaissance as they pertain to the fire building and the suppression activities taking place within and around it.
- explain proactive behaviors that can be accomplished by the RIT.
- explain and identify some of the ways a RIT can provide the means for rapid egress if needed by interior working companies during fire-suppression activities.
- explain how time plays a role in the outcome of a rapid intervention operation.
- identify the possible resources needed to carry out rapid intervention operations.

CASE STUDY

On January 28, 2004 at 2341 hours, the Hinsdale, Illinois Fire Department responded to the report of a house fire in a building under construction. Police officers had been patrolling the neighborhood on a report of a smell of smoke for approximately ten minutes prior to the neighbor's 9-1-1 call confirming a fire.

Units enroute included one engine with three personnel, a truck with three personnel, and an ambulance with one person. In addition, a still alarm was called out the door by the responding captain that included an additional engine, two trucks, an ambulance, an incident safety officer, a rapid intervention team (RIT) officer, and a chief officer. This timely still alarm request would prove invaluable later.

Upon arrival, the shift captain reported that sectors A and B were clear, but there was fire in a protrusion of the D sector of the home, including some exterior fire in that sector at the roofline. The captain ordered the crews to switch to fireground red (a tactical radio channel) and had the engine lead out, dropping both beds, which included a 3-inch supply line to a wye with an 1¾-inch preconnected line and a 2½-inch line. The other bed contained the 5-inch supply line. The truck took the front of the building and assumed command and accountability.

The first crew took the 1¾ to the A/D corner where there was a door that led into a library. From the doorway, the crew was able to knock almost 90 percent of the fire. In the meantime the 2½ was connected and the crew switched lines and began to enter the library to continue suppression efforts.

At this point, the first mutual aid company, a truck company, arrived on the scene and was assigned to a second line to work fire at the roofline. An upgrade of the alarm was requested per Command. The crews had now been on the scene for approximately ten minutes.

What occurred next is every fire chief's nightmare. The interior crew consisting of two firefighters had made entry with the 2½ line. They were approximately 25 feet into the library when the lead firefighter fell through a hole in the floor. At the time he went through the floor, both he and his partner were attempting to pull more line into the building. The firefighters were only 3–5 feet apart.

When the lead firefighter fell through the hole, he fell approximately 10–12 feet and landed on his hands and knees. There was fire evident at the ceiling line off to his right, but the fire had no direct contact with him. His partner did not initially know what had happened. However, the first thing the downed firefighter did was inform his partner to hold his position exactly where he was because there was a hole. He then radioed a Mayday call. The fallen firefighter communicated that he was capable of assisting in his own rescue and that there was fire at the basement ceiling but not in direct contact with him at that time.

His partner, due to the short distance they were into the structure, crawled back to the door to ensure that the captain had heard the Mayday call, which had been acknowledged by the shift lieutenant. The shift lieutenant relayed this information to the captain. An exterior hoseline was used to protect the downed firefighter. At this time, the deputy chief arrived on the scene and had all companies, except the rescue companies, switch to fireground white (a tactical radio channel) for operational communications.

As the mutual company came up the front walk to take their assignment as the second line, they were immediately reassigned to the RIT along with the remaining crews. A 20-foot roof ladder was taken into the structure and the plan was to get it into the hole as quickly as possible so the downed firefighter could self-rescue. At the same time, the shift Lieutenant was in search of another entrance to the basement from the exterior.

The visibility in the library at this point was zero and the fire was growing and threatening to cut off the access hole from below. The fallen firefighter's partner used voice contact to direct the RIT's effort to ladder the hole. The first couple of attempts were unsuccessful due to hitting debris or overshooting the fallen firefighter, who was struggling with limited visibility and whose low-air alarm was activating at this time.

Finally, the ladder made its way through the hole, and the fallen firefighter saw it go over his head and grabbed it. His first attempt to ascend the ladder was met with great heat and fire and he had to retreat while the area was cooled. A second attempt met with the same frustration. Finally, the third attempt was successful, and as the firefighter came though the hole, he was grabbed by the RIT and pulled from the building, where he was handed off to the mutual aid ambulance company on the scene.

The deputy chief called for a personnel accountability report (PAR) at that time and was able to confirm that all companies were out of the building and accounted for. The time from the Mayday call until the confirmation of the PAR was approximately six minutes.

—Case Study by Chief Patrick Kenny, Hinsdale Fire Department, Hinsdale, Illinois

Introduction

It goes without saying that the protection of life is and always will be the priority of the fire service. It is also common knowledge that the occupation of firefighting is one of the most dangerous, high-risk jobs available. When we define the word *dangerous* in the scope of an occupation, we are most often referring to the environment in which a person is trying to perform a task. In the case of rapid intervention, it is fighting fire and facilitating rescues under the most extreme conditions that causes us to place our own lives on the line.

Societal factors such as demographics, economics, and technology have made our jobs as firefighters increasingly more difficult and hazardous through the years. Historically, numerous firefighters have given the supreme sacrifice in efforts to protect and save the lives and property of the public. Unfortunately, it was not until the 1990s that the ideas of saving our own firefighters and rapid intervention began to take a prominent foothold in the profession. Some of the earliest concepts of self-survival techniques and rapid intervention were in departments that had formed rescue squads, such as the New York Fire Department (F.D.N.Y.). These rescue companies dedicated themselves to devising ways of rescuing civilians as well as other firefighters. Rapid intervention came into being as a result of the devastating losses of firefighters in precarious situations who were without properly trained, dedicated personnel readily available to help them. This led to the recognition that serious change was needed to better the odds for firefighters' well-being.

It should be realized that there are many degrees of danger on the fireground. The need for the eyes and ears of RITs is now, as it should be, a priority for every fireground situation. There are so many different risks present in a fire situation that it is improbable as well as impractical to think that one or even a handful of individuals can account for the safety of all. The permanent need for rapid intervention should now be a mainstay within the fire service in order for firefighters to survive the multiple dangers of firefighting.

Rapid Intervention Teams

Each year, an average of 100 firefighters die in the line of duty, and thousands succumb to other types of injuries **(Figure 1-1).** Collapse of fire buildings, disorientation within smoke-filled environments, getting into trouble when operating on the floors above the fire, flashovers, backdrafts or smoke explosions, fires involving below-grade applications, falling through floors, and many other unforeseen events are reasons why we must diligently apply rapid intervention concepts and the ability to rescue our own in the event of any operation going bad. It should be realized that it is important to be creative and open-minded in order to move forward to generate new ideas and concepts aimed at improving our ability to bring our members home safely at the end of their tour of duty.

It should be understood that the rapid intervention team is derived from many different terms and concepts. **Rapid intervention** encompasses specialized teams involving the use of rescue techniques under the most adverse conditions. These teams are referred to by many names and descriptions, such as **rapid intervention team (RIT), rapid intervention crew** (or **company**) **(RIC), firefighter assist and support team (FAST), rescue assist team (RAT),** and many other acronyms that have been given to them by various departments across the country. (To simplify, the term *RIT* will be used throughout this text.) The actions of these teams are summarized by the term **RICO,** which stands for **rapid intervention company operations.** The names and other related jargon are not important; it is the task and purpose of these teams on the fireground that is vitally important.

Rapid intervention operations can save firefighters' lives, but they should not give a false sense of security to those firefighters operating on the fireground. As firefighters, we should realize that we are not indestructible, and we should not take extreme chances just because we feel an extra cushion of safety exists in the form of a RIT.

A true understanding of what a rapid intervention operation consists of is critical for all involved. These operations are resource-intensive and will require multiple RITs to complete.

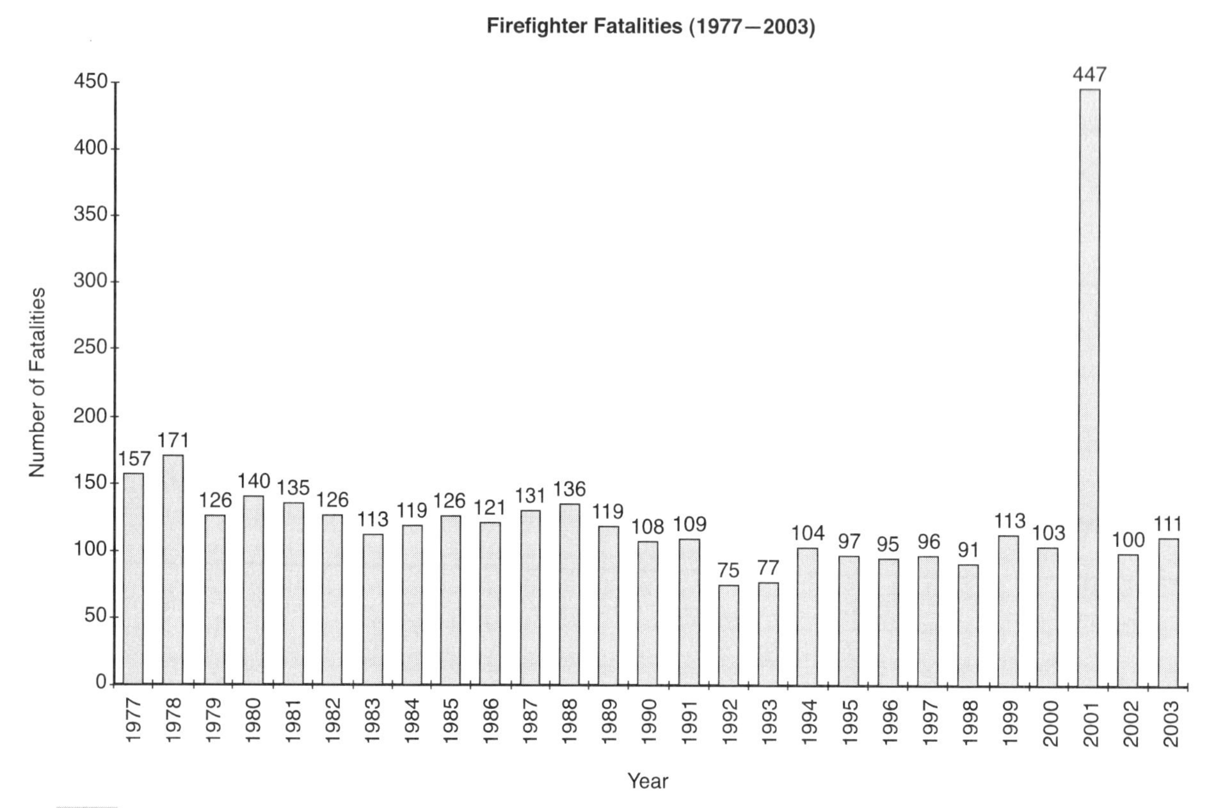

FIGURE 1-1

Each year an average of 100 firefighters die in the line of duty, and thousands succumb to other types of injuries. The number for 2001 is abnormally high due to the loss of 344 firefighters at the World Trade Center. *(Source: US Fire Administration)*

Command and company officers must prepare and think ahead when setting up rapid intervention operations at an incident.

Goals and Objectives of Rapid Intervention Teams

Though it may be possible for one team to successfully remove a distressed firefighter, it will more than likely take two or three teams to get the job done. Teams, however, must be made up of four to six well-trained firefighters. Two firefighters standing by as the RIT will not be adequate. Rescuing a downed firefighter is labor-intensive. Sufficient personnel must be on-scene and in position to act at any moment.

This realization is the first rule of thumb: Ensure that there are adequate resources available to dedicate to a RIT operation. If one team is ready and waiting, then there should be another one in staging. Any RIT members deployed to the scene should know their roles and responsibilities. Their initial mission is to prepare themselves by evaluating the fireground. They will also stage tools to make certain they are equipped for potential situations that may be presented. The RIT should be in a constant state of readiness. If a distressed firefighter needs assistance, the RIT should function with five goals in mind:

1. Locate the downed firefighter.
2. Assess the firefighter's condition and the environment.
3. Provide an emergency air supply.
4. Call for additional teams or resources.
5. Attempt to remove the firefighter to safety.

Once these goals have been accomplished, additional teams can concentrate on the rescue and removal effort.

Rapid Intervention Readiness

Training is an essential part of RIT operations. Training helps identify what works and what does not work. It identifies the tools and resources that will be needed and also demonstrates the need for solid procedures to deal with fireground emergencies. Training creates readiness.

Training should begin by introducing the basics and then proceed to advanced skills and techniques. It should also concentrate on equipment that may be used during these operations. Identifying tools and equipment that may need to be purchased increases the chances of success during an actual operation **(Figure 1-2).**

The RIT should be as self-sufficient as possible when preparing for action on the fireground. When the time comes for the team to go to work, members should not be looking for tools or resources to be used for the rescue. This will delay the overall operation and may affect the final outcome.

RITs should be established on the fireground and should assemble the equipment needed based on the particular type of incident. These tools should be exclusive and available only for use by the RIT. These tools might include:

- forcible entry tools
- RIT pack or spare breathing apparatus with mask
- saws
- fire extinguishers
- thermal imaging cameras
- search rope
- individual ropes and webbing
- pulleys and carabiners

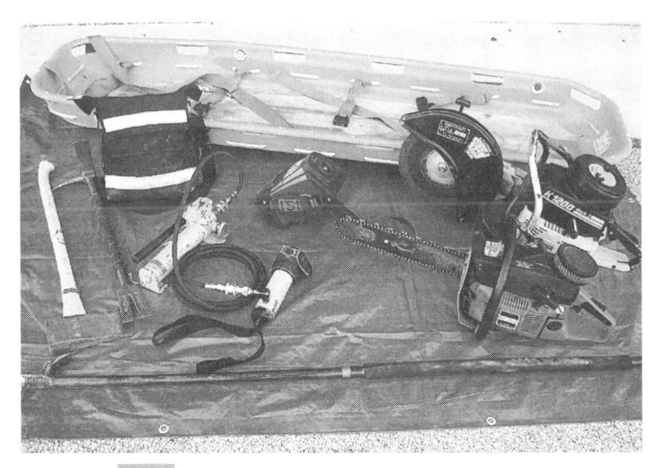

FIGURE 1-2

Specialized tools such as thermal imaging cameras, rescue litters, large-area search kits, RIT packs, and hydraulic forcible entry tools may be needed in addition to basic tools such as irons, hooks, and saws for RIT operations.

- hydraulic rescue tools
- ladders of various size
- charged hoseline

It is also important for team members to have portable radios and individual light sources.

The complexity and size of an incident will determine how many teams will be needed. This is especially true when dealing with larger incidents. As the situation escalates in complexity, it will become apparent that a separate branch in the **Incident Management System (IMS)** (organization system used to manage resources at an emergency incident) will have to be established in order to provide for the safety of everyone involved. Teams will work under a separate **RIT operations chief** or branch officer in the command system. During **Mayday** operations—operations in which a person is in a life-threatening situation—it is unreasonable and unsafe to expect one person to command both the fire suppression and rapid intervention operation simultaneously. It is very important for officers to adjust their standard operating procedures when these types of scenarios are presented. They must take control of communications and maintain accountability.

Note

A **safety officer** involved in the safety of the fire-suppression situation should not be expected to oversee a response to a fireground Mayday. An additional safety officer should be put in place to be exclusive to the RIT operation.

Proactive Behavior

When the RIT arrives on scene, the team leader should report to Command to collect information and determine a setup or staging area that will work to the team's advantage. This will allow the team to gain a clear view of what is taking place.

The RIT should begin to recognize significant hazards and dangers developing on the fireground and communicate those dangers to the proper divisions and Command personnel. The RIT should develop a solid rescue plan should one be needed.

RITs are often underutilized and held to one place on the fireground throughout an incident. Most of the time they are in the wrong place, have an improper vantage point, or are too close to the Command vehicle to be of any real assistance if an emergency occurs **(Figure 1-3).** This often occurs because the firefighters assigned as the RIT do not like the assignment because they want to be involved in the "action." It is easy for complacency to develop within the team when this occurs. The fact is, the RIT position is one of the most important positions on the fireground. The best-trained and best-equipped firefighters should make up the RIT.

The fireground operation of a RIT starts with the immediate surveillance and reconnaissance of the fire building and the ground around it. The RIT should prepare the fireground by taking measures that will increase firefighter safety and survival **(Figure 1-4** and **Figure 1-5).**

FIGURE 1-3

The RIT should be positioned in an advantageous place on the fireground and be ready to go to work at a moment's notice.

FIGURE 1-4

RIT teams must be in constant reconnaissance, monitoring fire conditions and egress hazards.

FIGURE 1-5

The RIT will need to clear all glass from windows that may be used by interior crews for escape. Coordination must always take place with interior crews before venting windows to prevent unwanted fire spread.

Firefighter Rescue—Safety and Survival

As stated previously, the primary philosophy of rapid intervention is the safety and survival of the firefighters working on the fireground. The RIT must be aware of the locations of the different fire companies and firefighters and must know the various assignments. The team should be responsible for monitoring all radio transmissions and necessary frequencies. Likewise, it is the responsibility of Command to disseminate as much information as possible to the RIT. The accountability system in place should be well understood by the **RIT leader** or **officer (RITLO).** (The RIT officer will be discussed in greater detail in Chapter 3.)

It is very important that the RITLO establish good "face-to-face" communications process with Command. The RIT should be aggressive in seeking out information while being careful not to interfere with Command's ability to run the fireground. A properly completed **tactical worksheet** can be used by the RITLO to facilitate this (**Table 1-1** on page 9). The RIT should take immediate action to eliminate or correct any hazards that may cause injuries or harm to firefighters operating on the fireground.

FIGURE 1-6

Smoke conditions can tell the RIT a great deal about what is taking place inside of the structure. Density, velocity, volume, and color are conditions that the RIT should pay particular attention to.

On the Outside

The RIT is responsible for continuous surveillance of the fireground to identify immediate hazards **(Figure 1-6).** This allows the team to estimate the size of the incident and to recognize any irregular fireground or construction features that may create potential problems **(Figures 1-7A** and **1-7B).**

A

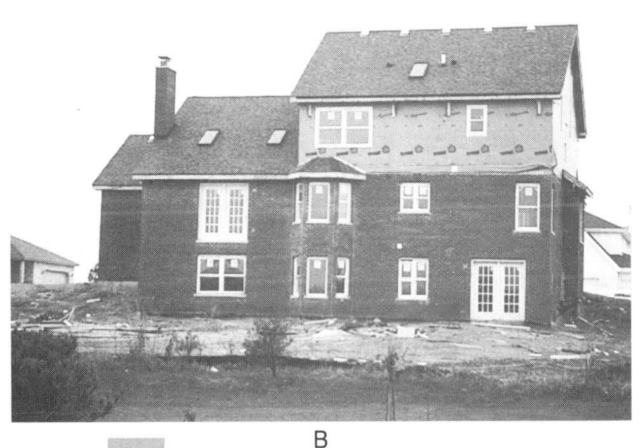

B

FIGURE 1-7

Getting a read of all four sides of a building is very important. How does size up change when all sides of this single family dwelling are viewed? What appears from the front (A) may be totally different from what is presented in the rear of the building (B).

Escape

The team should not hesitate to remove any obstructed means of escape or egress for firefighters working on the interior. They should make certain that all windows are clear of obstructions such as burglar bars, gates, plywood coverings, or any other materials **(Figure 1-8).** Glass should remain intact in windows unless removal is coordinated with interior companies to directly change fire conditions or aid in rescue (venting for life). The RIT should be proactive and raise ground ladders to windows for potential egress and escape **(Figure 1-9).**

FIGURE 1-8

This building is a RIT team's nightmare when it comes to rapid escape. In addition to the burglar bars, notice the boxes blocking the egress path from the door as well as the hazard that is presented by the air conditioning units placed in the windows. Access to some of the windows with ladders will also present a challenge due to limited space.

Construction

The RIT should recognize the main type of building construction and the impact the fire is having on the structure. Specifically, the RIT should recognize any type of weaknesses, misalignments, or faulty areas that exist **(Figure 1-10).**

Note

Understanding how loads are imposed and resisted by structural components will also aid in predicting how stable the building is.

FIGURE 

Ladders should be proactively placed at windows as secondary egress points for interior crews.

FIGURE 

It is important for the RIT to be able to point out flaws or renovations in construction that could affect operations on the fireground. The visible plates on this masonry wall signify that this wall may be compromised.

Tactical Objectives	
Size Up	
Call for Help (Upgrade Alarm)	
Save Lives (Search/Rescue)	
Cover and Contain ❑ Fire Attack ❑ Exposures	
Ventilation ❑ Horizontal ❑ Vertical	
Rapid Intervention Team ❑ IRIT ❑ RIT	
Extinguish ❑ Water Supply ❑ Back Up Line	
Overhaul	
Salvage	

Fireground Tactical Worksheet

Incident Location ______________________________ Time ___________

Box Card # ____________ Temperature ___________ Wind ___________

Strategic Priorities
1) Occupant Removal
2) Life Safety / Incident Stabilization
3) Conserve Property
4) Safety / Accountability of Personnel

Fire Flow
_____ GPM
L × W / 3 (per floor)

Company	Task / Assignment

Benchmarks
❑ All Clear
❑ Secondary Search
❑ Vent. Complete
❑ Loss Stopped

Additional
❑ Accountability
❑ Adequate EMS
❑ Rehab
❑ Staging Est.
❑ Utilities Cont.
❑ Police
❑ Investigator

PAR	Structural Stability Check
10 min. ___ 20 ___ 30 ___ 40 ___ 50 ___ 60 ___	10 min. ___ 20 ___ 25 ___ 30 ___ 35 ___ 40 ___ 45 ___ 50 ___ 55 ___ 60 ___

C

B D

A

TABLE 1-1

Tactical worksheet.

It is the responsibility of the RIT to report immediately anything suspicious in these areas to the sector to which they pertain or the safety officer.

Hazards

Another important area of service for the RIT is identifying different types of hazards that may exist on the fireground. Hazardous material information such as **NFPA 704 System** markings found on the building or auxiliary buildings tell of hazards or hazardous materials that may be used or stored on the premises **(Figure 1-11).** Teams should also identify natural gas and propane products that may exist.

It is a good idea for the RIT to establish a checklist to help organize the mentioned information. This can be accomplished by establishing some type of abbreviated method, printing it on a laminated card, and attaching it to a rapid intervention equipment bag. Information to record on this card should include that shown in **Table 1-2.**

FIGURE 1-11

NFPA 704 markings can provide information to the RIT that additional hazards may be present.

The Value of Time

Time is a major limiting factor during a fireground emergency. Valuable time can be gained by ensuring that all firefighters are trained in the techniques of self-survival. This training should include information about how to anticipate problems. Additional training in rapid intervention operations will allow crews in the immediate vicinity of a fireground emergency to attempt to solve the problem. This immediate action can reduce the overall time required by the RIT to deal with the emergency. Time affects all aspects

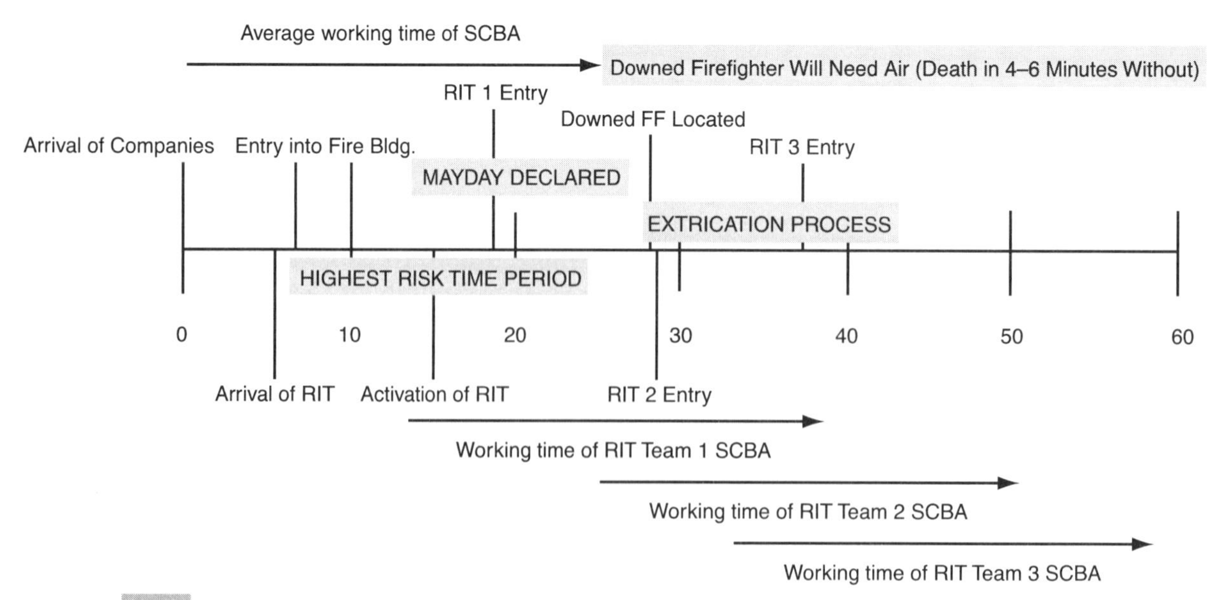

FIGURE 1-12

Rapid intervention timeline.

RIT Responsibilities Checklist

Tool Assignment

RIT Leader/Officer
Radio
Rope bag
Thermal imaging camera
Hand light

Firefighter 1
Radio
Forcible entry (irons)
Hand light

Firefighter 2
Radio
RIT pack
Hand light

Firefighter 3
Radio
Specialty tools
Hand light

Additional Tools
Webbing
Attic ladder
Rescue litter
Power saws
Tarps for tool staging
Extra SCBA cylinders
Handline
EMS jump kit

Rescue
- o Immediate location of companies
- o Identify hazards and safety issues

Escape
- o Open up and remove blocked, secured doors and windows
- o Raise ladders to windows and roofs

Circle Survey
- o Building size and special features
- o Dimensions
- o Construction and stability
- o Exposures within and without

Hazards
- o Information placards
- o Electrical power and downed wires
- o Hazardous materials and hazard conditions
- o Gases: natural/propane/other

TABLE 

RIT responsibilities checklist.

of fireground operations. Time components have a tremendous impact on the overall outcome of the incident **(Figure 1-12).**

Sample Fireground Timeline

1. First-arriving fire-suppression companies.
2. Establishing attack offense/defense/search.
3. First RIT arrival.
4. RIT reports to Command, sets up at division.
5. The communications process of a Mayday received.
6. Response of RIT.
7. Identifying who and where.
8. Locating/identifying resources needed.
9. Extrication and rescue process.
10. Complete removal, hand over to EMS.

The Case Study at the beginning of this chapter illustrates how quickly things can happen on the fireground and the benefits of having well-trained firefighters as part of the RIT. It also proves that yes, rapid intervention does work! But again, everything mentioned in the Case Study took place in a matter of only six minutes. It can not be stressed enough that time is of the essence when we have to save one of our own.

The value of time during RIT operations is further revealed in a series of studies that were conducted by the Phoenix, Arizona Fire Department and Dr. Ron Perry of Arizona State University. These studies were conducted as a result of the tragic loss of firefighter Bret Tarver at the Southwest Supermarket fire on March 14, 2001.

The studies consisted of over 200 rapid intervention drills that were held in three buildings sized between 5,000 and 7,500 square feet. A scenario of a hoseline extended into the structure 150 feet with two firefighters off the line 40 feet was presented to the participants.

Throughout the studies, it took an average of eight to nine minutes for the RIT to reach the downed firefighter from the time that the Mayday was called. This included briefing, entry preparation, and search. The average time to find, package, secure an air supply, and remove the downed firefighter was approximately twenty-two minutes. Also worth mentioning is the fact that it took an average of twelve firefighters to rescue each downed firefighter, while one in five, or 20 percent, of the rescuers experienced a Mayday of their own.

The times obtained in these studies were not under heat and smoke conditions as may be experienced in a real incident. An involved, lengthy extrication process was also not required for the downed firefighter (under a partial collapse, entanglement, etc.). The Phoenix studies illustrate the resources and time that may be involved in rescuing one of our own. If there is doubt in these numbers, set up a similar exercise at your department. In our experiences of conducting scenarios and drill sessions, even under different circumstances, we have consistently noticed time intervals that are very similar in nature.

Additional Issues Concerning Rapid Intervention

Fire departments should establish a rapid intervention plan for fireground operations to produce and activate true RITs. Many fire departments are forced to consider the "risks versus the benefits" of establishing rapid intervention capabilities. They must alter their standard operating procedures or rely on other fire departments to help them establish the capability. Other fire departments that have the manpower and apparatus available may have specific procedures that assign incoming units to perform RIT. Some fire departments may establish regional teams made up of individuals from many departments. In order for any of these configurations to work, cooperation among fire departments is needed. This entails training together, practicing skills, and attaining the competencies necessary to perform firefighter rescues.

Injuries and fatalities continue to occur on the fireground. Firefighters may become lost and disoriented. Firefighters may be injured and killed in collapses, especially those involving truss construction. Firefighters may get caught in backdrafts and flashover conditions. Fire departments and incident commanders must deal with inadequate resources. The aggressive nature of our business seems to have us continuously charging in when probably the best thing to do is to stay out. Protective clothing, which has improved tremendously over the years, hides the true dangers of the environment in which firefighters operate. Firefighters continue to go too deeply into structures without the use of safety lines. Repeated alarms with "*nothing showing*" create complacency. For these reasons and many more, rapid intervention must become a mainstay in the fire service. For those departments that rely on mutual aid, it may be better to withhold members when a fire attack is marginal to allow additional help to arrive and establish an acceptable safety margin.

Fire departments and incident commanders must exceed the **National Fire Protection Association (NFPA)** standards and **Occupational Safety and Health Administration (OSHA)** regulations pertaining to RIT and the 2-in-2-out rule in order to be successful. The minimum size of a rapid intervention team should be four members. Remember, the actual rescue of a downed firefighter will require the efforts of several teams.

Summary

Fire departments should stress training programs that emphasize basic fireground skills, self-survival techniques, and RIT competencies. Rapid intervention training should involve studying and reviewing incidents to learn from the lessons of others. After proper training, RIT members should be able to identify the fire main goals of rapid intervention, how to use surveillance and reconnaissance techniques to effect a positive outcome, and the proactive behaviors that will create a climate of success.

When a Mayday occurs, a firefighter is either lost or trapped and is running out of time. It is unfortunate that when this suddenly happens, many departments will painfully realize that they are not prepared to deal with the situation. Many resources, such as the proper tools, are needed for a successful RIT intervention; foremost among them is time. **Be aggressive, be proactive, be safe, and be trained; don't just simply be.**

■ KEY TERMS

Firefighter assist and support team (FAST)
Incident Management System (IMS)
Mayday
National Fire Protection Association (NFPA)
NFPA 704 System
Occupational Safety and Health Administration (OSHA)
Rapid intervention
Rapid intervention company operations (RICO)
Rapid intervention crew (or company) (RIC)
Rapid intervention team (RIT)
Rescue assist team (RAT)
RIT leader or officer (RITLO)
RIT operations chief
Safety officer
Tactical worksheet

■ REVIEW QUESTIONS

1. The initial step in the RIT's mission is to prepare themselves by
   a. staging.
   b. evaluating the fireground.
   c. communicating with interior companies.
   d. securing the tools they need from other companies.
2. The RIT should function with five goals in mind. They are:
   1.
   2.
   3.
   4.
   5.
3. The __________ and __________ will determine the number of RITs needed at an incident.
4. A well-trained incident commander should be able to effectively manage both the suppression and rapid intervention efforts.
   True False
5. A safety officer put into place to oversee fire-suppression operations can also be utilized to help supervise the fireground Mayday.
   True False
6. RITs are one of the most underutilized resources on the fireground. This is often true because the RIT
   a. is not equipped.
   b. does not have an aggressive officer.
   c. is positioned poorly.
   d. is assigned to poorly skilled firefighters.

7. The best communication process to allow the RIT officer to obtain information needed from Command is through
   a. a radio network.
   b. face-to-face communications.
   c. portable radios.
   d. summaries from Dispatch.
8. The minimum size of a RIT should be _____ members.
   a. two
   b. three
   c. four
   d. five
9. NFPA 704 markings can give the RIT information regarding to
   a. construction type.
   b. number of occupants.
   c. possible hazards.
   d. water supply.
10. Training identifies
   a. tools and resources needed.
   b. what works and what does not.
   c. the need for procedures.
   d. all of the above

ADDITIONAL RESOURCES

Cobb, R. (1996, July). Rapid intervention teams: They may be your only chance. *FireHouse,* pp. 54–57.

Cobb, R. (1998, May). Rapid interventions teams: A fireground safety factor. *FireHouse,* pp. 52–56.

Coleman, J., and Lasky, R. (2000, January). Managing the Mayday. *Fire Engineering,* pp. 51–62.

Dodson, D., *Fire department incident safety officer.* Clifton Park, NY: Thomson Delmar Learning, 1999.

Firefighter's handbook, 2nd ed. Clifton Park, NY: Thomson Delmar Learning, 2004.

Jakubowski, G., and Morton, M. *Rapid intervention teams.* Fire Protection Publications, Stillwater, OK: Fire Protection Publications, 2001.

Kreis, S. (2003, December). Rapid intervention isn't rapid. *Fire Engineering,* pp. 56–66.

Lamb, P. (2002, August). The rapid intervention time line and crew survivability. *Fire Engineering,* pp. 93–95.

Smith, M. (2001, August). Rapid intervention: What is it? *Firehouse,* pp. 18–19.

CHAPTER

2 Compliance and Standards

Learning Objectives

Upon completion of this chapter, you should be able to:

- explain the three basic principles of risk management when operating at an emergency incident.
- discuss the basic parameters of OSHA 29 CFR 1910.134 and how it relates to rapid intervention.
- define the NFPA standards that specifically relate to rapid intervention and how they may relate to your organization.
- identify areas of responsibility that RIT members may be involved with on the fireground that would allow them to still function as part of the RIT.
- explain the exemption regarding the 2-in-2-out rule in the OSHA Respiratory Standard when life is in jeopardy.
- explain what is meant by *consensus standards* and explain which NFPA standards address rapid intervention.
- identify the main objectives in NFPA 1710 and 1720 regarding a fire department's operations related to rapid intervention.
- identify some of the NFPA standards regarding the actions taken by fire department personnel at the initial stages of an incident.

Case Study

"On November 25, 2002, at approximately 1320 hours, occupants of an auto parts store returned from lunch to discover a light haze in the air and the smell of something burning. They searched for the source of the haze and burning smell and discovered what appeared to be the source of a fire. At 1351 hours they called 911. Units were immediately dispatched to the auto parts store with reports of smoke in the building. Firefighters advanced attack lines into the auto parts store and began their interior attack. Crews began opening up the ceiling and wall on the mezzanine, where they found fire in the rafters. Three of the eight firefighters operating on the mezzanine began running low on air. As they were exiting the building, the ventilation crews on the roof began opening the skylights and cutting holes in the roof. The stability of the roof was rapidly deteriorating, forcing everyone off the roof. The IC [incident commander] called for an evacuation of the building. Five firefighters were still operating in the building when the ceiling collapsed. Two firefighters escaped. Attempts were made to rescue the three firefighters while conditions quickly deteriorated. Numerous firefighters entered the building and removed one of the victims. The victim was transported to the area hospital and later pronounced dead. Approximately two hours later, conditions improved for crews to enter and locate the other two victims on the mezzanine."

—Case Study taken from NIOSH Firefighter Fatality Report # 2002-50, **"Structural Collapse at an Auto Parts Store Claims the Lives of One Career Lieutenant and Two Volunteer Firefighters—Oregon,"** ***full report available online at http:// www.cdc.gov/niosh/face200250.html.***

The importance of this Case Study as it relates to Chapter Two is that it is one of the first cases in which OSHA levied fines against a fire department for failure to comply with regulations related to rapid intervention. OSHA cited a total of sixteen violations while levying fines of $54,450 against the municipality. Violations considered the most serious included:

- a less than appropriate Incident Management System.
- the absence of a RIT.
- the lack of a personnel accountability system.
- various violations of the Respiratory Standard, including medical exams, fit testing, maintenance, and program administration.

Introduction

The need for rapid intervention is here to stay. The question of what defines an adequate RIT as it relates to the specific standards and regulations from the various governing bodies or agencies is open to individual interpretation. No matter what the definition is, it must be applied to the overall fireground operation in order to provide for the safety of the firefighters. Firefighter safety is the main objective **(Figure 2-1)**.

FIGURE  2-1

RIT operations need to consider every emergency incident, not only structure fires.

Rapid Intervention Standards and Regulations

Both the NFPA (National Fire Protection Association) and OSHA (Occupational Health and Safety Administration) indicate that RIT members shall be properly equipped and trained for the operations likely to be encountered. Simply stated, the purpose of RIT operations is to rescue lost, trapped, or injured firefighters. In order to meet the objectives of the standards and regulations, RITs should always be comprised of individuals trained in the techniques of rapid intervention.

Inexperienced personnel should never be placed on a RIT because it reduces the team's ability to meet its purpose.

All members of the fire department should be aware of the standards and regulations that apply to them. Firefighters should make certain that their department provides the appropriate instruction, along with policies and procedures, to allow them to operate within the standards. With respect to rapid intervention, this means that departments should provide a solid RIT policy and provide all members with the rapid intervention training necessary to be able to act as a member of a RIT.

Not complying with or following accepted standards and regulations can present serious consequences for fire departments. Losing a firefighter is an ultimate price to pay for noncompliance. Severe monetary fines and civil liability could also result. Fines and suits levied in situations where firefighters perished will never pay for their loss. Sharing lessons learned should be encouraged both as a tribute to those fallen firefighters and as a protective measure to prevent similar future situations. Specific OSHA and NFPA standards will be discussed later in the chapter.

FIGURE 2-2

Proper risk management principles are essential for firefighter safety. Situations where life and property are beyond saving should always warrant an exterior fire attack.

Rapid Intervention and Risk Management

Risk management at an emergency scene is based on three principles:

1. Significant risk can be taken to save a life.
2. Minimal risk should be taken to save property.
3. No risk should be taken to save lives or property that cannot be saved.

The risk management responsibilities of RIT operations are very difficult and can cause further injury or even death if not objectively analyzed. RIT operations involve going into the same areas that have caused the injury, . . . entrapment, or possible death of a firefighter. By acknowledging this, we must properly assess the operational risks and recognize when a rescue has turned into a recovery, no matter how hard this may seem **(Figure 2-2).**

The Occupational Safety and Health Administration (OSHA)

The Occupational Safety and Health Administration was created by the William Steiger Act, which was signed into law by President Nixon in 1970. It was established to ensure the safety of and health of people inside the workplace. OSHA is regulated by the U.S. Department of Labor. Violations of OSHA standards can carry fines up to $70,000 for each violation.

Public employees in state and local governments are not covered by OSHA, but most states have their own occupational safety and health plans to cover them. State plans covering private-sector employees must also cover state and local government employees. Twenty-two states and territories have state occupational safety and health plans that are OSHA approved. An additional four—Connecticut, New Jersey, New York and the Virgin Islands—have state plans that cover public employees only **(Figure 2-3).** These plans may be even more stringent than the regulations set forth by the federal OSHA regulations. The federal program is the minimum set of standards that must be addressed in the individual state plans. The individual state has control for enforcement of the regulations in these states. Unless a local government in a state without an OSHA-approved plan for public safety employees adopts OSHA provisions, federal OSHA has no authority in enforcement. In these instances, enforcement will be carried out by the individual state's Department of Labor.

One of the most relevant regulations regarding firefighting operations is the federal OSHA Respiratory Protection Standard, **OSHA 29 CFR 1910.134.** The Respiratory Protection Standard was adopted in 1971 and revisions were made in 1998 to include a provision known as the **2-in-2-out rule.** In short, OSHA 29 CFR 1910.134, Section (g)(4), specifically references deployment of first-arriving fire personnel and the provision of personnel available for their rescue if needed. It states that if personnel are to enter a hazardous atmosphere, at least two individuals are to remain in visual or voice contact

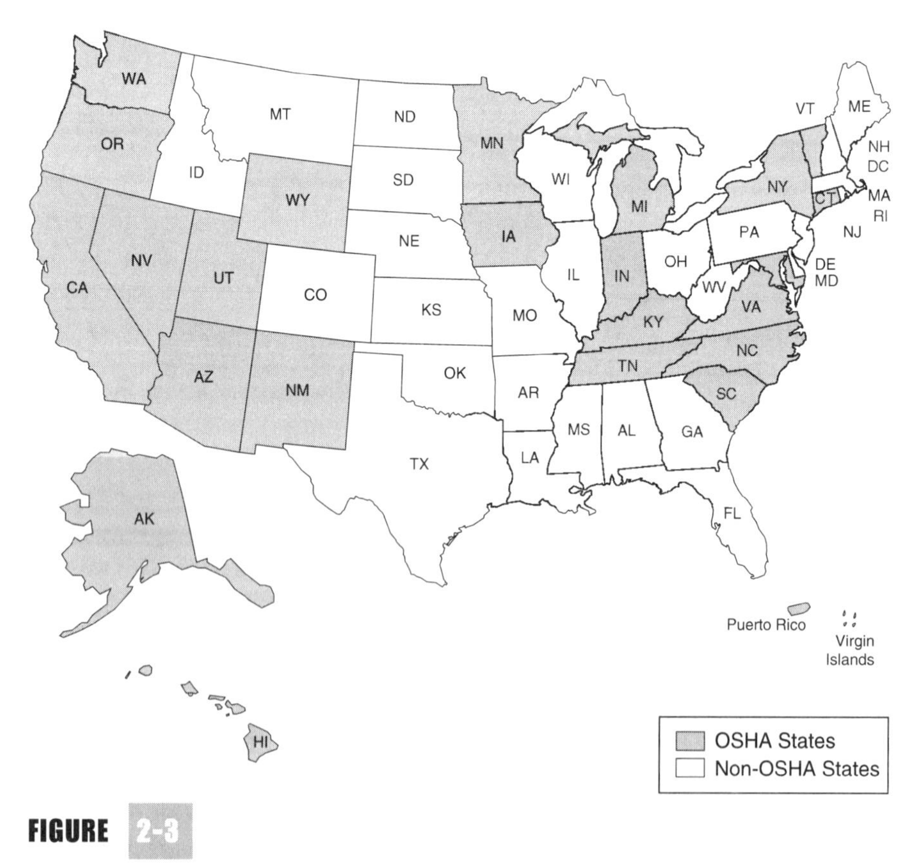

FIGURE 2-3

Twenty-five states are currently authorized by OSHA to run their own safety and health plans. All OSHA regulations apply to both public and private employees in these states.

with them at all times. It also mandates that these two additional personnel be located on the outside of this hazardous atmosphere and that both pairs of individuals be required at all times to use and wear self-contained breathing apparatus. Additional paragraphs of the regulation describe the roles and function of the two individuals who are located on the outside of the hazardous atmosphere.

If we look at the regulation and its additional paragraphs in more detail we can see how the fire service might misinterpret, and at times even manipulate, the regulation beyond its intentions. We have to remember that the regulation specifically applies to two individuals on the inside of the hazardous atmosphere and two individuals on the outside, as a form of backup and accountability if something should go wrong. By no means is the regulation stating that only two personnel are adequate for the rescue of a firefighter in trouble; staffing issues are not defined in the standard. The regulation provides the beginning stages of accountability and safety for firefighters involved in any suppression activities.

Of the two personnel that OSHA requires being on standby outside the **immediately dangerous to life and health (IDLH)** environment, one of them must not be assigned any additional responsibilities that would be vital to the safety of others working on the scene. The second member of the outside team can be involved in specific fireground operations as long as these assignments do not interfere with the ability to perform rescue with the other member if deemed immediately necessary.

The bottom line is to establish a full RIT of at least four to six members specifically assigned to that responsibility and no other. Being proactive and not reactive is the key to success in these matters.

There are situations where a fire department will assemble on the scene of a structure fire and find they have to perform an emergency rescue prior to meeting the OSHA 2-in-2-out requirement. The OSHA regulation does allow an exception for a known life hazard within a structure. The firefighters would immediately react and conduct the rescue in order for the victim to be viable and have a favorable outcome.

Points to Remember

- OSHA 29 CFR 1910.134 states that if firefighters must enter an area that is immediately dangerous to life and health (IDLH), two members should enter together and remain in visual or voice contact with one another at all times **(Figure 2-4).** Radio contact only has been deemed unacceptable due to the possibility of radio failure and possible chance of inability to operate the radio by the affected firefighter in an extreme situation.
- OSHA states that at least two properly equipped, trained firefighters must be positioned outside the IDLH atmosphere and will account for the interior teams while remaining capable of rapid rescue of those interior teams if needed.
- OSHA states that the incident commander has the flexibility to determine whether more than two outside firefighters are necessary when more than two firefighters are operating on the interior of structures while relating it to the size and scope of the incident.
- OSHA states that there may be an explicit exemption in the Respiratory Protection Standard if a life is in jeopardy. The 2-in-2-out rule can be waived in this instance.

FIGURE 2-4

OSHA 29 CFR 1910.134 states that if firefighters must enter an area that is immediately dangerous to life and health (ILDH), two members should enter together and remain in visual or voice contact with one another at all times.

Firefighters at the scene must decide whether the risks posed by entering the interior structural fire prior to the assembly of at least four firefighters is outweighed by the need to rescue victims who are at risk of death or serious physical harm. There is no violation of the standard under rescue circumstances.

National Fire Protection (NFPA) Standards

The NFPA standards are classified as **consensus standards,** which means that they have been developed by members of a specific trade, industry, or profession. OSHA considers NFPA standards as reference when an issue is not addressed specifically in an OSHA regulation.

The NFPA addresses rapid intervention in several standards including: **NFPA 1500** (Fire Department Health and Safety Programs), **1710,** and **1720** (Organization and Deployment of Fire Suppression Operations, Emergency Medical Operations, and Special Operations to the Public by [1710] Career Fire Departments and [1720] Volunteer Fire Departments).

NFPA 1500 makes reference to the 2-in-2-out provision in OSHA 1910.134 and provides guidelines for decisions pertaining to risk management.

One of the main objectives of NFPA 1710 and 1720 is to define how a fire department will incorporate different levels of service, response capabilities, and staffing levels. NFPA 1710 describes rapid intervention on two levels. The first level refers to the **initial rapid intervention crew (IRIC)** and states that two members of the initial attack crew must be assigned as a rapid deployment rescue team for the purpose of rescuing lost or trapped firefighters. (NFPA standards 1710 and 1720 use the term *rapid intervention crew* or *RIC,* which has the same connotation as *rapid intervention team* or *RIT.*) The second level refers to the RIT comprised of firefighters that are specifically "assigned for immediate deployment to rescue lost or trapped members."

What the NFPA standard is saying is that the first-arriving company must have the ability to establish an IRIC consisting of two members that are properly trained as well as equipped. This is easily described on paper, but in the realities of firefighting it is not always possible. Another important point of NFPA 1710 is that the IRIC will be replaced with what can be interpreted as a permanent RIT of at least four members when the incident escalates beyond the initial alarm assignment. It also states that when the magnitude of a given incident is exposing firefighters to an increased level of risk, the incident commander will upgrade the two-member initial rapid intervention crew to a full RIT that is comprised of a minimum of four fully equipped and trained members. NFPA 1710 specifically states that the initial rapid intervention crew is not to take the place of or be confused with establishing a dedicated RIT for the incident.

NFPA 1720 addresses the difficulties faced by volunteer fire departments in regard to how they provide and deploy their firefighting and emergency medical service (EMS) personnel to a specific incident. Due to a wide range of deployment factors, mutual aid agreements, and lengthy response times, NFPA 1720 defines rapid intervention strictly in regard to establishing a RIT rather than an initial rapid intervention crew because of the problems experienced with minimum staffing and deployment procedures. Instead, NFPA 1720 simply indicates that volunteer departments provide sufficient numbers to establish a safe working environment when an initial fire attack begins. NFPA 1720 goes on to describe what the initial attack is and how it will relate to the OSHA standard of 2-in-2-out, which mandates that there will be two individuals involved in the hazardous atmosphere or area while two additional members will be outside the hazardous area to be ready to assist or rescue the team operating on the interior. OSHA's 2-in-2-out rule also says that one of the members on the outside of the structure can be involved or engaged in other activities as long as that member is capable and able to leave those activities immediately if a rescue is necessary.

Both NFPA 1710 and 1720 describe how the RITs will be incorporated into the IMS and **accountability system.** They also indicate that departments should provide some form of dispatching procedures that will provide additional companies to be assigned as RITs.

Points to Remember

- Both NFPA 1710 and 1720 state that members operating in hazardous areas at emergency incidents shall operate in teams of two or more.
- NFPA 1710 and 1720 state that in the initial stages of an incident, where only one team is operating in the hazardous area at a working structural fire, a minimum of four individuals is required. This consists of two individuals working as a team in the hazardous area with two individuals present outside this hazard area for the purpose of assistance or rescue.
- No one shall be permitted to serve as a standby member for a firefighting team when the other activities in which they are engaged compromise their ability to assist in and perform a rescue if necessary.
- NFPA 1500 states that firefighters shall abide by the 2-in-2-out rule during operations in the hazard zones at fire incidents.
- The outside team must be ready with proper turnout gear and SCBA as well as the appropriate equipment to protect the entry team **(Figure 2-5).**
- Two personnel must be available at all times on the outside to help interior crews in the case of a Mayday. Of these two, only one can be engaged in another activity.
- A first-in company engaged in structural firefighting should have a dedicated RIT in place once additional units arrive. Accountability should be maintained to guarantee the location of all personnel in IDLH environments as well as in the immediate hazard zones.
- When arriving at incidents requiring a life rescue, firefighters should quickly analyze the risk involved and determine their actions. They should also realize that a RIT may not be in place for their needs.
- NFPA states that the composition and structure of a RIT shall be permitted to be flexible based on the type of incident and the size and complexity of operations.

FIGURE 2-5

OSHA and NFPA standards specify that a minimum of at least two rescuers be made available outside of the hazard area. They should be in a position with the proper tools to respond at a moment's notice. It is strongly suggested that the RIT be made up of at least four firefighters, if possible.

- NFPA 1720 states that on-scene members performing rapid intervention company operations who become involved with other functions should be ready to redeploy to perform rapid intervention duties if needed. "The assignment of any personnel shall not be permitted as members of the rapid intervention team if by abandoning their critical tasks to perform a rescue clearly jeopardizes the safety and health of a member operating at the incident."
- NFPA states that resources beyond the fire department's initial attack assignment are to establish a RIT through one of the following: on-scene members designated and dedicated as RITs or on-scene members or companies located for rapid deployment for the purposes of being dedicated to a RIT.
- Both NFPA 1710 and 1720 state that at least one RIT shall be standing by with equipment to provide for the rescue of members that are performing special operations, or for members that are in positions that present a high hazard for injury in the case of a catastrophic event.

Summary

It does not matter what rule, standard, or regulation a fire department is trying to comply with. The bottom line is that firefighter safety must be the main objective. The RIT must be established as soon as possible and be made up of properly trained and equipped firefighters.

Firefighters must understand local standards and regulations as well as those mandated by the NFPA and OSHA, such as NFPA 1500, NFPA 1710, NFPA 1720, and OSHA's 2-in-2-out rule.

Firefighters should act according to the three basic principles of risk management during an emergency incident and should know what fireground activities they can be involved in while still functioning as part of the RIT.

NFPA standards and OSHA regulations have come a long way in helping to make the firefighter's job safer. It is important that these standards and regulations be considered and adhered to. Additional information on these standards and regulations can be obtained by contacting the organizations directly. In addition, many technical reports on this subject matter are available through these organizations and the U.S. Fire Administration. They all should be looked into thoroughly prior to establishing rapid intervention policies for any given fire department or organization.

■ KEY TERMS

2-in-2-out rule
Accountability system
Consensus standards
Immediately dangerous to life and health (IDLH)
Initial rapid intervention crew (IRIC)
NFPA 1500
NFPA 1710
NFPA 1720
OSHA 29 CFR 1910.134

■ REVIEW QUESTIONS

1. Risk management at emergency incidents is based on three basic principles. They are:
    1.
    2.
    3.
2. OSHA is a result of the ___________ signed by President Nixon in 1970.
    a. Taft–Hartley Act
    b. William Steiger Act
    c. OSHA Act
    d. Federal Safety Act
3. OSHA is operated under the direct supervision of
    a. the U.S. president.
    b. the U.S. Congress.
    c. the Department of Labor.
    d. the Office of Standards and Professional Qualifications.
4. States that have OSHA-approved occupational safety and health plans in place have the authority and responsibility for enforcement of the regulations.

    True False
5. OSHA-approved state occupational safety and health plans will be less stringent than federal OSHA regulations.

    True False
6. According to OSHA 29 CFR 1910.134, at least two individuals are to remain in ________ contact with personnel inside the hazardous atmosphere.
    a. physical
    b. voice
    c. visual
    d. both b and c

7. The 2-in-2-out rule allows exceptions to the regulation when
   a. life hazards are present.
   b. the response is elevated.
   c. portable radios are used.
   d. the fire is extinguished.
8. NFPA 1710 addresses rapid intervention on two levels: the first level, or the ________, and the RIT that is specifically assigned for the purpose of rescue.
   a. first team
   b. "A" team
   c. IRIC
   d. immediate company
9. __________ refers specifically to volunteer fire departments and rapid intervention.
   a. OSHA 29 CFR 1910.134
   b. NFPA 1710
   c. NFPA 1720
   d. OSHA 29 CFR 1910.146
10. NFPA ________ makes reference to the necessary guidelines pertaining to risk management on the fireground.
   a. 1500
   b. 1710
   c. 1720
   d. 1901

ADDITIONAL RESOURCES

Dodson, D. *Fire department incident safety officer.* Clifton Park, NY: Thomson Delmar Learning, 1999.

Firefighter's Handbook, 2nd ed. Clifton Park, NY: Thomson Delmar Learning, 2004.

Jakubowski, G., and Morton, M. *Rapid intervention teams.* Stillwater, OK: Fire Protection Publications, 2001.

NFPA 1500 Fire Department Occupational Safety and Health Program. Quincy, MA: National Fire Protection Association, 1997.

NFPA 1710 Organization and Deployment of Fire Suppression Operations, Emergency Medical Operations, and Special Operations to the Public by Career Fire Departments. Quincy, MA: National Fire Protection Association, 2001.

NFPA 1720 Organization and Deployment of Fire Suppression Operations, Emergency Medical Operations, and Special Operations to the Public by Volunteer Fire Departments. Quincy, MA: National Fire Protection Association, 2001.

United States Fire Administration. *Rapid intervention teams and how to avoid needing them.* Technical Report Number 123. Available online from the USFA Web page at http://www.usfa.fema.gov.

U.S. Department of Labor. *All about OSHA.* Government Publication Number OSHA 2056-07R, 2003.

CHAPTER

3 Mayday Management, Operations, and Communications Preparedness

Learning Objectives

Upon completion of this chapter, you should be able to:

- explain the benefits of pre-fire planning.
- identify proactive actions that can be taken by the RIT on the fireground to promote firefighter survival.
- name the benefits, importance, and components of well-written and well-understood RIT procedures.
- explain the importance of having procedures for communication on the fireground.
- list procedures for how a RIT should be placed into action.
- define basic actions that must be taken by an incident commander (IC) when a Mayday is initiated.
- clarify the roles and responsibilities of the RIT leader or officer (RITLO).
- describe the advantages of having a separate branch in the IMS system led by a rescue branch officer or RIT operations chief.
- tell the duties and responsibilities of the safety section officer (SSO).
- explain the purpose of acquiring a personnel accountability report (PAR) during Mayday communications.

CASE STUDY

"On December 3, 1999, six career firefighters died after they became lost in a six-floor, maze-like, cold-storage and warehouse building while searching for two homeless people and fire extension. It is presumed that the homeless people had accidentally started the fire on the second floor sometime between 1630 and 1745 hours and then left the building. An off-duty police officer who was driving by called Central Dispatch and reported that smoke was coming from the top of the building. When the first alarm was struck at 1815 hours, the fire had been in progress for about 30 to 90 minutes. Beginning with the first alarm, a total of five alarms were struck over a span of 1 hour and 13 minutes, with the fifth called in at 1928 hours. Responding were 16 apparatus, including 11 engines, 3 ladders, 1 rescue, and 1 aerial scope, and a total of 73 firefighters. Two incident commanders (IC#1 and IC#2) in two separate cars also responded.

Firefighters from the apparatus responding on the first alarm were ordered to search the building for homeless people and fire extension. During the search efforts, two firefighters (Victims 1 and 2) became lost, and at 1847 hours, one of them sounded an emergency message. A head count ordered by Interior Command confirmed which firefighters were missing.

Firefighters who had responded on the first and third alarms were then ordered to conduct search-and-rescue operations for Victims 1 and 2 and the homeless people. During these efforts, four more firefighters became lost. Two firefighters (Victims 3 and 4) became disoriented and could not locate their way out of the building. At 1910 hours, one of the firefighters radioed Command that they needed help finding their way out and that they were running out of air. Four minutes later he radioed again for help. Two other firefighters (Victims 5 and 6) did not make initial contact with Command nor anyone at the scene, and were not seen entering the building. However, according to the Central Dispatch transcripts, they may have joined Victims 3 and 4 on the fifth floor. At 1924 hours, IC#2 called for a head count and determined that six firefighters were now missing. At 1949 hours, the crew from Engine 8 radioed that they were on the fourth floor and that the structural integrity of the building had been compromised. At 1952 hours, a member from the Fire Investigations Unit reported to the Chief that heavy fire had just vented through the roof on the C side. At 2000 hours, Interior Command ordered all companies out of the building, and a series of short horn blasts were sounded to signal the evacuation. Firefighting operations changed from an offensive attack, including search and rescue, to a defensive attack with the use of heavy-stream appliances. After the fire had been knocked down, search-and-recovery operations commenced until recall of the box alarm 8 days later on December 11, 1999, at 2227 hours, when all six firefighters' bodies had been recovered."

—Case Study taken from NIOSH Firefighter Fatality Report # 1999-47,* "Six Firefighters Killed in Cold Storage and Warehouse Building Fire—Massachusetts," *full report available online at http://www.cdc.gov/niosh/face9947.html.

Some important key points for consideration were mentioned in the National Institute for Occupational Safety and Health (NIOSH) report related to this incident and are very relevant to the material contained in this chapter. They are listed in the report as:

- "Ensure that inspections of vacant buildings and pre-fire planning are conducted which cover all potential hazards, structural building materials (type and age), and renovations that may be encountered during a fire, so that the Incident Commander will have the necessary structural information to make informed decisions and implement an appropriate plan of attack."
- "Ensure that the incident command system is fully implemented at the fire scene."
- "Ensure that a separate Incident Safety Officer, independent from the Incident Commander, is appointed when activities, size of fire, or need occurs, such as during multiple alarm fires, or responds automatically to pre-designated fires."
- "Ensure that standard operating procedures (SOPs) and equipment are adequate and sufficient to support the volume of radio traffic at multiple-alarm fires."

CASE STUDY (CONTINUED)

- "Ensure that Incident Command always maintains close accountability for all personnel at the fire scene."
- "Ensure that a Rapid Intervention Team is established and in position upon their arrival at the fire scene."
- "Identify dangerous vacant buildings by affixing warning placards to entrance doorways or other openings where firefighters may enter."

Introduction

It is important for firefighters to understand the principles of risk management when dealing with rescuing firefighters in distress. Failure to do this has contributed to firefighters' lives being lost. Risk management begins with well-written and well-understood policies, procedures, and guidelines. Policies, procedures, and guidelines should be in place to cover all aspects of an incident, from pre-incident planning to debriefing and post-incident analysis.

Five common factors in firefighter fatalities are:

- breakdowns in the Incident Management System
- lack of or inadequate standard operating procedures (SOPs)
- a breakdown in the accountability of personnel on the fireground
- breakdowns in communications on the fireground
- lack of recognition of key aspects related to building construction

All of these factors can be controlled to some extent by taking the proper steps prior to an incident.

Pre-Fire Planning

Pre-fire knowledge of buildings goes a long way toward improving the odds of success when something goes wrong on the fireground. Firefighters and commanders should understand the risks they are likely to encounter in each type of scenario. Developing solid pre-fire plans is the place to start. By preplanning and analyzing possible fire situations that may cause problems, we increase our success rate while diminishing the possibility of injury or death.

Pre-fire planning allows us to formulate and organize information in a way that can be retrieved easily at the time of a fire incident. Pre-fire planning also identifies potential threats to firefighters before they are exposed to them under emergency conditions **(Figure 3-1).**

Lightweight truss construction contributes significantly to firefighter fatalities. One of the biggest problems with lightweight truss construction is the connectors (sometimes called **gusset plates** or **gang nails**) that hold the components of the assembly together **(Figure 3-2).** Temperatures in the area of 1200°F will cause these metal connectors to expand, twist, and pull away **(Figure 3-3).** Each truss assembly is dependent on each other. When one fails, the others will fail also. Building construction should be a major component of fire department preplanning.

During fireground operations, combining preplanning information with actual fireground conditions allows firefighters and commanders to anticipate problems that are likely to occur. For instance, fire impinging on truss components for an extended amount of time should indicate potential collapse. Delayed or inadequate ventilation may increase the chances of flashover. Excessive water flow with little or no runoff should indicate potential collapse conditions due to the increased weight load on the structure. These are just some of the conditions that should be observed by all who are involved on the fireground, especially the company that is assigned to rapid intervention.

FIGURE 

Will you or your company know at 3 AM that the roof of this building is really a bowstring truss? Even with a circle survey an officer may not pick up this critical information due to not being able to see the roof from a close distance or poor lighting. The only remedy is to know the buildings in your company's response district.

One of the biggest problems with lightweight truss construction is the connectors (gusset plates) that hold the components of the assembly together. Firefighters need to be aware of the buildings in their district that utilize this type of construction.

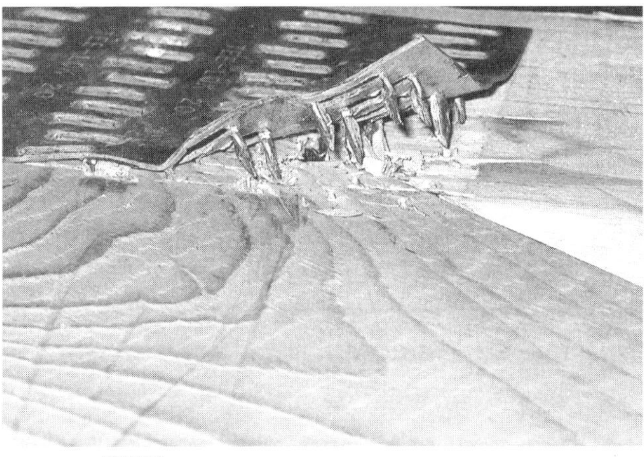

FIGURE 

Exposure to high heat will cause the metal connectors in lightweight truss construction to expand and pull away, causing failure of the truss assembly.

Training and Proactive Behavior

Another area of preparedness for success involving Mayday incidents is in the area of firefighter survival training. It is often said that the first rule in managing a Mayday on the fireground is to do everything possible to avoid a Mayday. The majority of a firefighter's training should focus on how to stay out of trouble. Strong basic skills will serve as a foundation to this statement. A department that is well trained in self-survival skills will also be able to generate a prediction of what a trapped, lost, or disoriented firefighter will do in extreme situations. This allows the RIT, whose members understand these skills, an increased chance of success in affecting a rescue. This will also diminish the time it takes to locate, assess, and extract the victim.

It should be understood that a well-prepared fireground involving proactive behavior is very important. Without proactive RICO there would be less likelihood of the downed firefighter managing any secondary means of escape or egress from a structure. Minimum proactive behavior for RITs should incorporate the use of ladders at secondary egress points around the building to provide for immediate access as well as escape. RITs should provide lighting at every entry point that a fire company has penetrated. This will provide a point of orientation for a possible lost firefighter or crew. The RIT should try to provide an additional hoseline for possible backup if conditions deteriorate at these entry points. Finally, remove all barriers, especially boarded or barred windows, chain gates, window gates, accordion gates, and any exterior items that may block escape. This proactive action also allows the RIT to gain quick access at various points around the structure.

Operational Guidelines and Procedures

In order to increase operational success on the fireground, fire departments should develop well-written operational procedures. This enhances firefighter safety and facilitates the decisions that are made by all personnel on the fireground. Having good **standard operating procedures (SOPs)** provides both control and accountability.

There are still many fire departments that do not have written procedures to manage their incidents. This exposes them to liability and also increases the possibility of a firefighter injury or fatality on the fireground. It should be realized that a firefighter injury or fatality starts as a Mayday, whether it is communicated on the fireground or not. Departments without SOPs will be limited in their ability to successfully manage a firefighter Mayday.

Nowhere is the importance of operational procedures more evident than with the fireground implementation of an Incident Management System (IMS). It should be the responsibility of the first-arriving unit to establish a command presence by initiating Incident Command. Failure to establish an early Command presence will ultimately lead to chaos and disaster. It is the actions of Command that will establish order and assign firefighters to work towards a common goal. A framework that can be adapted for a written policy for rapid intervention is shown in **Table 3-1.**

Radio Communications

Fireground communications should occur on a tactical channel and department-wide dispatch should occur over the main frequency. Firefighters should be well versed in the proper use and format of communications. It is important for the RIT to listen carefully to all radio communications involving fire companies and the tactics that are being employed.

Rescue Tip

All fire departments should establish written SOPs dealing with communications. Without these, the fireground can be lost and ruled by chaos from the early stages of an incident.

Mayday Communications

Any firefighter in distress, whether conscious or unconscious, has a limited amount of time to survive. The first line of defense is to immediately communicate a radio message alerting others that you need assistance. Firefighters must be trained to call a Mayday the moment that they even think that they may be in trouble. As stated in Chapter 1, time is the most valuable resource! Waiting to realize that you are actually in a bad situation may be too late. Firehouse badgering and teasing as well as individual pride will often prevent us from calling for help. These attitudes must be overcome. Help may not be there unless you ask for it! A firefighter may communicate a Mayday for himself or for any other member on the fireground that is in distress. When a Mayday is given, it should receive top priority over the radio.

Procedures must be put into place for calling a Mayday, and firefighters must be trained to follow such procedures. A sample procedure that can be modified is as follows:

1. Remain calm and call for the Mayday, "Mayday, Mayday, Mayday." (Make certain that members understand the meaning and use of the term.) Activate the emergency button on the radio, if equipped, at this time.

OPERATING GUIDELINES

Rapid Intervention Team (RIT)

Purpose

To provide for the establishment of a rapid intervention team (RIT) within the Incident Management System

Scope

The primary function of RIT is to provide personnel for the safety and rescue of firefighters as needed. The XXX Fire Department recognizes that RIT operations are labor intensive and require a substantial commitment of firefighters. To ensure adequate resources for RIT operations the XXX Fire Department has increased the number of responding companies to all general alarm incidents as well as those requiring multiple alarms through mutual aid.

The objectives of this guideline is to:

A. Establish a system of effectively assisting members suddenly threatened by a dangerous fireground situation or IDLH condition.
B. Ensure that adequate resources are maintained on the scene to continue extinguishment efforts while simultaneously rescuing the member/members in need.

Dispatch of Apparatus (RIT)

A. An initial rapid intervention team (IRIT) team shall be established at all working fires, general alarms or higher, and incidents requiring entry into IDLH environments.
B. An IRIT shall be established immediately upon the arrival of the first two XXX FD companies.
   1. From those two companies, the two apparatus operators shall remain outside the fire building to establish accountability and provide for immediate rescue if needed.
   2. The IRIT remains in effect until replaced by the establishment of the RIT.
C. The third XXX FD engine and the second ambulance due to the scene shall be assigned to RIT. The company officer will assume the role of RIT officer.
D. As an incident expands in size or complexity, the number of RITs teams shall increase with the demands of the incident as determined by the incident commander (IC) and the RIT operations chief.

Reporting

A. The third XXX FD engine and the second ambulance shall bypass staging, positioning their apparatus near the incident while remaining clear of initial suppression operations. The RIT leader or officer (RITLO) will then report to command.
B. RIT teams shall assemble near the point of entry of the working suppression companies, in full personal protective equipment (PPE), with SCBA and bring equipment needed to complete their task (see RIT equipment checklist).
C. The RITLO will ensure that all appropriate tactical objectives of the XXX FD RITLO checklist have been met (see RITLO/RIT operations chief responsibilities checklist).

Communications

A. A RIT will be assigned to XX FD fireground channel #2, frequency ####, unless otherwise specified (RIT will only operate on this frequency in a Mayday situation, otherwise they will remain on primary fireground, ######).
   1. All communications within RIT and with safety section officer (SSO) shall be on this frequency.
   2. The RITLO is responsible for monitoring both the fireground operations channel and the RIT radio channel.
B. A Mayday can be declared by any fire department member and/or personnel monitoring the radio frequencies (i.e., Dispatch).
   1. Command will issue an *Emergency Traffic Declaration* over every fire channel being used at the incident, communicating that a Mayday has been called.
   2. Command will conduct a PAR (personal accountability report) on the primary fireground channel with emphasis on members operating in the effected area.
   3. Command will deploy RIT.
   4. Safety Section Officer will take over RIT operations.
   5. Command shall move nonaffected companies to a secondary fireground channel.
   6. Command will then request additional alarms and specialty teams as warranted.

TABLE 3-1

Sample Operating Guidelines for Rapid Intervention Team.

7. When available, support aides and/or support officers shall be assigned to monitor individual radio frequencies.
8. When possible, Command will move operations into the mobile command center.
9. Dispatch will notify Command at ten-minute intervals of the need for additional PARs.

Tactical Objectives (RIT Team)

A. In order to avoid freelancing, each member on the scene will be responsible for exercising appropriate control when implementing this procedure.
B. The RIT shall maintain a "ready" state at all times to facilitate a rapid response if needed.
C. The RITLO shall report to Incident Command for a briefing.
D. The RIT operations chief and the SSO will be responsible for monitoring all assigned radio frequencies.
E. RIT shall conduct a continual size up of the incident, reporting findings and safety concerns to Command.
F. RIT shall survey apparatus for additional equipment that may be needed. This shall be assembled in a RIT staging area. The need for additional resources shall be communicated to the SSO, or Command if this position is not yet filled.
G. The RITLO is expected to ensure that all items on the RIT officer's checklist are completed or provided for.
H. RIT shall perform proactive emergency operations. Examples may include:
   - Additional ladders for secondary egress
   - Establish backup/safety line
   - RIT shall provide forcible entry to ensure secondary means of egress
   - Additional lighting
I. Mayday search-and-rescue efforts shall begin at the last reported location of the missing/injured firefighter/firefighters.

Strategic Objectives (Command)

A. **Rescue**
   - ☐ Deployment of RIT
   - ☐ Assignment of SSO
   - ☐ Second RIT pulled from tactically staged companies

B. **Fire Control**
   At location of RIT operations
   - ☐ Consider water source
   - ☐ Is additional engine needed for protection of rescue efforts?

   Rest of Structure
   - ☐ Is alarm level sufficient?

   Strategy
   - ☐ Offensive
   - ☐ Defensive

C. **Supporting the Rescue Effort**
   - ☐ Monitor the structure for fire control, stability, air quality, and IDLH conditions.
   - ☐ Ensure continued ventilation efforts to maintain tenability during the rescue.
   - ☐ Coordinate and control the search efforts.
   - ☐ Emergency Medical Service (EMS) available for rescuers and the downed firefighter.
   - ☐ Request representatives from code enforcement, public works, and the fire prevention bureau.

D. **Post-Rescue**
   - ☐ Call out Critical Incident Stress Debriefing (CISD) team for all members.

TABLE 3-1

Sample Operating Guidelines for Rapid Intervention Team. *(continued)*

2. Transmit your unit number and name.
3. Give your last known location or assignment.
4. Tell what your emergency is and the status of your crew.
5. Secure acknowledgement by Command.
6. Activate **personal accountability safety system (PASS)** device (a PASS device detects motion—if a firefighter has been immobile for a set length of time, the device will activate an alarm).

The accountability of firefighters as well as the action and progress of the fire is a high priority for the IC. To achieve accountability, fire departments should have well-established procedures that specify fireground assignments. An assignment example would be to have a RIT in place and ready to deploy in the event that a Mayday is transmitted over the radio.

Discipline versus Chaos

The actual deployment of a RIT can create chaos. The response to deployment is usually not orderly and incorporates gut feelings of well-intentioned firefighters. Fireground commanders must immediately gain control of the event and establish order and discipline to ensure all aspects of the fireground (RIT deployment, fire suppression) are handled.

When the RIT is deployed to rescue a downed firefighter, members will be taxed both mentally and physically. All individuals will become so focused on rescuing their fallen brother that they will not realize or admit their own limitations. This may ultimately expose RIT members to possible injury, making them part of the problem. The importance of RIT training can not be overemphasized as it relates to decision making and discipline by members of the RIT.

Firefighters on the fireground may want to become involved with the fire incident operations. It is understood that if a Mayday were to occur, members on the fireground might have an uncontrollable urge to move towards the emergency. Command officers must maintain control of their personnel and ensure that suppression activities continue simultaneous to any rescue attempt. This is important for two reasons:

- It limits possible exposure to the RIT.
- It improves accountability in the original divisions or groups that were performing suppression activities.

It is important for fire departments to provide guidelines and build the confidence of their members so that discipline will be present in any situation.

Mayday Operations

Whenever a Mayday is communicated, the IC should immediately call for a **personnel accountability report (PAR).** Calling a PAR allows the IC to confirm the missing member and that member's potential location. Until the exact information is known it is impossible to properly initiate an effective rescue plan. A PAR should also be utilized for other fireground events such as collapse or change of strategy (offensive to defensive mode). The PAR should begin with companies nearest to the fireground event (i.e., collapse, flashover, etc.) reporting first and working away from that area.

One of the most important areas for the IC is to make sure that the fire fight is continued during rapid intervention operations. The IC will have to realize that the Command system may need to be expanded. If nothing else, the IC should assign a personal aide in order to help in the communications process. An IC will not be able to effectively manage both the rescue operation and the incident itself. When ICs are faced with a Mayday communication they will have to expand the Command system and apply a high priority to the rescue. A separate rescue branch officer or RIT operations chief (the term *RIT operations chief* will be used for this text) will have to be assigned by Command to the efforts of the RIT **(Figure 3-4).** The situation can become very emotional and cause quick reactions that may lead to delays, misinformation, and disorganization when trying to institute an appropriate plan of action. ICs must concentrate on making sure that accurate information is obtained and that the RIT operations chief is acting upon the information. RIT operations must be flexible. An ever-changing rescue plan may be necessary. It is important to remember that flow charts and checklists, however helpful, do not make the problem go away. The people plugged into these positions in the IMS system must be disciplined and show a strong leadership presence.

The idea that one RIT, or even two, can resolve a fireground Mayday is somewhat naive. The complications found in large buildings, including disorientation, collapse, and other unknown hazards, create situations that will likely require resources and manpower far beyond the

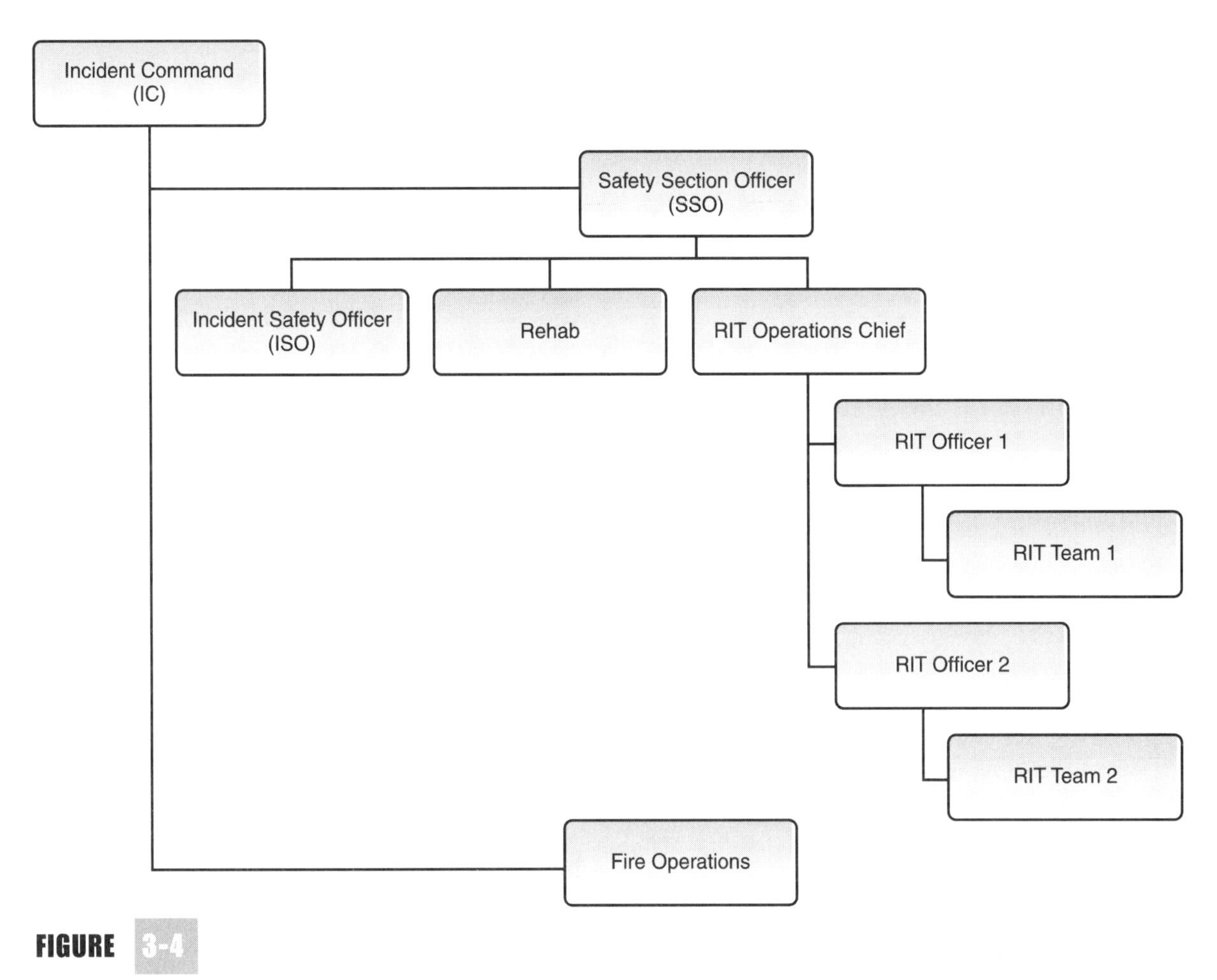

FIGURE 3-4

When ICs are faced with a Mayday communication they will have to expand the Command system and apply a high priority to the rescue. A separate SSO and RIT operations chief will have to be assigned by Command to the efforts of the RIT. An expanded IMS may look like example shown.

capabilities of a single RIT. Because of this, the IC should immediately request additional alarms or mutual aid companies when faced with a fireground Mayday. Additional alarms or the request of additional resources can be done in a proactive manner as well. If there is any question as to whether help will be needed, it should be requested. Unused help can always be sent back. This proactive approach ensures the safety of the fireground and increases the probability of a successful rescue. Other areas that ICs should address are emergency medical service (EMS) support and technical rescue teams.

Technical rescue teams may require additional equipment and resources depending on the situation presented. Early notification will allow these resources to be assembled before they are needed. EMS support is also essential. A number of medical personnel will be required to treat the downed firefighter as well as the rescuers themselves. It has been repeatedly documented that these types of incidents create additional injuries involving firefighters involved with the rescue.

RIT Deployment Operations

By now it is obvious that the IC should have a RIT, or RITs, on-scene and in a proactive mode whenever fire-suppression activities are taking place. The team should be properly prepared with the necessary equipment and should monitor all radio traffic to interpret fireground actions as they relate to the tactical plan. If a Mayday is communicated the RIT can thus be deployed immediately. An additional RIT should be put in place immediately. Maintaining control and keeping things organized is of prime importance to the IC and those under the IC's command in order to ensure accountability.

Fire departments should institute a dispatching system that permanently assigns a RIT, whether it be an engine company, truck company, or heavy rescue squad, to every working structure fire. If this RIT is not needed, the IC can return it. Anytime an incident requires multiple alarms, the IC should consider multiple RITs as well. It is important for the IC to understand the capabilities of a RIT. The IC should know that in order for any rescue to be achieved, the RIT must know the last known location of the firefighter, if there is more than one firefighter involved, and what caused the problem or situation.

FIGURE  3-5

A strong Command presence needs to be established to maintain control and accountability on the fireground. The IC should be focused and ready to run the incident.

Division of Labor

When a Mayday event occurs it is important for the IC to immediately seek the help of others. An additional branch under the IMS system will be required. The importance of establishing a rescue or RIT branch, and assigning strong leadership to it, can not be overemphasized. Command presence of the entire fireground operation, as well as the RIT operation, must be established and maintained **(Figure 3-5)**. The size of the Command presence is determined by the complexity of the incident.

Note

It is also paramount to remember that the original IC will not be able to effectively manage both the suppression efforts and the RIT operation.

The RITLO should be in place with the RIT when it arrives. The main function of the RITLO is to lead and direct the operations of the RIT on the interior or site of the rescue **(Figure 3-6)**. So many activities will be taking place simultaneously that it will be difficult for the RITLO to successfully handle any additional responsibilities once the RIT is deployed. As stated before, because the RIT will become so focused on rescuing the fallen firefighter, members may not realize or admit their own limitations; this includes the RITLO. This may ultimately expose RIT members to possible injury, making them an additional part of the problem. For this reason, the utilization of a RIT operations chief is highly recommended.

The RIT operations chief position does not necessarily have to be filled by a chief officer; it requires a disciplined individual that is knowledgeable and can demonstrate a strong leadership presence. This position will directly support the RIT operation and will provide several advantages in handling a fireground Mayday. The presence of an RIT operations chief provides the following advantages:

- allows the RITLO to focus on the rescue efforts

FIGURE 3-6

The RITLO must possess strong leadership skills and keep constant vigilance of changing conditions on the fireground.
(Photo by D. Tasso)

- supports the RIT with manpower, equipment, and special needs
- controls freelancing in the rescue effort
- provides direct accountability for the deployed RIT
- constantly monitors fireground conditions
- ensures that backup RITs are in place and briefed

All communications from the RITLO should be directed to the RIT operations chief because this position will be responsible for providing resources and making any decisions related to the rescue. The RIT operations chief should be mobile, to a degree, from the interior to the RIT's point of entry so that he can get a true indication of the operations taking place in the rescue effort.

The RIT operations chief and RITLO should work together to make certain that all necessary tasks and considerations are addressed prior to a Mayday. A RIT operations chief/RITLO checklist such as that shown in **Table 3-2** can prove to be beneficial in this situation.

RITLO/RIT Chief Responsibilities

- ☐ Monitor Radio While Enroute
- ☐ Consult/Review Preplan if Available

Report to SSO or IC (if SSO not established) upon Arrival for Briefing Ascertain:

- ☐ Location to Stage Equipment (Usually Primary Entry Point)
- ☐ Confirm Fireground Operations Radio Channel and RIT Radio Channel
- ☐ Location of Fire and Volume
- ☐ Expected Fire Extension
- ☐ Which Fire Companies Are Operating and Where
- ☐ How Long Have They Been Operating
- ☐ What Progress Have They Made
- ☐ Are Sufficient Resources Available?
- ☐ What Type of Accountability System Is in Place?

Conduct Size Up

- ☐ Identify Building Construction and Features
- ☐ Locate All Doors and Windows, Removing any Barriers
- ☐ Establish Access to Sides or Rear of Building
- ☐ Establish Second Way Off of Roof
- ☐ Raise Additional Ladders
- ☐ Look for Signs of Collapse

- ☐ Report Findings and Actions to SSO
- ☐ Monitor Fireground Radio for Signs of Trouble and Progress

Survey Closest Engine

- ☐ Size and Amount of Hose Available
- ☐ Master Stream Devices Available
- ☐ Portable Ladders Available

Survey Closest Ladder Truck

- ☐ Ensure Crew Can Operate Aerial Apparatus
- ☐ Portable Ladders Available?
- ☐ Specialized Equipment?

- ☐ Ensure EMS Is Available and Establish Communications with Them

TABLE 3-2

Example of a Responsibility Check for the RITLO/RIT Chief.

Dedicating a **safety section officer (SSO)** to the Command team can also help control and evaluate risks in and around the actual rescue operation. This position has proven to be a valuable part of the IMS. The SSO is separate from the incident safety officer (ISO). The SSO is an aide to the IC who is in charge of the ISO, the RIT operations chief, on-scene accountability, and **Rehab** efforts (the Rehab branch of the IMS provides for the health and medical concerns of firefighters in the fireground, such as the need for water and food). The person in this role will be located at the Command post or inside the Command van close to the IC. The main purpose of the SSO is to take command of the RIT operation when a Mayday is called **(Table 3-3).** This will allow the IC to concentrate on the firefighting efforts of the incident.

Safety Section Officer (SSO) Operational Duties
☐ Confirm Incident Safety Officer (ISO) is in place
☐ Assign RIT
○ Assign RITLO
○ Assign RIT Operations Chief
☐ Establish Communications on Secondary Tactical Channel
☐ Establish Accountability for All On-Scene Companies
☐ Perform Hazard Evaluation of Scene
○ Utilities Controlled
○ Collapse Zones Established
○ Water Supply Secured
○ Back Up Lines in Place
○ Scene Lighting Provided
☐ Prepare Emergency Operations Plan of Action
○ Evacuation Plan Set Up
○ PAR
☐ EMS Available for Responders
☐ Establish Rehab Sector

Evacuation Plan
☐ Initiate evacuation order
☐ Activate radio emergency evacuation tones
☐ Have on-scene units manually sound evacuation—5 airhorn blasts
☐ Inform ISO
☐ Inform RIT operations chief
☐ Initiate PAR
☐ Confirm missing companies—attempt contact

PAR Benchmarks—Log					
All Clear	Loss Stopped	Mode Change (Off to Def)	Event (Collapse etc)	Mayday	Evacuation
______	______	______	______	______	______

Company	# personnel	PAR 1	PAR 2	PAR 3	PAR 4	PAR 5

TABLE 3-3

Safety Section Officer Checklist.

A breakdown of the IMS for handling a Mayday is illustrated in Figure 3-4.

When a Mayday is called, the following should be considered by the Command team:

1. If a Mayday is received, IC should immediately perform a PAR in order to determine the number of firefighters involved, their location, and what caused their situation. Are they trapped, lost, disoriented, injured, running out of air, etc.?
2. Switch fire-suppression operations to an alternative secondary tactical channel. This will help the affected firefighter communicate with the RIT and not get walked over by other traffic. Only the RIT, RIT operations chief, and SSO should be on the same channel as the downed firefighter. It is strongly recommended that the RIT, RIT operations chief, and SSO also have a third tactical channel available to relay progress reports and other important information. This keeps the downed firefighter on an open channel for communication and also can prevent the downed firefighter from hearing a progress report that may not be favorable. If alternate tactical channels are to be used, though, it is imperative that the RIT operations chief has a second radio available to use. The channel that the downed firefighter is calling the Mayday on must never be abandoned. To do so may cause valuable information being given by the downed firefighter to be missed.
3. Immediately deploy the RIT utilizing search-and-rescue techniques that can be rapidly carried out.

Once the RIT is deployed, the SSO should take control of rescue operations. It will be necessary for the SSO to collect certain information and employ certain procedures or actions. For the purposes of simplicity we will indicate these actions in the following order so that they are remembered easily. The considerations are of highest importance:

4. The RIT should identify the needs of the victim and communicate this to the RIT operations chief, especially in the areas of air supply, entrapment with extrication needs, and any possible fire threat that may be approaching.
5. The RIT should provide progress reports to the RIT operations chief while in the process of removing the downed firefighter.
6. The SSO shall provide EMS with mobile intensive care units ready to receive the downed firefighter or possibly other injured rescue members.
7. Upon the removal or successful completion of the rescue, the SSO should perform another PAR to account for all members on the fireground.

A checklist such as that shown in **Table 3-4** may help the Command team remain focused on the tasks that must be performed.

Command Team Checklist
(Report of a Lost or Downed Firefighter)

- ☐ Emergency Traffic Declaration to Alert Fireground That a Mayday Has Been Declared
- ☐ Move Nonaffected Units to a Secondary Fireground Channel
- ☐ PAR Conducted by Operations on Secondary Channel
- ☐ Immediately Request Additional Alarm(s)
- ☐ Commit the Rapid Intervention Team
- ☐ Change Plan to a High-Priority Rescue Effort
- ☐ Withdraw Companies from Affected Areas (as Needed)
- ☐ Reinforce Firefighting Positions
- ☐ Open/Unlock All Doors
- ☐ Ventilate—Maintain Tenability
- ☐ Provide Additional Lighting
- ☐ Closely Coordinate and Control Search Efforts
- ☐ Special Call for TRT Teams if needed
- ☐ Monitor Structural Stability
- ☐ Maintain Strong Supervision and Control of Crews
- ☐ Control the Media
- ☐ Assign Aides at Command Post to Monitor Each Radio Channel
- ☐ Special Callback for Additional Chief Officers

TABLE 

Command Team Checklist. (*Source: Adapted from Tempe, Arizona Fire Department. Policies and Procedures 205.02*)

Communication and Operational Dispatch Centers

When a firefighter becomes lost or trapped he will, if conscious, communicate a Mayday. The Mayday should include information describing the firefighter's situation. There may be so much fireground noise that a Mayday may barely be audible. It is the responsibility of all members on the fireground to be tuned in to the possibility of a Mayday. It is an essential responsibility of the RIT to monitor the tactical channels for a possible Mayday. The dispatch center must also constantly monitor for a possible Mayday transmission. This is an excellent addition to the safety protections provided to all fireground members and is also beneficial to the IC **(Figure 3-7).**

Dispatch centers handle Mayday events in a variety of ways. In some situations the dispatch center is required to sound an emergency traffic tone on all tactical channels to advise everyone that a Mayday is in progress. The IC should also realize that the RIT has been monitoring all tactical channels on the fireground and more than likely will have immediate and accurate information regarding the specifics of a given Mayday.

The IC can direct the dispatch center to advise all agencies on the fireground of the Mayday. This will allow them to be alert to any radio communications that take place. Once the RIT is on the downed firefighter's frequency, it should be understood that all other fireground radio communications need to be shifted to another channel. Communications regarding progress reports between the RIT and IC should be performed on a channel separate from the downed firefighter.

In some fire departments the use of portable radios with built-in emergency traffic tones provides an excellent method of identifying the distressed member. The use of these types of systems necessitates different procedures in order for them to be effective and understood by the members on the fireground. The principle behind the use of portable radio emergency activation is that the portable radio is supplied with a signal built in so that when pressed it will transmit an audible signal to a communications center that is capable of identifying what unit, company, or radio the signal is coming from. The communications center or dispatch will then attempt to make contact with the radio to determine if there is an emergency. Any information that is acquired through dispatch from this type of alerting system will be immediately relayed to the IC. It will then be the IC's responsibility to contact the radio to determine the emergency and what it is. If radios are equipped with features such as this, it is crucial that all members understand how the feature actually operates. Operations of the feature will be dependent on the programming of the radio itself.

FIGURE 3-7

An alert dispatcher monitoring radio traffic may be able to pick up a firefighter calling for help or even initiate a Mayday in instances where responders on-scene are not able to hear the firefighter in trouble.

Rescue Tip

All members operating on the fireground should also be aware that a Mayday communication might be given on a channel other than the main fireground channel. This is another reason why all channels should be monitored (by the RIT and dispatch center) for potential Mayday events. The IC should be made aware of any such Mayday event and determine a course of action. A downed firefighter may not have a second opportunity to establish a second communication, so it is critical that all messages are acted upon.

Additional Considerations

- When a Mayday occurs, the responsibilities of the IC, the SSO, branch officers, and chief officers involve many actions and decisions that affect the fireground. When structural firefighting operations are complicated by the sudden addition of an injured or distressed firefighter the IC must aggressively acquire the necessary information and determine the nature of the specific problem. By doing so, the IC will be qualified to implement a specific action plan that will address the needed resources and expansion of the Command system. Even though the IC is not specifically involved with the RIT, the IC must understand what the operation will require. At the same time, the IC must also direct the existing fire-suppression activities to help indirectly support the RIT's rescue plan.
- An additional area that the IC must consider is having medical response personnel ready to treat and transport any injured firefighters. This may require another branch under the IMS system.
- The IC should provide for additional ventilation to improve interior conditions that may be slowing the RIT operation.
- RITLOs should develop a search plan that systematically covers the search area while avoiding duplication of effort.
- All personnel on the fireground must consider the possibility of structural collapse. A technical rescue team that can integrate with the RIT will enhance the knowledge and success of any rescue effort that involves collapse. The IC should have a means of providing for a technical rescue team in the event that one is needed. When involved in a collapse rescue operation, the RIT should be prepared to take a supportive role in the overall rescue effort, assisting the technical rescue team. It is essential that the technical rescue team and the RIT work hand-in-hand to accomplish the rescue.
- One final area that should be addressed is the IC's responsibility in establishing a Rehab unit **(Figure 3-8).** These types of incidents push firefighters to their limits and can result in additional injuries that will complicate an already stressed fireground. The IC should remain aware of the physical limitations of companies and provide for relief before it is necessary. The rescue effort may be so exhausting for personnel that the IC may be faced with the unfortunate circumstance of terminating the rescue effort. A good IC will know when this has to be done for the protection and safety of the firefighters. It is understood that this is an unbearable decision but one that is sometimes ultimately the best decision.

FIGURE 3-8

Firefighters involved in RIT operations will require extensive rehab after being pushed to their limits. Rehab should be a priority at all incidents.

Summary

Understanding risk management on the fireground begins with well-written and well-understood policies, procedures, and guidelines. Discipline and knowing what to do can make the difference in saving the life of a firefighter who is in trouble. Policies, procedures, and guidelines should include all entities that may be involved in an incident. Training, review, and evaluation should take place on a continual basis to ensure optimal understanding and performance.

Proactive actions as well as pre-fire planning can be as important to firefighter survival as actions that take place at the fireground. Procedures for communication and RIT deployment at the fireground are also critical. The Mayday initiation procedure must be well understood, as well as the roles of the various members of the IMS, such as the IC, the SSO, the RITLO, and the RIT chief.

KEY TERMS

Gang nails
Gusset plates
Lightweight truss construction
Personnel accountability report (PAR)
Personal accountability safety system (PASS)
Pre-fire planning
Rehab
Safety section officer (SSO)
Standard operating procedures (SOPs)

REVIEW QUESTIONS

1. Five common factors of firefighter fatalities are:
   1.
   2.
   3.
   4.
   5.
2. The connectors used in lightweight truss construction are sometimes referred to as
   a. spike nails.
   b. turnbuckles.
   c. plate nails.
   d. gusset plates.
3. The connectors used in lightweight construction will begin to fail at __________ °F.
   a. 300
   b. 800
   c. 1,200
   d. 2,000
4. The advantage(s) of self-survival training is that it
   a. generates predictability.
   b. increases chance of success.
   c. diminishes time required for rescue.
   d. all of the above
5. The two ramifications of not having well-written and well-understood procedures are:
   1.
   2.
6. Incident Command should be established by
   a. the first unit on scene.
   b. the first chief on scene.
   c. dispatch.
   d. predesignated policies and procedures.
7. Command and sector officers must maintain control of their personnel and ensure that suppression activities continue simultaneous to any rescue attempt. This is important for two reasons:
   1.
   2.
8. It is recommended that only the affected firefighter call for a Mayday because that firefighter knows the situation better than other people on the fireground.

   True False

9. PARs should be conducted in four instances. They are:
   1.
   2.
   3.
   4.

10. When a Mayday is initiated, fire-suppression operations should
    a. remain on the same tactical radio channel.
    b. change to a different tactical radio channel.
    c. no longer use radios for communication.
    d. none of the above

11. Communications between the RIT and the RIT operations chief should be performed on the same radio channel as the downed firefighter's communications so that the downed firefighter knows what progress is being made.

    True False

12. The advantages of having a RIT operations chief and SSO when dealing with a Mayday situation are:
    1.
    2.
    3.
    4.
    5.
    6.

ADDITIONAL RESOURCES

Brannigan, F. *Building construction for the fire service,* 3rd ed. Quincy, MA: National Fire Protection Association, 1992.

Brunacini, A. *Fire command,* 2nd ed. Quincy, MA: National Fire Protection Association, 2002.

Dodson, D. *Fire department incident safety officer.* Clifton Park, NY: Thomson Delmar Learning, 1999.

Firefighter's handbook, 2nd ed. Clifton Park, NY: Thomson Delmar Learning, 2004.

Hoff, Robert, and Kolomay, R. *Firefighter rescue and survival.* Tulsa, OK: PennWell Publishing, 2003.

Rapid intervention teams and how to avoid needing them. Technical Report Series, Report #123. Washington D.C.: U.S. Fire Administration, 2003.

CHAPTER

4 Firefighter Self-Survival

Learning Objectives

Upon completion of this chapter, you should be able to:

- explain the importance of developing programs that deal with training firefighters to handle their own emergencies on the fireground.
- identify the three directions of fire within a structure that a firefighter should be aware of when entering the fire building.
- identify personal equipment that firefighters should carry in their possession.
- identify measures to be taken to ensure preparedness for a fireground Mayday.
- define considerations for a proper fireground size-up.
- explain disorientation and the steps that firefighters can take to increase their chances of overcoming or avoiding it.

CASE STUDY

"On October 13, 2001, a 40-year-old captain (the victim) died and another captain was injured while fighting a fifth-floor high-rise apartment fire. At 0448 hours, units were dispatched to a fire alarm. Units arrived on the scene at 0453 hours and reported heavy fire showing from the exterior of the building. Crews made immediate entry and attack, but after running low on air the victim and the other captain decided to exit. In the process, the victim apparently became disoriented and lost, whereas the other captain was able to escape. Rescue crews were sent to the fifth floor where the victim was located in the elevator common area. The victim was transported to an area hospital where he was pronounced dead at 0615 hours. NIOSH investigators concluded that to minimize the risk of similar occurrences, fire departments should

- *ensure that the department's high-rise standard operating procedures (SOPs) are followed and refresher training is provided.*
- *ensure that team continuity is maintained.*
- *ensure that personnel are in position to maintain an offensive attack.*
- *ensure that a lifeline is in place to guide firefighters to an emergency stairwell.*
- *instruct and train firefighters on initiating emergency traffic (Mayday-Mayday) when they become lost, disoriented, or trapped.*
- *ensure that a rapid intervention team (RIT) is established and in position.*
- *ensure that a backup line is manned and in position to protect exit routes.*
- *ensure that an adequate number of staff are available to immediately respond to emergency incidents.*
- *ensure that the incident commander (IC) continuously evaluates the present weather conditions (i.e., high winds) during high-rise fire operations."*

—Case Study taken from NIOSH Firefighter Fatality Report # 2001-33,* "High-Rise Apartment Fire Claims the Life of One Career FireFighter (Captain) and Injures Another Career Firefighter (Captain)—Texas," *full report available on-line at http:// www.cdc.gov/niosh/ face200133.html.

Introduction

Strong basic skills will prevent a firefighter from getting into trouble. This can only be accomplished through prior training and experience. When performing on the fireground it is essential that you develop a sound philosophy with regard to your safety. When performing a certain function or task, you should understand the reason you are performing the task and the potential consequences of performing it.

A RIT will need to know the abilities of a well-trained firefighter who is knowledgeable in self-survival skills in order to understand and predict what that firefighter will do when dealing with his own Mayday. This knowledge increases the RIT's probabilities of successfully completing its mission.

Basic Skills

Firefighters should have some basic instincts in order to increase their chances of avoiding injury or death. These basic instincts should start being developed at the beginning of the firefighter's career and should be continually reinforced through training. Being constantly aware and focused on the fireground—being oriented to where you are, what you are doing, and where you are going—is the basic instinct that should be continuously developed. Always try to stay oriented and avoid getting lost inside structures and have a plan to exit quickly if needed. Know your position within the fire building (front, rear, sides) along with what floor you are on. Try to orientate yourself with regard to the type of room you are in (bathroom, living room, hallway, bedroom). It is very difficult for firefighters to keep track of their location when going from one room to another.

Firefighters should always maintain awareness of their position within a room, along with the location of objects and furniture encountered within the room. If a search pattern starts to the left, continue to the left. Do not change to a right-handed pattern once started. Actions such as this will lead to disorientation. Doors and windows are survival landmarks that are important to every firefighter operating inside structures. Conditions can deteriorate rapidly, sometimes without warning—knowing the location of the closest door or window is a self-survival basic.

All firefighters involved in search and rescue operations should stay as close to the floor as possible. Remaining low provides better visibility and is the coolest environment in the room. Monitoring fireground conditions is another basic instinct that firefighters should develop. This includes being aware of any changes in fire, visibility, intensity, and heat conditions. Firefighters need to be cognizant that conditions can be very different from one floor to the next as well as from room to room. Know where the fire has been, where it is at, and where it is going. Firefighters should track the conditions and color of smoke and must know how to interpret these findings. Firefighters should constantly look around themselves—above, below, and to their sides—to monitor fire and heat conditions. Monitoring heat conditions within the fire building is extremely important in order to recognize sudden changes in conditions that would require an emergency exit **(Figure 4-1)**. Techniques such as **penciling,** in which short bursts of water are directed toward the ceiling, may need to be utilized to determine heat levels at the ceiling. If the water directed at the ceiling above a search team does not return down, conditions are obviously getting too hot, indicating possible flashover.

Firefighters must also be aware of communications on the fireground. A portable radio is an essential tool for every firefighter performing fireground operations. This may be the only link to fellow firefighters, officers, and the IC. The radio can be used to report the progress of fire-suppression activities, changes in fire development, and, most important, to communicate a Mayday if needed. By utilizing a portable radio, the firefighter in distress increases the chances of surviving a Mayday. A portable radio gives a firefighter the ability to call for immediate assistance, no matter what the situation or problem. Radios must be able to be operated while wearing firefighting gloves.

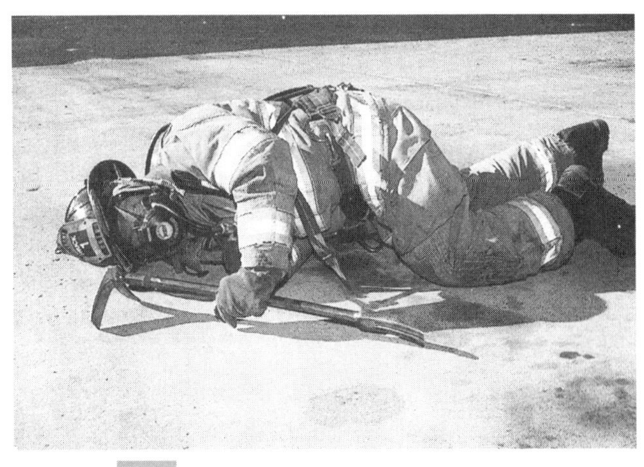

FIGURE 4-1

Firefighters should stay low and constantly monitor conditions above them when operating inside of a hostile environment. It may be necessary to turn to one's side to allow full view of the area overhead due to the brim of the firefighter's helmet and SCBA bottle preventing full head movement.

One of the hardest skills for firefighters to develop is to try to remain in control of their situation and stay focused on what they are doing. This is extremely difficult during emergency situations. Individuals can increase their chances of successfully handling an emergency through progressive training involving self-survival skills. If the firefighter loses control, all other basic instincts, along with the firefighter's ability to prioritize and act in an appropriate manner, will be forgotten.

Firefighter Self-Survival Training

The development of a fireground survival program is important when developing and training firefighters to handle emergency situations. By incorporating firefighter survival training we improve the confidence of a firefighter when dealing with emergencies. Lack of focus, including such things as not wearing your SCBA properly, can lead to disaster **(Figure 4-2).**

Because of improvements in public awareness and technology, there are fewer and fewer structure fires each year. This diminishes our experience level. This also causes an increased chance of running into problems on the fireground. Firefighter fatalities have remained at a relatively constant level year after year even with the decrease in fires. This inexperience with structural firefighting requires ongoing training programs that provide realistic scenarios. This inexperience is also the reason that firefighters must understand the importance of self-survival training. Firefighter self-survival training should be continuous and repetitive in order to maintain proficiency of the skills involved. Even with continuous and diligent training, anything can happen on the fireground that might require the use of self-survival skills and RIT operations. Firefighter survival training should lead into rapid intervention training. It should be understood that during an individual firefighter Mayday, the firefighter begins to solve his fireground emergency while the RIT is searching for his location.

FIGURE 4-2

Wearing equipment improperly is asking for trouble. The straps on this firefighter's SCBA are not fastened, thus creating an additional entanglement hazard. In addition, it increases the firefighter's chances for back injuries because the weight of the unit is not properly dispersed without the straps being fastened.

Training in self-survival and RIT operations will produce the instincts required to deal with a fireground emergency while reducing the amount of critical thinking that must take place under stressful conditions. Departments should take the time to train in self-survival techniques and to incorporate standard operating procedures that reflect emergency conditions—before a true emergency occurs.

Self-Preparedness

There are a number of reasons why firefighters end up in distress, and we cannot plan for every possible situation. Some of the personal equipment that a firefighter should carry includes:

- full protective clothing, including bunkers, firefighting gloves, nomex hood, helmet, and fire boots
- SCBA and mask that fits properly
- flexible emergency breathing tube with 1-inch diameter and 3/4-inch interior diameter, approximately 3 ft. long
- truckman's/utility belt
- 25-foot, 1-inch tubular or flat webbing
- medium-size handheld wire cutters
- two hand lights
- two carabiners
- 30 to 50 feet of personal escape rope in bag
- search line

- Halligan, axe, or pike pole
- two door wedges

Vehicle Preparedness

All firefighters should have a thorough knowledge of the apparatus that they are assigned to **(Figure 4-3).** They should have a clear understanding of their roles and responsibilities in relationship to the apparatus, as well as to the other members of the crew. If it is an engine company, the firefighter needs to know and understand the layout of hose lengths and diameters as well as the nozzles that will be used for fire attack. Firefighters should make sure to look in all compartments to know the location of additional equipment. For a truck company, members should know the location of all tools as well as the length of available ladders. Specific seat assignments provide responsibilities for tasks to be accomplished on the fireground while promoting accountability **(Table 4-1).**

FIGURE 4-3

Firefighters need to thoroughly check apparatus and personal equipment, such as SCBA, each time they report for duty.

Communications

All firefighters should be able to communicate with other members on the fireground. Each firefighter should have a properly working portable radio. Every member should be able to locate and establish the tactical channels that will be utilized during that tour of duty. This includes regular radio communications for alarms as well as emergency traffic and fireground tactical channels. If the portable radios are equipped with emergency tones they should be tested on a regular basis.

Other Preparedness Activities

All firefighters should be knowledgeable in fireground size-up procedures. Proper size-up provides a solid foundation for the many basic instincts for survival that should be developed. Proper size-up should consider:

- general conditions upon arrival
- the location and extent of fire
- type of structure (single-family, high-rise)
- number of floors and location of fire floor
- entry and exit points
- types of stairwells
- mode of fire attack
- progress of the fire attack
- monitoring radio communications
- wind direction
- hazardous conditions
- where are you (at all times)?

These are all very important observations that every firefighter should be gathering while working on the fireground. This information is needed in order to establish a proper self-survival plan.

Truck 5 Riding Assignments Date ____________ Shift ____________		
Position	**Responsibility**	**Tools**
Officer ____________	Safety/Team Leader Interior	Thermal Imaging Camera, Halligan, Search Rope, Radio
Driver/Operator ____________	Aerial Main/Roof	Hook, Flat Head Axe, Halligan, Rope, Radio
Jumpseat 1 ____________	Roof	Saw, Hook, Radio
Jumpseat 2 ____________	Outside Vent/Interior	Hook, Flat Head Axe, Halligan, Radio

TABLE 4-1

Example of a riding assignment card.

Disorientation

Firefighter **disorientation** occurs when a firefighter becomes lost while working in a hazardous atmosphere involving fire, smoke, heat, and darkness. Disorientation causes stress and quickly leads to exhaustion. When disorientation occurs it affects the firefighter's ability to gain a sense of direction. This condition may affect any firefighter, experienced or inexperienced, and it can occur suddenly without warning. Its occurrence has been documented countless times in firefighter fatalities. Firefighters who perform searches before initial hoselines are in place, above the fire, or without the use of personal search ropes are at an increased risk for disorientation.

Training in self-survival techniques, search techniques, SCBA use, preplanning, size-up, and communications, along with the proper use of tools and personal equipment will increase the firefighter's ability to overcome situations involving disorientation. A firefighter who becomes lost within a structure is involved in a true emergency.

Gaining Control of Disorientation

Disorientation will allow fear and panic to set in to create further difficulties and confusion. A disoriented firefighter may allow panic to gain so much control that he will be unable to save himself or even communicate his problem to others. The first action firefighters should take when dealing with disorientation is to stop what they are doing. They should then take a few seconds to calm down and gain control of their breathing. Once they gain control of this, they will be able to size-up what is happening. The important thing is to remember to call for a Mayday immediately. It is not the time to let personal pride stand in the way. Establishing communication as soon as possible will help calm the firefighter. Once communication has been established the

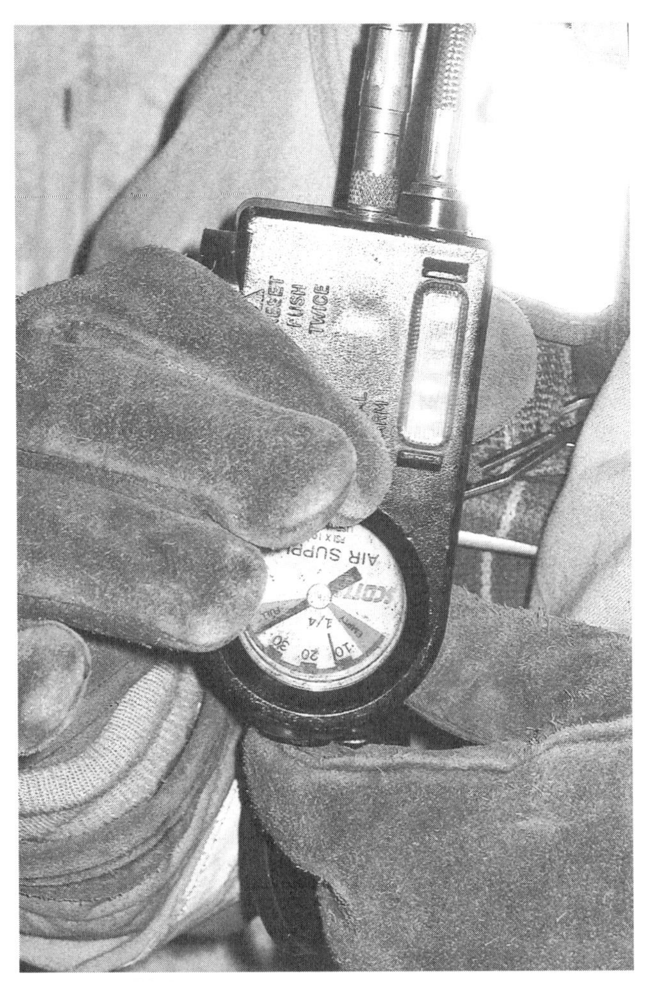

FIGURE 4-4

PASS devices should be activated whenever a firefighter is in trouble.

firefighter should activate the **PASS device (Figure 4-4).** The next important step is to confirm the remaining air supply.

It is important for firefighters to mentally note the size and type of structure that they are going into. One of the main objectives for firefighters who are lost in structures is to try to find a wall. Walls provide orientation by providing a point of reference to move throughout the structure. Walls can lead to windows and doorways (it is important to sweep high enough on the wall to locate the door handle or window sill) and possibly out of the building all together **(Figure 4-5).** Firefighters should understand that gaining control of their behavior helps prevent them from just wandering throughout the building without a plan. Firefighters should try to establish a point of reference as soon as possible and proceed in a systematic fashion.

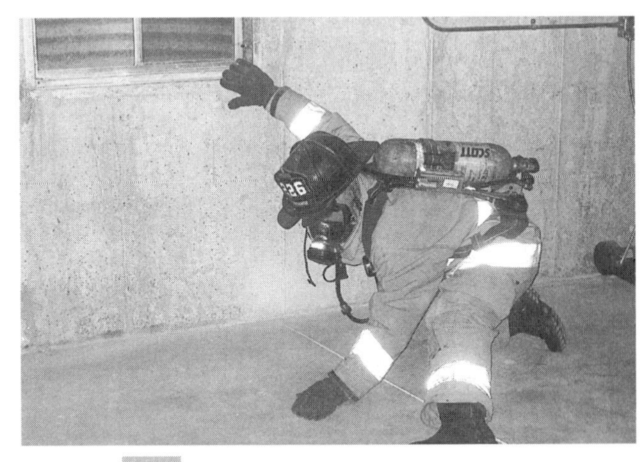

FIGURE 

Firefighters need to make certain that they sweep high enough on walls to feel window exits and door handles.

Another means of finding your way out of the building or structure when you are lost or disoriented is by locating a hoseline and following it to safety. When firefighters enter structures, whether for fire suppression or search, they should try to be aware of the number of hoselines (and lengths) that are going into the structure. The difficulty for firefighters that are disoriented within a structure is in finding the hoseline and then deciding which direction leads to safety.

When a firefighter encounters a hose coupling it should be examined to establish which direction leads to the outside and which leads to the fire (nozzle). It is important to establish the male and female coupling in order to determine which way to proceed. Encountering a female coupling first will lead a firefighter out to safety. Once the firefighter has located a hoseline and determined which direction to travel, every attempt should be made to stay with that line until it leads the firefighter to safety **(Figure 4-6).**

Firefighters that understand basic building construction can use this knowledge to assist them if they become lost or separated from their company. When working in large areas involving commercial spaces, the floors are usually made of concrete. All concrete floors or slabs require **expansion joints** to allow them to expand and contract to prevent cracking **(Figure 4-7).** Following these lines will eventually lead to a wall,

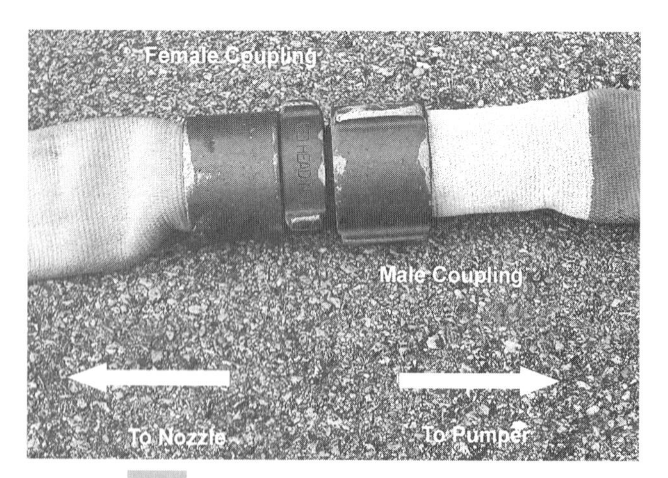

FIGURE 4-6

Lugs on the female coupling are much smaller than the lugs on the male ones. Firefighters should be able to tell the difference and map their orientation by feel if they cannot see.

FIGURE 4-7

Expansion joints in concrete floors can lead firefighters to exterior walls.

or column, which provides a point of reference to help become orientated.

During the physical activity of trying to become orientated, the firefighter must also be aware of his air supply. At some point a decision will have to be made with regard to ceasing all activities in order to conserve the remaining supply of air. A firefighter in trouble should never give up. Self-survival is largely dependent on attitude. At some point a firefighter may not be able to save himself. As the air supply diminishes, it is imperative that the firefighter assume a resting position by lying on the floor, preferably choosing a location on an exterior wall near a doorway or in a hallway, to increase the chances of being found by the RIT. The PASS device should remain activated and in a position that provides the loudest audible sound.

When a firefighter has decided to wait for rescuers to arrive at his position, he can use his personal equipment to attract attention to himself and make himself visible to approaching rescuers. Every firefighter should have two hand lights. When waiting for rescuers, the hand lights can be utilized by pointing them toward the ceiling or shining them low to the floor beneath the smoke. This will assist the RIT as it performs the search. The distressed firefighter should also make noise in order to draw attention to his position. Metal on metal provides the loudest noise. Consider the following summary of events when dealing with a Mayday situation:

- The problem occurs.
- Communicate a Mayday.
- Utilize portable radio emergency button.
- Communicate who you are, where you are, nature of your problem.
- Be prepared to communicate on any channel.
- Activate PASS device between communications.
- Stay with your partner or company.
- Find a hoseline.
  - Feel the coupling.
  - Female to safety.
  - Follow the line out.
- Continue to search for a way out.
- Move to a safe area of refuge when needed.
- Control yourself.
  - Control your breathing.
  - Stay calm.
  - Conserve air.
- Take a resting position near a wall, doorway, or hallway.
- Use flashlights and tools to indicate position.

Summary

A strong foundation in basic skills will prevent a firefighter from getting into trouble. This can only be accomplished through experience and solid training. Being prepared mentally as well as physically is imperative for all firefighters and Command personnel. Maintaining control and focus could mean the difference between life and death in a Mayday situation.

Safety skills include basic skills such as maintaining orientation inside the fire structure, maintaining communications, wearing and carrying proper equipment, knowing the location of additional equipment and materials, understanding size-up techniques, and gaining control when disoriented.

In all fireground situations and especially Mayday operations, self-preparedness is the key to firefighter survival.

■ KEY TERMS

Disorientation
Expansion joints
PASS device
Penciling

■ REVIEW QUESTIONS

1. The three directions of fire travel within a structure that a firefighter should be aware of when entering the fire building are:
   1.
   2.
   3.
2. Two advantages of training in fireground self-survival and RIT operations are:
   1.
   2.
3. Two advantages that result from specific assignments for riding positions are:
   1.
   2.
4. Name five considerations of size-up.
   1.
   2.
   3.
   4.
   5.
5. Firefighters who perform searches before initial hoselines are in place, above the fire, or without the use of personal search ropes are at an increased risk for disorientation.

   True False
6. Name four areas of training that will increase a firefighter's ability to overcome situations involving disorientation.
   1.
   2.
   3.
   4.
7. Encountering a _______ coupling first will lead a firefighter out to safety.
   a. male
   b. female
8. Lugs on the female hose coupling are _______ than the lugs on the male hose coupling.
   a. larger
   b. smaller

ADDITIONAL RESOURCES

Dodson, D., *Fire Department Incident Safety Officer.* Clifton Park, NY: Thomson Delmar Learning, 1999.

Firefighter's Handbook (2nd Ed). Clifton Park, NY: Thomson Delmar Learning, 2004.

McCormack, J., *Firefighter Rescue and Rapid Intervention Teams—Saving One of Our Own.* Indianapolis, IN: Fire Department Training Network, 2003.

McCormack, J., and Pressler, R., *Firefighter Survival.* Indianapolis, IN: Fire Department Training Network, 2002.

Rapid Intervention Teams and How to Avoid Needing Them. Technical Report Series, Report #123. Washington, D.C.: U.S. Fire Administration, 2003.

CHAPTER

5 SCBA Emergencies and Survival Skills

Learning Objectives

Upon completion of this chapter, you should be able to:

- discuss various techniques utilized in prolonging air supply duration.
- determine SCBA air consumption rate by employing the methods outlined.
- demonstrate and explain the emergency check procedure when a problem is experienced while wearing an SCBA.
- explain various techniques utilized for sharing air between two firefighters in an emergency.
- explain various techniques that can be utilized for self-survival when a firefighter runs out of air.
- demonstrate the various techniques utilized for escaping from restrictive areas.
- demonstrate the proper method or technique for escaping from an entanglement hazard.

Case Study

"On April 11, 1994, at 0205 hours, a possible fire was reported on the ninth floor of a high-rise apartment building. This building had been the scene of numerous false alarms in the past. An engine company and a snorkel company were the first responders and arrived at the apartment building at 0208 hours. The engine company was the first on the scene and assumed command.

"Five firefighters from the two companies entered the building through the main lobby. They were aware that the annunciator board showed possible fires on the ninth and tenth floors. Lobby command radioed one firefighter that smoke was showing from a ninth-floor window. All five firefighters used the lobby elevator and proceeded to the ninth floor.

"When the doors of the elevator opened on the ninth floor, the hall was filled with thick black smoke. Four of the firefighters stepped off the elevator. The fifth firefighter, who was carrying the hotel pack, stayed on the elevator (which was not equipped with firefighter control) and held the door open with his foot as he struggled to don his SCBA. His foot slipped off the elevator door, allowing the door to close and the elevator to return with him to the ground floor.

"The remaining four firefighters entered the small ninth-floor lobby directly in front of the elevator. One firefighter stated that he was having difficulty with his SCBA and asked for the location of the stairwell. Another firefighter said, "I've got him," and proceeded with him into the hallway, turning right. Later, one of the four firefighters stated that he had heard air leaking from the SCBA of the firefighter having difficulty and had heard him cough.

"The remaining two firefighters entered the hallway and turned left, reporting zero visibility because of thick black smoke. Excessive heat forced them to retreat after they had gone 15 to 20 feet. They proceeded back down the hall past the elevator lobby. There they encountered a male resident, who attacked one of the firefighters, knocking him to the floor and forcibly removing his facepiece. The two firefighters moved with the resident through the doorway of an apartment, where they were able to subdue him. One firefighter broke a window to provide fresh air to calm the resident. At about the same time, the low-air alarm on his SCBA sounded. The other firefighter was unable to close the apartment door because of excessive heat from the hallway. Both firefighters and the resident had to be rescued from the ninth-floor apartment window by a ladder truck.

"Firefighters from a second engine company arrived on the scene at 0209 hours. They observed a blown-out window on the ninth floor and proceeded up the west-end stairwell to the ninth floor carrying a hotel pack and extra SCBA cylinders. These firefighters entered the ninth floor with a charged fire hose and crawled down the smoke-filled hall for approximately 60 feet (the hallway was 104 feet long) before extreme heat forced them to retreat. As they retreated, they crawled over something they thought was a piece of furniture. They did not remember encountering any furniture when they entered the hallway. In the dense smoke, neither firefighter could see the exit door 6 feet away, and both became disoriented.

"After the firefighter from the first company rode the elevator to the ground floor lobby, he obtained a replacement SCBA and climbed the west-end stairs to the ninth floor. When he opened the ninth-floor exit door, he saw the two firefighters from the second engine company in trouble. He pulled both into the stairwell.

"When a rescue squad arrived at the scene at 0224 hours, lobby command could not tell them the location of the firefighters from the first company. They proceeded up the west-end stairs to the ninth floor.

"The rescue squad opened the ninth-floor exit door and spotted a downed fireman approximately 9 feet from the door. He was tangled in television cable wires that had fallen to the floor as a result of the extreme heat. The downed fireman was from the first engine company; his body may have been what the firefighters from the second engine company encountered in the hallway. He was still wearing his SCBA, but he was unresponsive. The rescue squad carried him down the stairs to the eighth floor, where advanced life support was started immediately.

"The rescue squad then entered the first apartment to the left of the exit door and found a second firefighter from the first engine company kneeling into a corner and holding his mask to his face. He was unresponsive. The rescue squad carried the firefighter down the stairs to the eighth floor where advanced life support was started.

CASE STUDY (CONTINUED)

Both firefighters were removed within minutes and taken to a local hospital, where advanced life support was continued; but neither responded. Both victims died from smoke and carbon monoxide inhalation.

"Both victims wore PASS devices; but because the devices were not activated, no alarm sounded when the firefighters became motionless."

—Case study taken from NIOSH Health Hazard Evaluation Report No. 94-0244-2431, full report available online at http://www.cdc.gov./niosh/hhe/reports.

Introduction

The SCBA is the most important and widely used tool in the fire service today. Its use has greatly expanded the capacities of firefighters when performing aggressive interior searches and fire attack. Every firefighter needs to be thoroughly familiar with the specific piece of breathing apparatus that they will use. The training that firefighters receive must include aspects involving the use of breathing apparatus, maintenance training, and most important, emergency procedures in crisis situations.

Self-Contained Breathing Apparatus Training

SCBA training should begin at a basic skill level and rapidly work toward a more difficult, refined level. Repetition of skills needs to be emphasized as skills must be practiced numerous times before proficiency is to be expected, and review must also be periodic to maintain them **(Figure 5-1).**

The only boundaries that exist when training on SCBA are safety on the training ground and the physical abilities of the participants. It is important to stress that the procedures introduced are to be utilized only under extreme circumstances. The procedures are meant to be utilized as last resorts in saving your own life or that of another firefighter.

Air Consumption for Survival

Basic physiology tells us that different firefighters will consume the air in their SCBA at different rates. The working air supply will depend on the firefighter's training, physical condition, activity, and mental state as experienced under the stressful conditions encountered during firefighting.

FIGURE  5-1

Basic skills such as SCBA donning should be practiced on a continual basis to ensure that a firefighter will be ready and capable to handle an emergency on the fireground.

Note

A question that every firefighter should be able to answer without hesitation is, "How long can I work while wearing an SCBA?" The fireground is not the place to figure out that answer.

Each firefighter is responsible for determining his individual **point of no return** (end of air supply) when entering a hazardous atmosphere. In most residential fires, the low-pressure warning device is adequate to signal a firefighter's point of no return because a firefighter should be able to exit a single-family structure easily after the alarm sounds. However, in larger structures, a firefighter may need more time to exit than would be allowed if the firefighter relied solely on a low-pressure warning device.

A simple method for determining a point of no return is to check the pressure gauge before entering a contaminated atmosphere and again on arrival at the objective. The amount of air used to reach this point is the minimum amount needed to get back to a nonhazardous atmosphere.

Calculating a consumption rate from a consumption test is another way to figure an individual's point of no return. It has an advantage because the firefighter is working at a constant rate when determining it. Consumption testing involves a firefighter working on an obstacle course while experimenting with different breathing rates and techniques. To get a true indication, the obstacle course will need to be set up in exactly the same way each time a consumption test is performed **(Figure 5-2).**

The test should be performed until the firefighter cannot draw any additional air from the SCBA. This will give the individual a true indication of exactly how much time he has left to operate in the event that he becomes trapped or lost. It will also demonstrate to the firefighter how much air is still left in an SCBA even after the **low-pressure/25% alarm** (alarm indicating

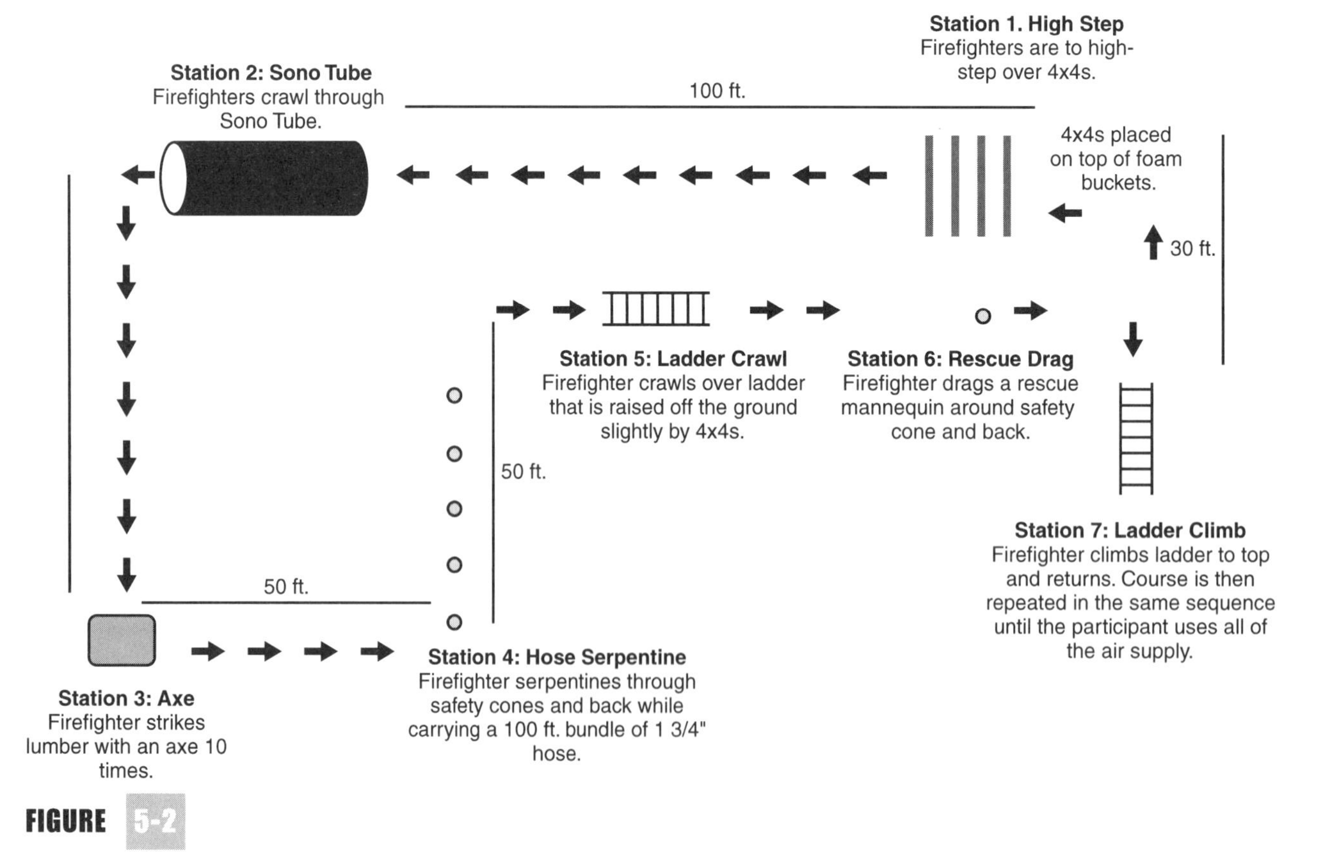

FIGURE 5-2

Typical course used for consumption test. Distances between stations need to be the same each time the test is performed to obtain accurate results. Firefighters should also have a partner to monitor their safety and record information at all times while on the course.

low tank pressure and only 25 percent of air capacity remaining) stops operating. Knowing this information can help keep a firefighter calm and possibly enable self-survival in an emergency situation **(Figure 5-3; Table 5-1).**

There are various methods of breathing that may help in reducing a firefighter's air consumption rate. It will take experimentation on the firefighter's part to find which one works the best. When using any method, it is important to take normal breaths and exhale slowly to keep the CO_2 in the lungs within proper balance. Firefighters should *never* hold their breath in an attempt to save air. Due to the body's release of adrenaline when firefighting, oxygen is being consumed at a higher rate and holding the breath could cause unconsciousness.

A controlled breathing method can provide for the most efficient use of air. Breathing only through the mouth or nose is not the answer to conserving air. Breathing only through the mouth results in an increased respiratory rate while also preventing the body from utilizing all available oxygen before exhalation. Breathing exclusively through the nose will result in short breaths that do not fill the lungs to their full capacity. However, inhaling through the mouth and exhaling through the nose is a method that provides more than adequate air exchange and can be beneficial while engaged in a heavy work load.

To reverse that pattern (inhale through the nose and exhale through the mouth) is another

FIGURE  5-3

Consumption courses should consist of tasks that will simulate the level of work to be performed on the fireground.

Air Consumption Test Record

Name ________________________

Date ________________________

Starting Cylinder Pressure _______ psi

Starting Time _________

Low-Pressure Warning Device (25%) Activation Time ________

Minutes / Seconds until Low-Pressure Warning Device Sounded

_________ Minutes _________ Seconds

Air Cylinder Empty Time (No more air to mask) ________

Time Worked after Low-Pressure Warning Device Sounded

_________ Minutes _________ Seconds

Total Time Worked from Start to Empty Cylinder

_________ Minutes _________ Seconds

TABLE  5-1

Air consumption test record.

method that provides for good air exchange and is easy to remember. Slow, deliberate exhalation is the key to both of these and any other breathing technique in helping to conserve air.

In an extreme emergency, being proficient in a technique such as **skip breathing** can mean the difference between life and death for a firefighter. Skip breathing allows for the maximum use of air that may be contained in an SCBA. A rated 30-minute SCBA that normally lasts 15 to 20 minutes can be extended well over an hour with this breathing technique. This extra time can be the deciding factor necessary for a RIT to be successful in removing the firefighter in trouble. To perform skip breathing:

1. The firefighter inhales fully and holds this breath for the duration that a normal exhalation would take.
2. At this point, the firefighter shall take an additional breath and begin to slowly exhale. This cycle is then repeated.

Whatever breathing techniques are used by a firefighter, it is paramount that they are performed calmly and efficiently. Training exercises such as regular consumption tests will allow a firefighter to practice and perfect breathing techniques and will also help promote emotional stability of firefighters while wearing SCBA. A regular physical fitness program will also help firefighters wear SCBA while experiencing less fatigue and maximizing their consumption rate.

Emergency Procedures Check

It is imperative that firefighters become thoroughly familiar with their breathing apparatus and possess a basic knowledge about preventative field maintenance for the particular unit that they are using. Minor failures such as free flow of air or improper connections are very common on the fireground. These are often the result of operator error or improper preventative maintenance. Upon reporting for duty, a firefighter should make it a point to thoroughly inspect and test the functions of the assigned SCBA. This should also be repeated every time the unit is put into use.

When a failure occurs in a hostile environment, a firefighter who is familiar with his SCBA unit will be able to remain calm and provide a remedy to the situation while exiting the area. The most important rule for a firefighter to remember in the case of a malfunction or depletion of air is to never remove the facepiece of the SCBA. The facepiece itself will afford the firefighter some protection for the face, eyes, and respiratory area while leaving the hazardous conditions.

To find and remedy a failure, a standardized **emergency procedure check** to find, locate, and remedy a malfunction in an SCBA is stressed. This ensures that firefighters will find the problem and execute the proper procedure that will enable them to leave the hostile environment. The procedures listed can be modified to accommodate the different types of SCBA that may be in use **(Skill 5-1)**.

To perform an emergency procedure check (see Figure 5-4):

1. Determine need. Is there a problem?
2. Place left hand on facepiece **(Skill 5-1A)**.
3. Slide hand down mask—check regulator **(Skill 5-1B)**.
4. Check air saver switch.
5. Check bypass or purge valve—is it open or closed?
6. Follow line from regulator to pressure reducer—check for problems **(Skill 5-1C)**.
7. Check if cylinder valve is in open position.
8. Check if cylinder is securely connected to high-pressure line.
9. Correct any problems found in check as you find them.
10. If not able to correct problem, leave area at once with assistance to safe area (consider **buddy breathing**—sharing a single air source).

Buddy Breathing and Other Emergency Escape Procedures

With the exception of SCBAs using an **emergency breathing support system (EBSS)**, a system that can connect a second tank of air without removal of the facepiece, buddy breathing is not a recommended or approved procedure as far as NIOSH, OSHA, or any manufacturer are concerned. Thus, these agencies do not have any guidelines or certification procedures for SCBAs when they are used to support two users. However, buddy breathing is a very important skill that all firefighters should become familiar with because it may be the only option that they have when they are faced with a life-or-death situation. Training on buddy breathing skills is very difficult, and proficiency in the basic fundamentals of SCBA use should be stressed prior to introducing a firefighter to them. Most problems with buddy breathing take place while firefighters are attempting to move toward safety and for this reason the skills should be practiced while moving. As firefighters become more skilled in the procedures, the difficulty level can be increased by obscuring

Safety

Buddy breathing should only be practiced in a nontoxic environment.

SKILL 5-1

How to Perform an Emergency Procedure Check

A Place left hand on facepiece to determine if there is a problem with the lens of the mask. If cracked, cover it with a gloved hand and proceed to exit with your partner.

B The mask-mounted regulator should be checked next, including the air saver (on/off) switch. Activating the bypass or purge valve may provide the air necessary for an escape.

C The high-pressure line leading to the cylinder valve should also be checked. Is the line torn? Is the line not secured to the cylinder valve? Is the cylinder open?

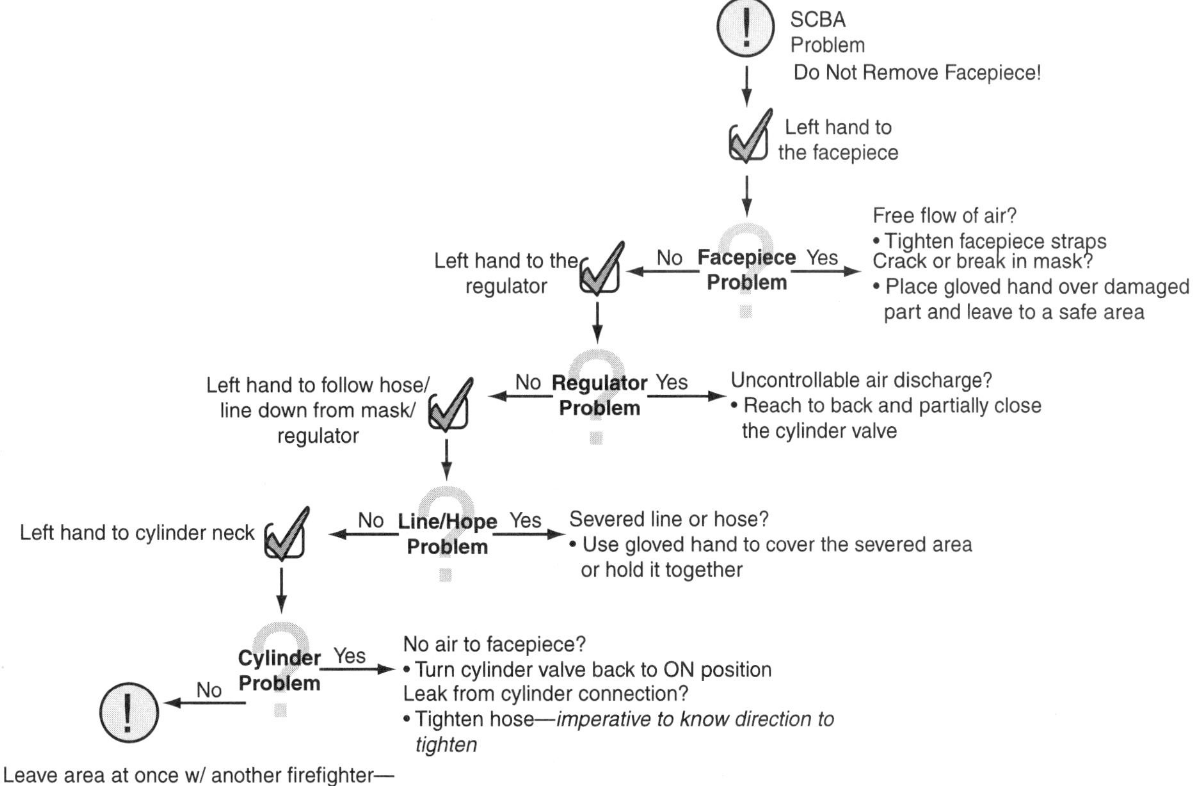

FIGURE  5-4

SCBA emergency procedure checklist.

their vision through the SCBA facepiece while practicing.

Emergency Breathing Support Systems

Most SCBA units on the market offer some form of an EBSS. The advantage to such a system is that a firefighter never has to remove the facepiece. Each particular manufacturer will have specific instructions for using the system that is equipped on their unit. In general terms for all units, the following points need to be recognized by firefighters:

1. Become thoroughly familiar with the system and how it works. This includes all couplings that may need to be connected or disconnected. It is under adverse conditions that the system will need to be utilized. Firefighters should be able to operate the system while wearing gloves and with obscured vision.
2. The rescuer must take charge and be in control. This is imperative because the rescuer is not the one in trouble and therefore will be calmer and more focused when handling the situation.
3. Before hooking into a support system, the rescuer should make certain that there is a sufficient air supply to do so. If not, another means of escape must be considered. Does the system equalize the air cylinders of both firefighters or are both firefighters now breathing from one tank of air? Most manufacturers will not advise utilizing a system unless the rescuer has at least one-half the rated capacity of air in his SCBA cylinder. Whatever the case, both firefighters must realize that they will have to be calm and in control of their breathing while operating on a very limited air supply.
4. Once connected, the movement of the two firefighters must be coordinated. The firefighter being rescued should follow the lead of the firefighter supplying air. If connected by facepiece-mounted regulators, the firefighters must be careful not to dislodge one another's mask while moving. Firefighters must also make certain that they do not get their connection hoses entangled in objects and must move in a coordinated fashion so that they do not cause each other to be off balance. As mentioned earlier, moving while buddy breathing is a very difficult task and must be practiced **(Figure 5-5).**

Not all departments have SCBAs that are equipped with an EBSS, and not all departments have SCBAs that are produced by the same manufacturer as those that may be used in neighboring departments, which could pose a problem during mutual aid situations. This does not mean that firefighters cannot perform buddy breathing to save their lives. It only means that they must now use or modify one of the following procedures to fit their situation.

Facepiece-to-Facepiece Method

This method of buddy breathing can be performed *without* the firefighters removing their facepieces. The only requirement is that one of the firefighters must carry a short piece of flexible plumbing tubing **(Figure 5-6).** This tubing should be at least 3 feet in length and needs to have walls thick enough to prevent it from collapsing when breathing through it.

FIGURE  5-5

An EBSS will limit the mobility of both firefighters. As with all buddy breathing procedures, moving will be very difficult. For this reason, firefighters should train on a continuous basis to ensure that they are confident in its use.

FIGURE 5-6

A piece of flexible plumbing tubing carried as part of a firefighter's personal equipment can be a valuable survival tool. Tubing with a 3/4-inch inside diameter fits over most cylinder threads. It should measure at least 36 inches long.

The following procedure should be followed to perform facepiece-to-facepiece buddy breathing **(Skill 5-2)**:

1. The rescuer is to place one end of the flexible plumbing tubing inside of his facepiece. If a nosecup is utilized inside the facepiece, the tubing must be placed into the nosecup **(Skill 5-2A)**.
2. The opposite end now needs to be inserted into the other firefighter's facepiece in the same fashion. Both firefighters will be able to breathe off the rescuer's unit at this point.
3. The gaps produced in both facepiece seals should be covered by the firefighter's gloved hand. There will still be a slight gap but the positive pressure from the functioning SCBA will be enough to keep out smoke as the firefighters exit the building. The functioning unit's purge valve or bypass can be utilized if more air is needed **(Skill 5-2B)**.

If a firefighter is alone for some unforeseen reason, this tubing can also be utilized as a last resort for escape. By having one end of the tubing placed inside an inner pocket of a turnout coat, the other end can be inserted into the facepiece. The firefighter will not be supplied with clean air but this will filter the smoke to some degree and provide a last chance for escape **(Figure 5-7)**.

SKILL 5-2

Facepiece-to-Facepiece Buddy Breathing

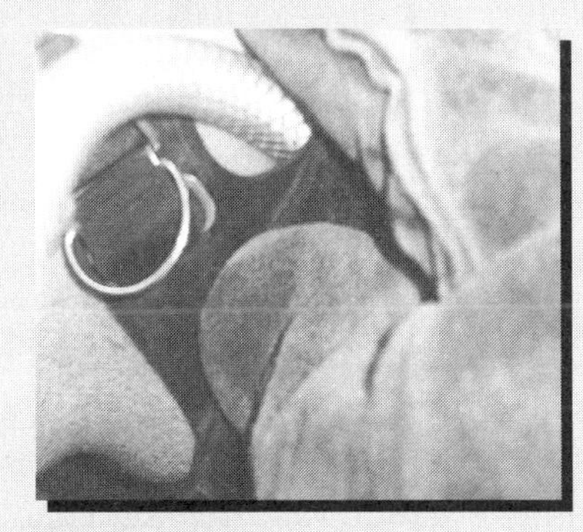

A The rescuer is to place one end of the flexible plumbing tubing inside of his facepiece and the other into the affected firefighter's mask.

B Firefighters demonstrating facepiece-to-facepiece buddy breathing. Their gloved hands are utilized to seal the gaps where the tubing enters their masks.

FIGURE 

Although it will not provide a fresh air supply, flexible tubing placed inside of the turnout coat as a filter may provide a firefighter with the time needed to successfully escape.

"Robbing the Bottle"

One of the most frightening experiences that exists in any firefighter's mind is the thought of becoming disoriented in a fire building and running out of air. From air consumption testing, it has been demonstrated that a firefighter can still have a good number of breaths left from the breathing apparatus after the low-pressure alarm ceases. Air in addition to this can still be available to a firefighter even after air stops being delivered to the firefighter's facepiece. Remaining air in the cylinder will not have the required pressure to make it through the SCBA components to be delivered through the facepiece. A firefighter will have to get this air directly from the cylinder itself. This air will only be a small amount but it may be all that a firefighter needs to make it to a window, door, stairwell, or other means of escape.

This method is only utilized as a definite last resort.

One method in which a firefighter can access this air, if wearing a unit with a mask mounted regulator, is **(Skill 5-3):**

1. Remove the SCBA unit from the back, leaving the facepiece in place.
2. The cylinder valve is then turned to the off position and the cylinder is removed from the high-pressure connection **(Skill 5-3A).**
3. Flexible plumbing tubing is then pushed over the threads of the air cylinder.
4. The firefighter removes the mask-mounted regulator from the facepiece and inserts the opposite end of the plumbing tubing into the facepiece and his mouth **(Skill 5-3B).**
5. The firefighter can now open the cylinder valve slightly to allow some air into the tubing. The firefighter should inhale this air through the mouth and slowly exhale through the nose, closing the cylinder valve in between breaths.

Restrictive Area Techniques

Firefighters caught in a collapse or cut off from their means of egress may have to fit through a tight spot to get themselves to safety while wearing their SCBA **(Figure 5-8).** Whenever going through obstacles, it is paramount that firefighters make certain that conditions on the other side are safe. It is also highly recommended that firefighters train on these techniques with their firefighting gloves in place so that they become confident in working with the different components of their SCBA in less than ideal conditions.

A conventional way for firefighters to get through an obstacle is to simply shift their SCBA to the left side, allowing protection of the regulator or pressure reducer. It will also allow more freedom of movement for the firefighter because

SKILL 5-3
Robbing the Bottle

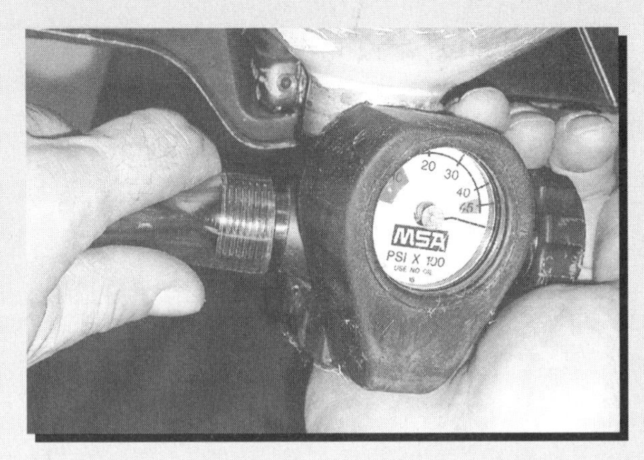

A The cylinder valve is turned off and flexible tubing is placed over the cylinder threads.

B The firefighter removes the mask-mounted regulator from the facepiece and inserts the opposite end of the plumbing tubing into the facepiece and his mouth. Air flow is controlled by turning the cylinder valve on and off.

lines from these devices to the facepiece are located on the left side of the SCBA on most units.

To shift the SCBA (Skill 5-4):

1. The right-side shoulder strap is loosened first, followed by the waist strap being extended **(Skill 5-4A).**
2. The SCBA is now shifted over to the left as far as possible, allowing the firefighter's profile to be reduced **(Skill 5-4B).** It may be necessary for a larger-framed firefighter to remove the right arm completely from the shoulder strap to accomplish this maneuver **(Skill 5-4C).**
3. After the obstruction is cleared, the SCBA is shifted back into the center of the back and the straps are tightened once again.

FIGURE 

With proper technique, a firefighter can easily fit between two wall studs. Notice the placement of the firefighter's left hand over the neck of the cylinder valve. This placement allows for protection of the valve when proceeding through obstacles.

SKILL 5-4
To Shift the SCBA

A The right-shoulder strap is loosened first, followed by the waist strap being extended.

B The SCBA is now shifted over to the left as far as possible, allowing the firefighter's profile to be reduced.

C It may be necessary for a larger-framed firefighter to remove his right arm completely from the shoulder strap to accomplish this maneuver.

Another method of a firefighter fitting through a tight space such as wall studs with a SCBA is to "swim" through the obstacle backwards. Advantages of this technique are that a firefighter is not required to remove or loosen any part of the SCBA, which can slow an escape considerably.

To employ the backwards "swim" (Skill 5-5):

1. The firefighter seats himself with his back against the wall and with the SCBA centered between two obstacles. The firefighter's feet should be positioned out in front of the body as far as possible while the buttocks are raised off the ground **(Skills 5-5A and 5-5B).** This is the key to proper use of this technique.
2. The firefighter then pulls his shoulders and elbows inward toward the center of their body while pushing backwards through the space.
3. Once the neck of the SCBA cylinder is clear of the wall stud, the firefighter should rotate his body and employ a "swim"-type maneuver to bring the rest of

SKILL 5-5

To Employ the Backwards "Swim"

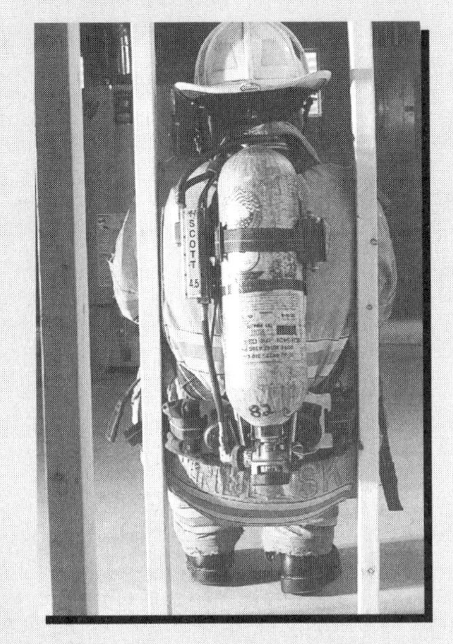

A To perform the backward "swim" through a wall, the firefighter sits with his back against the wall with the SCBA bottle centered between the stud space.

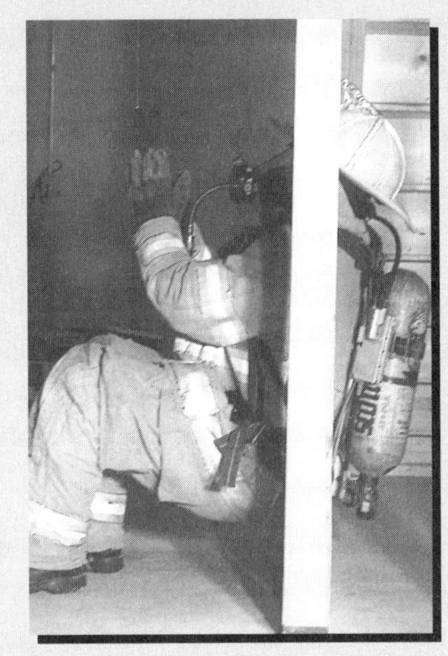

B The firefighter's feet should be out in front and the buttocks should be raised off the ground.

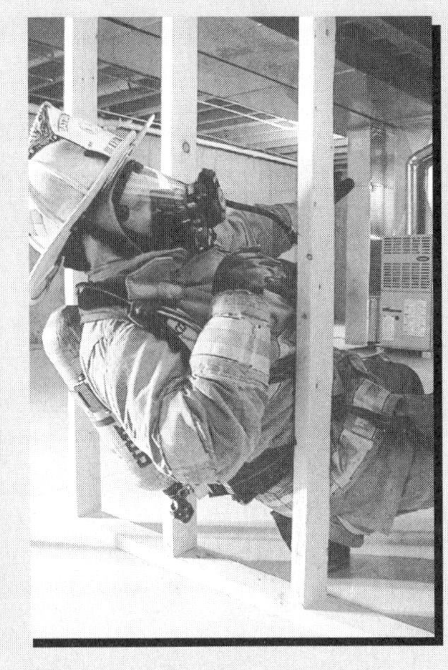

C As the firefighter moves backwards, he will begin to slightly rotate his body.

D As the firefighter rotates his body, he will reach his arm over in a swimming motion to pull him through the obstacle.

his body through the obstacle **(Skills 5-5C and 5-5D).**

Another simple method for a firefighter to reduce his profile to clear an obstacle is to employ a *forward dive* technique.

To execute the forward dive (Skill 5-6):

1. Centering himself between the wall studs or obstacle, the firefighter places both arms in front through the obstacle.

SKILL 5-6

To Execute the Forward Dive

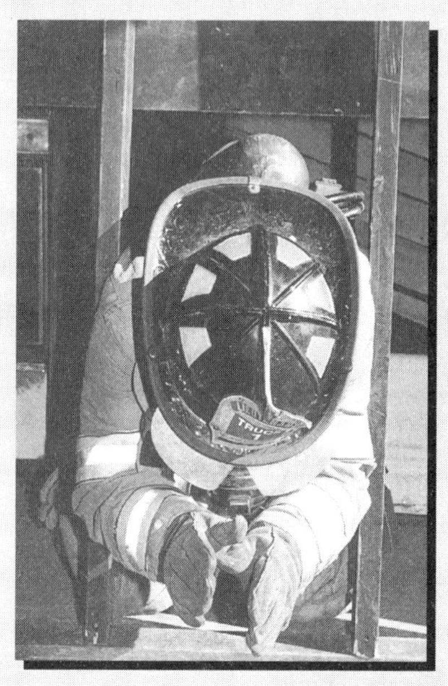

A As the firefighter exhales, he pulls his arms inward toward the center of his body, allowing them to fall forward through the obstacle when utilizing the forward dive.

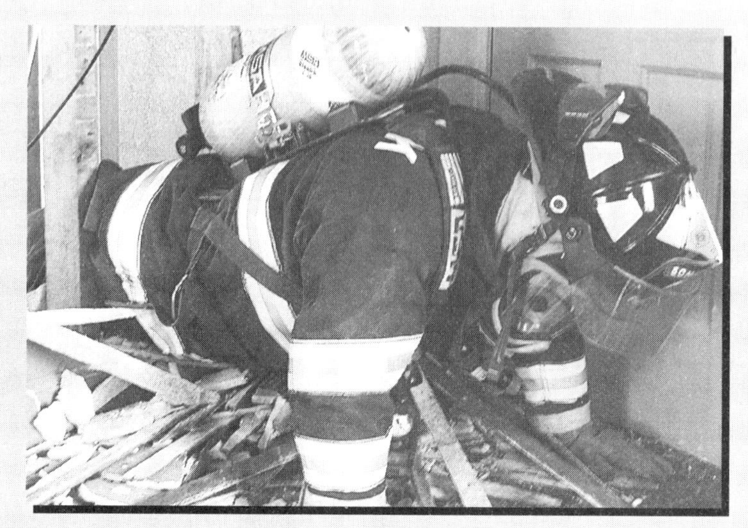

B Once the SCBA is clear of the obstacle, the firefighter can use his arms to pull himself the rest of the way through.

2. As the firefighter exhales, he pulls his arms inward toward the center of his body, allowing them to fall forward through the obstacle **(Skill 5-6A).**
3. Once the SCBA is clear of the obstacle, he can use his arms to pull himself the rest of the way through **(Skill 5-6B).**

Firefighters may have to get beneath an obstacle to facilitate their escape, which may also require them to lower their profile **(Figure 5-9).** The SCBA can be shifted to the left side by loosening the shoulder straps just as if the firefighter were advancing in an upright position **(Figure 5-10).**

Safety

The SCBA should not be removed from the firefighter's back unless absolutely necessary.

FIGURE  5-9

Firefighters may have to navigate through tight spots that may not allow them to be off the floor in their escape efforts.

FIGURE 5-10

The SCBA should be shifted to the left side just the same as the upright position when a firefighter must go underneath an obstacle to escape.

In a very extreme circumstance, a firefighter may have to resort to removing the SCBA from the body to facilitate clearing an obstacle. A firefighter must be very cautious when removing the SCBA as it further complicates the nature of the situation that the firefighter is presented with. To remove the SCBA in a constricted area or if heat conditions dictate, a firefighter should be as low as possible and the following steps can be utilized.

Removing the SCBA (Skill 5-7):

1. While lying to the side, the firefighter will need to loosen the shoulder straps and remove the waist belt of the SCBA.
2. The firefighter should next "roll" out of the pack by rolling over to the left. Going to the left will allow the firefighter increased freedom of movement. **(Skill 5-7A).**
3. To further facilitate freedom of movement, the SCBA should be rotated so that the cylinder valve is facing away from the firefighter. All straps will need to be placed in a neat, organized manner on top of the SCBA to facilitate an easy redonning **(Skill 5-7B).**

The firefighter should then move with the SCBA in front but keeping it close to the body to protect it and prevent the facepiece from being pulled off. It is imperative that the firefighter makes certain that no holes or elevation changes

SKILL 5-7

Removing the SCBA

A This firefighter is rolling out of the SCBA with it to the left. The straps are loosened and the firefighter rolls out of them, leaving the cylinder on his left side. Note how the straps of the SCBA are fully extended in a ready position.

B This firefighter is clearing the obstacle after removing the SCBA. He is moving the pack in front of him but keeping it close enough that it can be protected. The SCBA straps are placed neatly on top of the pack to facilitate a quick and efficient redonning.

exist in the floor when moving forward. When clear of the obstacle, the firefighter can redonn the SCBA by laying out the straps and rolling back into the pack.

Entanglement Hazards

With the increasing technological advances in communication systems and networking in today's world also comes an increased hazard to firefighters. Numerous cables and wires are run through **false** or **drop ceilings** to facilitate these various systems **(Figure 5-11).** Many case studies have documented firefighters becoming trapped or entangled in these types of wires or cables when these ceilings collapse. The entanglement often occurs in the area of the firefighter's SCBA. A **V-type** (failure of an interior support resulting in void spaces on either side of the collapse) or **lean-to collapse** (collapse in which supports fall to one side) will often occur when a drop ceiling collapses. For this reason, most wires and cables will be located away from the walls toward the center of the room. If possible, it is recommended that firefighters position themselves face up with their SCBAs against a wall to navigate through an entanglement obstacle. This causes the body to protect the SCBA from entanglement while enabling firefighters to use their hands and allows them to see the obstacle as they go through it **(Figure 5-12).**

If a firefighter becomes entangled it is imperative to stop forward movement—trying to "muscle" forward will only pull an entanglement tighter on the SCBA and might wedge it into a position on the SCBA that is unreachable. It is strongly recommended that firefighters carry a knife or wire cutters to free themselves from an entanglement **(Figure 5-13).** When cutting wires

FIGURE 5-12

Firefighters should navigate through entanglement hazards face up with the SCBA protected to avoid wires being entangled in a spot that may not be visible or reachable.

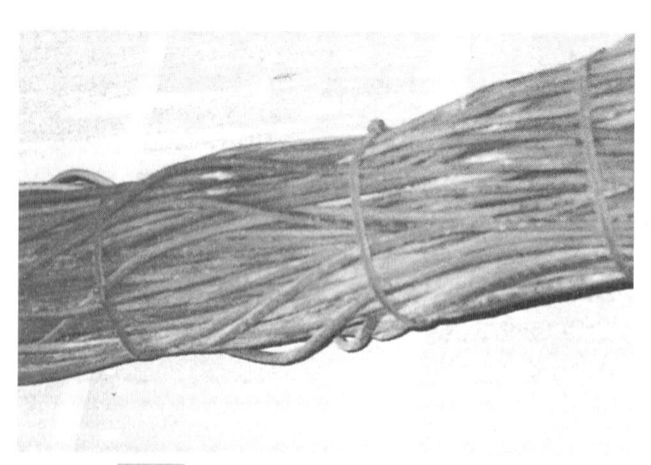

FIGURE 5-11

False or drop ceilings are often covering large amounts of wires used for communication and television. The wire hangers and gridwork also pose a serious entanglement hazard if the ceiling collapses.

FIGURE 5-13

All firefighters should carry a pair of wire cutters as part of their personal equipment to cut their way out of entanglement hazards. Cutters should be large enough to be manipulated with a gloved hand and durable enough to cut through the materials encountered.

during an entanglement, it is important that only one wire at a time is cut to avoid possible injury in the case that utilities have not been fully controlled. For this same reason, firefighters should brush wires out of their path with an open hand or the back of their hand **(Figure 5-14).** A firefighter may be able to navigate through an entanglement by employing a "swimming" technique or a "football" carry technique.

To perform the swim method for entanglement (Skill 5-8):

1. Identify the entanglement. Stop forward movement. Try to get low and possibly back out of the hazard **(Skill 5-8A).**
2. A firefighter should position himself face up with his SCBA against a wall if at all possible.
3. While slowly moving forward, the firefighter should reach one arm across his body and slide an open hand across the floor surface, catching any entanglement hazards. This arm should then be lifted to hold the hazard above the firefighter while moving under it **(Skill 5-8B).**
4. The firefighter's opposite arm should now be slid across the floor, repeating the same motion, holding any additional entanglement hazard above the firefighter **(Skill 5-8C).**
5. The first arm can now be brought down to the waist and the technique can be repeated. Firefighters should work together as a team to navigate the hazard if possible **(Figure 5-15).**

If already entangled, a firefighter can attempt to free himself by rotating and executing a variation of the "swim" to reach and remove a hazard.

To perform the reach and "swim" (Skill 5-9):

1. Once the firefighter has identified the entanglement and stopped forward motion, he can initiate a full forward "swim" motion by dropping an arm down to his waist and bringing it across his back as far as he can reach **(Skill 5-9A).**
2. If unsuccessful, he should try the same technique with the opposite arm.
3. If this is still not successful, he should rotate his body about one-quarter of the way to its side and attempt these same motions again. It is important that the firefighter does not roll too far in one

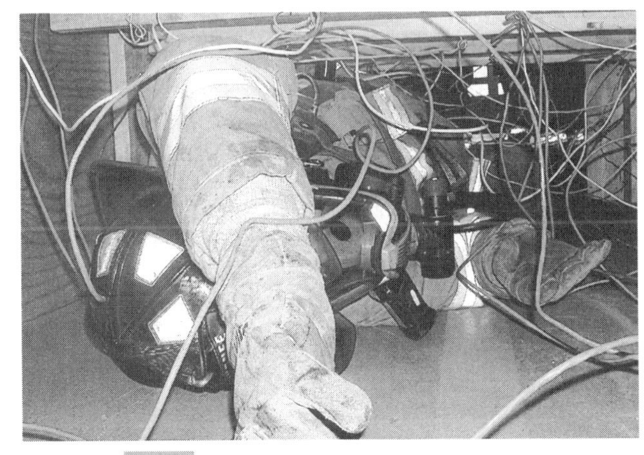

FIGURE 5-14

Firefighters should brush wires or cables out of their path with an open hand and avoid grasping them to avoid harm in case the utilities are not controlled.

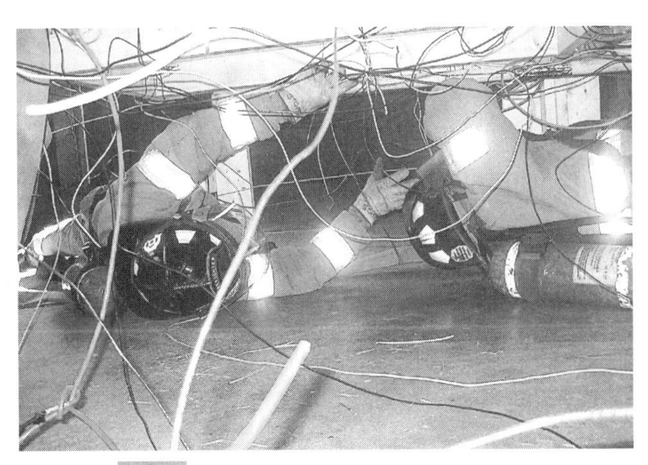

FIGURE 

Firefighters can work as a team to navigate their way through an entanglement by holding the hazard up so that the other firefighter can maneuver beneath it. A tool such as a pike pole, halligan bar, or even a hoseline can be also utilized to help control the entanglement hazard by lifting it up.

SKILL 5-8

To Perform the Swim Method for Entanglement

A Stop forward movement. Try to get low and possibly back out of the hazard.

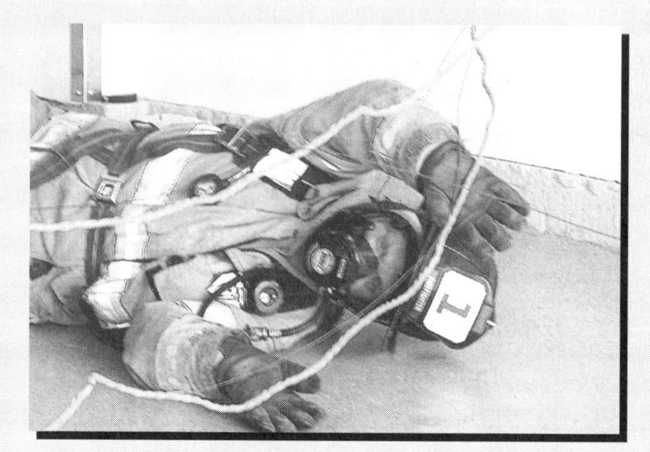

B While slowly moving forward, the firefighter should reach one arm across the body and slide an open hand across the floor surface, catching any entanglement hazards. This arm should then be lifted to hold the hazard above the firefighter while moving under it.

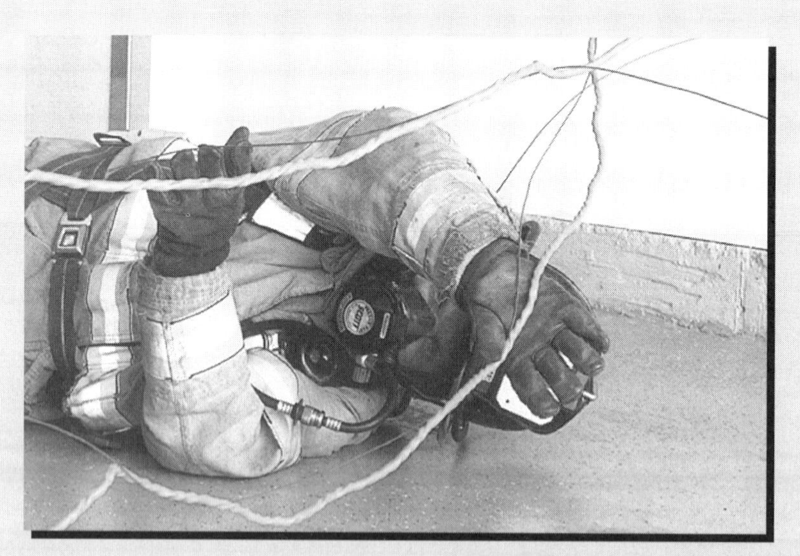

C The firefighter's opposite arm should now be slid across the floor, repeating the same motion, holding any additional entanglement hazard above the firefighter.

SKILL 5-9

To Perform the Reach and "Swim"

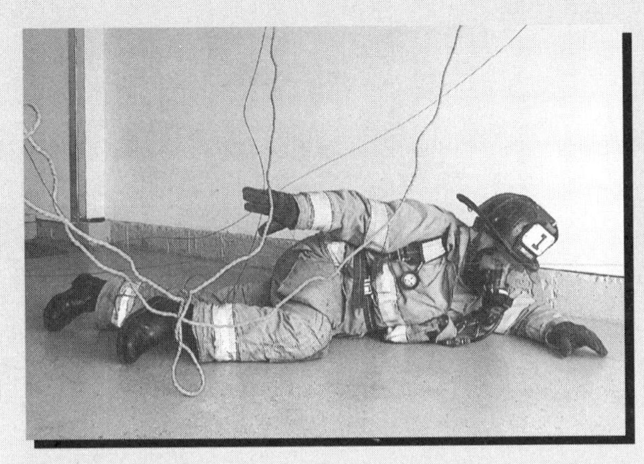

A Once the firefighter has identified the entanglement and stopped forward motion, he can initiate a full forward "swim" motion by dropping an arm down to his waist and bringing it across his back as far as he can reach.

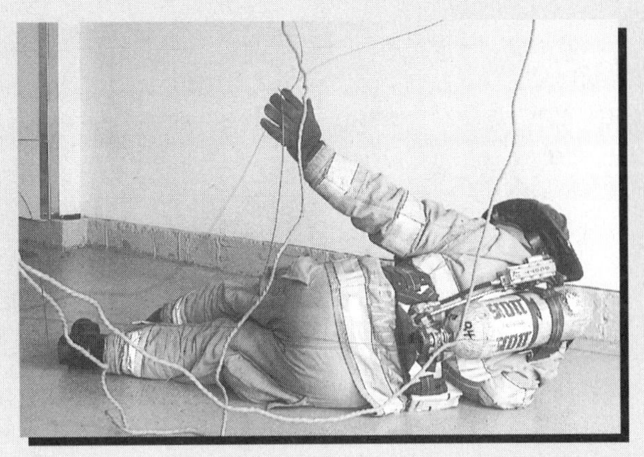

B If this is still not successful, he should rotate his body about one-quarter of the way to its side and attempt these same motions again. It is important that the firefighter does not roll too far in one direction, as this could cause the entanglement to wrap even tighter around the firefighter.

direction as this could cause the entanglement to wrap even tighter around the firefighter. This sequence should be repeated until contact is made with the entanglement and it is able to be removed **(Skill 5-9B)**.

Performing the "football" carry technique for entanglement (Skill 5-10):

1. Identify the entanglement hazard.
2. Stop forward motion. Sometimes just stopping forward motion and backing up slightly may disengage an encountered entanglement.
3. The firefighter loosens up the shoulder straps of the SCBA, removes the pack, and places his arms back into the shoulder straps that were opposite. The SCBA will be on in front of his torso like a papoose **(Skill 5-10A)**.
4. The firefighter lies on his left side, tucking the SCBA into his body as much as possible and proceeds through the entanglement with his arms crossed over, protecting the SCBA. If needed, one arm can be used to "swim" hazards up and out of the way **(Skill 5-10B)**.

An advantage of the "football" carry technique is that it allows a firefighter the ability to protect the areas that are vulnerable to getting tangled. It also gives the firefighter the ability to reach the high-probability entanglement areas that are quite difficult to reach with the SCBA on the back.

Toxic or "Hot Bottle" Change

A toxic bottle change consists of a firefighter doffing the SCBA unit from the back with the facepiece still in place and changing over to a full SCBA cylinder. Changing an air cylinder inside a toxic environment is not a practice that should be routinely done. If a situation does arise, such as a firefighter becoming trapped in a collapse where he cannot be immediately removed, a firefighter should be able to save himself with this

SKILL 5-10

Performing the "Football" Carry Technique for Entanglement

A The firefighter loosens the shoulder straps of the SCBA, removes the pack, and places his arms back into the shoulder straps that were opposite. The SCBA will be on in front of the torso like a papoose.

B The firefighter lies on his side, tucking the SCBA into his body as much as possible and proceeds through the entanglement with his arms crossed over, protecting the SCBA.

maneuver if a good cylinder can be accessed. Practicing toxic bottle changes is a great drill for firefighters to gain confidence in their SCBA and become familiar with its working components. It is not an easy skill to master and it must be repeated from time to time to maintain proficiency. Firefighters will have to experiment with their particular SCBA to find the procedure that works best with the units that they use.

To perform the toxic or "hot bottle" change (Skill 5-11):

1. Secure a site for the bottle change that is as safe as possible.
2. The firefighter should double-check that the bottle he will be changing over to is in fact a good bottle **(Skill 5-11A).**
3. Remove the SCBA while keeping the harness straps in an order that will facilitate easy redonning of the apparatus without getting it tangled **(Skill 5-11B).**
4. With the unit in front and the new bottle secured, the firefighter should attempt to calm himself down and gain control of his breathing. If advantageous, the firefighter can undo any latches or straps that hold the cylinder to the backpack assembly **(Skill 5-11C).**
5. When ready, the firefighter should turn off the cylinder valve and breathe down any remaining air inside of the high-pressure line of the unit. Once this is done the cylinder can be disconnected and the new

SKILL 5-11

To Perform the Toxic or "Hot Bottle" Change

A The firefighter should double-check that the bottle to be changed over to is in fact a good bottle.

B Remove the SCBA while keeping the harness straps in an order that will facilitate easy redonning of the apparatus without getting it tangled.

C With the unit in front and the new bottle secured, the firefighter should attempt to calm himself down and gain control of his breathing. If advantageous, the firefighter can undo any latches or straps that hold the cylinder to the backpack assembly.

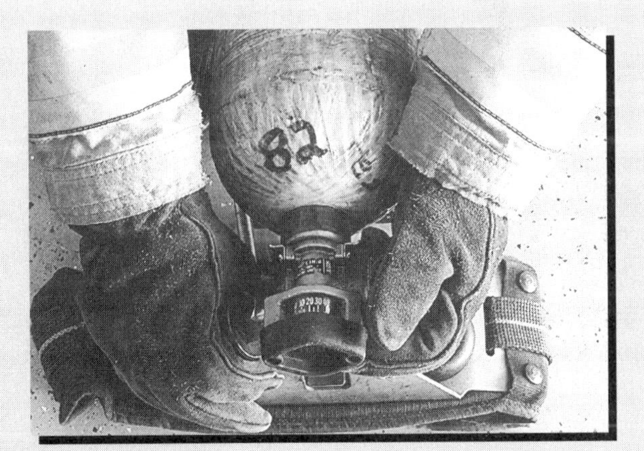

D When ready, the firefighter should turn off the cylinder valve and breathe down any remaining air inside of the high-pressure line of the unit. Once this is done the cylinder can be disconnected and the new cylinder can be threaded on to the high-pressure line.

cylinder can be threaded on to the high-pressure line **(Skill 5-11D)**. It is very important that the high-pressure line be completely threaded on prior to opening the new bottle cylinder valve. If it is not, the possibility exists that the O-ring on the high pressure line can get blown and render the SCBA useless until repaired.

As firefighters get proficient at this skill, they can increase the difficulty level by performing it with their vision obscured. This procedure may prove to be easier to perform under stressful conditions if a firefighter's partner is able to change the bottle for him. This will allow the firefighter that needs air to relax and to lie flat on the floor **(Figure 5-16)**. By lying flat on the floor, the firefighter's chest will be able to expand, allowing a greater volume of air on the last breath before the air supply is shut down. For this same reason, a firefighter should try to keep the upper body in an upright position as much as possible when performing this maneuver alone.

FIGURE 5-16

A two-firefighter "toxic" bottle change may prove to be easier to perform under stressful conditions if a firefighter's partner is able to change the bottle.

A system of communicating will need to also be worked out between the firefighters. This should be done in training prior to having to perform this maneuver in an emergency. Another way to perform this task with two firefighters is to employ buddy breathing, such as the emergency breathing support system (if equipped) or the facepiece-to-facepiece method, while the firefighter's bottle is changed. This eliminates the need for the firefighter to be without air at any time.

Summary

The fireground is a very dynamic and dangerous environment. Every firefighter, from the most seasoned veteran to the greenest rookie, can experience a problem with their SCBA. Emergency procedures involving SCBAs should be second nature to all firefighters. As with all aspects of firefighting, practice and training prior to an emergency will increase a firefighter's chance for survival.

Firefighters should be trained in and familiar with techniques used to prolong air supply duration, their own personal air consumption rate, emergency procedures check for SCBA equipment, techniques used to share air between two firefighters, and what to do if the SCBA runs out of air.

In addition to these skills, firefighters should also be trained in techniques used to escape from restricted areas and overcome obstacles and entanglements.

As noted, self-preparedness is key to firefighter survival. During an emergency is not the time to develop self-survival skills.

KEY TERMS

Buddy breathing
Emergency breathing support system (EBSS)
Emergency procedure check
False or drop ceiling
Lean-to collapse
Low-pressure/25% alarm
Point of no return
Skip breathing
V-type collapse

REVIEW QUESTIONS

1. What are the four factors that a firefighter's working air supply will depend on?
    1.
    2.
    3.
    4.
2. A simple method for determining a point of no return is to ______________________________ ______________________________ ______________________________.
3. To get a true indication of air consumption, the obstacle course should be set up differently each time.
   True False
4. Inhaling through the mouth and exhaling through the nose while wearing an SCBA will help provide adequate air exchange.
   True False
5. Skip breathing in an emergency can extend a rated 30-minute SCBA to over an hour if employed properly.
   True False
6. Minor failures of SCBAs are often the result of
    a. operator error.
    b. improper maintenance.
    c. make or manufacturer of the unit.
    d. both (a) and (b)
7. Firefighters should remove their facepiece immediately upon running out of air to prevent suffocation.
   True False
8. It is recommended to shift the SCBA to the left side when navigating through obstacles to allow __________ and __________.
9. In a V-type or lean-to collapse, most entanglement hazards will be located
    a. against the walls.
    b. toward the center of the room.
    c. both (a) and (b)
    d. neither (a) nor (b)
10. If trapped in an entanglement, a firefighter should
    a. grasp the wires and move them out of the way.
    b. cut as many wires at once to free himself quickly.
    c. pull his way through to break the wires as he advances.
    d. stop forward movement.

ADDITIONAL RESOURCES

Firefighter's Handbook (2nd ed.). Clifton, NY: Thomson Delmar Learning, 2004.

McCormack, J., and Pressler, R., *Firefighter Survival.* Indianapolis, IN: Fire Department Training Network, 2002.

Rapid Intervention Teams and How to Avoid Needing Them. Technical Report Series, Report #123. Washington, D.C.: U.S. Fire Administration, 2003.

Thiel, A., *Prevention of Self- Contained Breathing Apparatus Failures.* Special Report, United States Fire Administration, March 1998.

CHAPTER

6 Ropes, Knots, Anchoring, and Basic Mechanical Advantage for Rapid Interventions

Learning Objectives

Upon completion of this chapter, you should be able to:

- identify, tie, and explain the use of basic knots and hitches such as:
  - the figure eight on a bight
  - the tensionless hitch
  - the butterfly knot
  - the handcuff knot
  - the bowline knot
  - the Prussik or rescue loop
  - the Munter hitch
  - the water or tape knot
- assemble the modified diaper harness to be used for rescue.
- identify and establish suitable anchor points for rapid intervention company use such as:
  - a wall stud anchor without a tool
  - a wall stud anchor with a tool
  - a corner window anchor
  - a halligan floor/roof anchor
  - fixed and nonfixed object anchors
  - a body anchor
- explain the basic principles of a mechanical advantage system.
- demonstrate three different methods of assembling a 2-to-1 mechanical advantage system for rapid intervention company use.

Case Study

"On March 8, 2001, a 38-year-old male career firefighter (the victim) fell through the floor while fighting a structure fire, and died 12 days later from his injuries. At 1231 hours, Central Dispatch notified the career department of a structure fire with reports of the occupants still inside. The assistant chief arrived on the scene along with Engine 70 and assumed incident command (IC). The IC immediately called for the second alarm, began conducting the initial size-up of the structure, and confirmed heavy fire in the left front section. At that time, the neighbors approached the IC and informed him that the occupants were trapped inside. The IC ordered the firefighters on scene to commence search and rescue efforts, and then verified the stability of the structure through radio and face-to-face communications.

"Engine 68 arrived on the scene at approximately 1250 hours with an assistant chief and the victim. The assistant chief provided tactical command of the fireground, and along with the victim, conducted search and rescue operations. Other crews conducted searches with a thermal imaging camera of the first floor and basement level of the residence with no sign of any occupants. During these searches, the stability of the structure was diminishing due to the intense fire that was now venting through the roof.

"Firefighter #3 and the victim were at the front entrance conducting a defensive attack as the third emergency evacuation signal was sounded. The neighbors were still insisting to the IC and firefighters that the occupants were trapped inside, and one of the occupants was handicapped. The victim and one other firefighter conducted another search of the structure. The heat and flames were now extending from the basement level to the first floor when the firefighter's low-air alarm sounded. The victim and the firefighter were backing out of the structure when the floor beneath the victim gave way, causing him to fall through the floor and become trapped in the basement. Attempts were made from the first floor to rescue the victim by utilizing a handline and an attic ladder, but they were unsuccessful due to the intense heat and flames. Two rapid intervention teams were deployed simultaneously from separate entrances into the basement to perform a search and rescue operation for the downed firefighter. The RITs were able to locate and remove the victim on their initial entry. He sustained third degree burns to over half of his body and died 12 days later."

—Case Study taken from NIOSH Firefighter Fatality Report #2001-16,* "Career Firefighter Dies after Falling through the Floor Fighting a Structure Fire at a Local Residence—Ohio," *full report available online at http://www.cdc.gov/niosh/face200116.html.

Introduction

A RIT will be required to know and understand basic knots and concepts of mechanical advantage systems to be able to be successful in the performance of its duties. These concepts and knots will have to be kept as simple as possible due to the fact that they will be utilized under some of the most extreme conditions and time constraints. It must be realized that a quick retrieval of a downed firefighter is not a technical rescue and must not be treated as such—if a firefighter is down and conditions are deteriorating, he must be removed quickly to preserve his life. A lot of firefighters are intimidated by this subject area; this intimidation is unwarranted. Knot tying and basic mechanical advantage concepts is an essential subject area that all firefighters should be trained in. Before moving into the more complex areas and techniques associated with rapid intervention, it is paramount that the firefighter has a thorough understanding of this area as well as other essential basic firefighting skills.

Knot Tying and Basic Mechanical Advantage

Although there are numerous knots that are acceptable for use by the RIT, we will introduce what we feel are the most versatile, simplest to tie, and easiest to remember. It is also worthwhile to mention that the techniques introduced are not the only acceptable ways of tying these knots—some knots may have as many as five different ways of tying them. Find the technique that is easiest for you to perform. The important thing is, make certain that the finished knot is indeed correct, set, dressed, and always backed up with a safety to prevent slipping. Any rope, harnesses, or hardware utilized to support a firefighter in a life-safety situation must meet the requirements set forth by **NFPA 1983** (Standard on Fire Service Life Safety Rope and System Components). This chapter is not meant to be all-inclusive of the aspects associated with the NFPA standard or performing a high-angle rescue. It is strongly suggested that readers become familiar with the NFPA standard as well as basics such as rope types, construction, characteristics, uses, and maintenance if they are to function as part of the RIT.

Figure Eight on a Bight

The **figure eight on a bight** is a simple knot that is used extensively in many of the maneuvers that a RIT may have to carry out. It is advantageous because it allows a loop to be placed into a rope at any point and it will not slip.

Tying the figure eight on a bight (Skill 6-1):

1. Form a bight in the rope and cross it over the standing part of the rope **(Skill 6-1A).**
2. The bight is then wrapped back over the standing part from behind **(Skill 6-1B).**
3. The bight is then fed through the "eye" that was formed in step 1. The knot should then be set and properly dressed **(Skill 6-1C).**
4. Set the knot and back up the figure eight on a bight with a safety **(Skill 6-1D).**

Tensionless Hitch

The **tensionless hitch** can be used to establish anchor points **(Skill 6-2).** It consists of a figure eight on a bight wrapped around the object to be used as an anchor. It is recommended that it be wrapped around the object a minimum of four times. The figure eight is left to hang

SKILL 6-1

Tying the Figure Eight on a Bight

A Form a bight in the rope and cross it over the standing part of the rope.

B The bight is then wrapped back over the standing part from behind.

(continued)

SKILL 6-1 (CONTINUED)

Tying the Figure Eight on a Bight

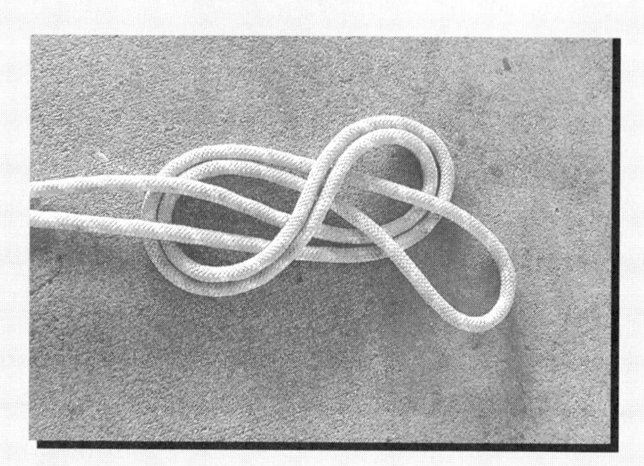

C The bight is then fed through the "eye" that was formed in step 1. The knot should then be set and properly dressed.

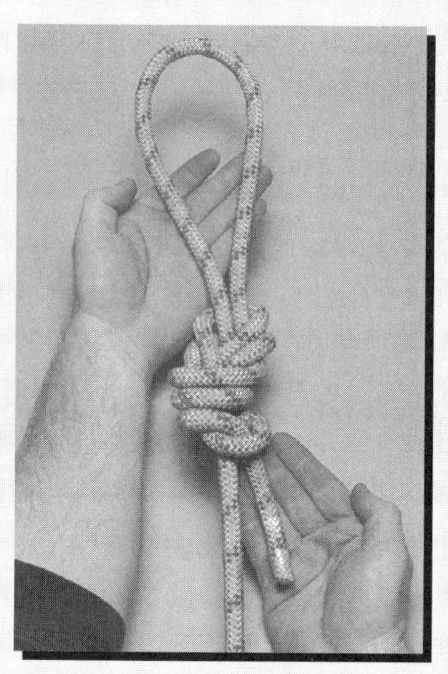

D Set the knot and back up the figure eight on a bight with a safety.

SKILL 6-2

Tensionless Hitch

A Tensionless hitch around a ladder rung.

freely, where a carabiner will be attached to it **(Skill 6-2A)**.

Butterfly Knot

The **butterfly knot** is used when a three-directional pull or attachment point is needed in a rope. This may be encountered when conducting large-area searches within the rapid intervention environment.

Tying the butterfly knot (Skill 6-3):

1. Lay a loop of the rope loosely over the hand **(Skill 6-3A)**.
2. Lay a second loop over the hand, crossing over the first loop **(Skill 6-3B)**.
3. A third loop is formed and positioned parallel to the second loop **(Skill 6-3C)**.
4. Slack is pulled from the second loop, forming a bight **(Skill 6-3D)**.
5. The bight is then pulled beneath the other two remaining loops **(Skill 6-3E)**.
6. The running and standing parts of the rope are pulled to set the knot **(Skill 6-3F)**.

SKILL 6-3

Tying the Butterfly Knot

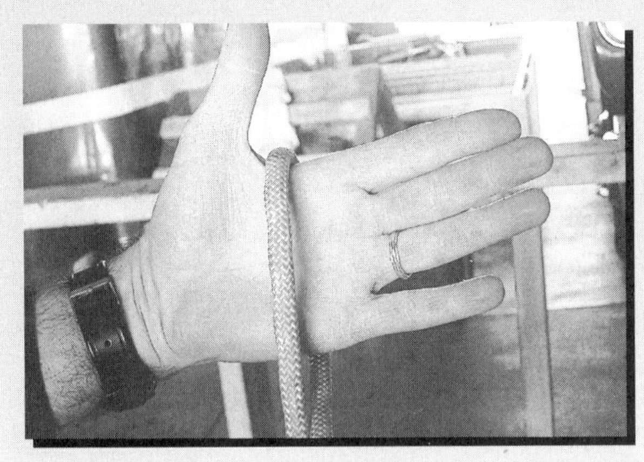

A Lay a loop of the rope loosely over the hand

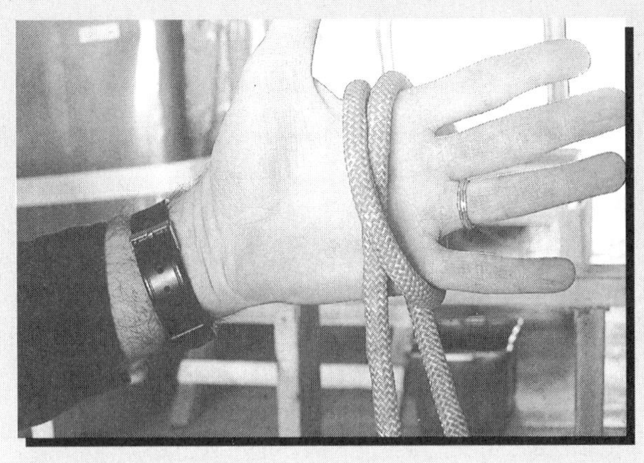

B Lay a second loop over the hand crossing over the first loop.

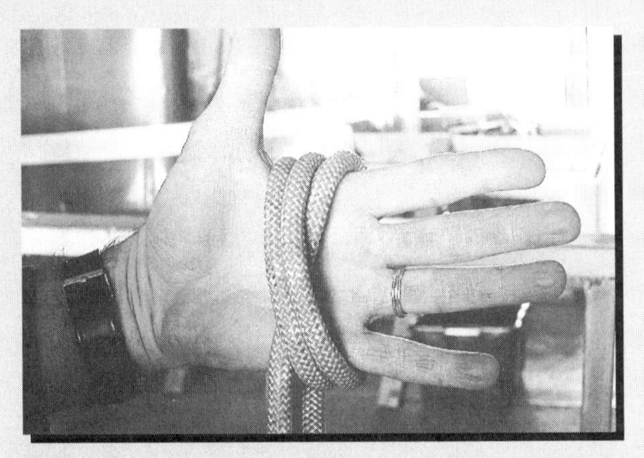

C A third loop is formed and positioned parallel to the second loop.

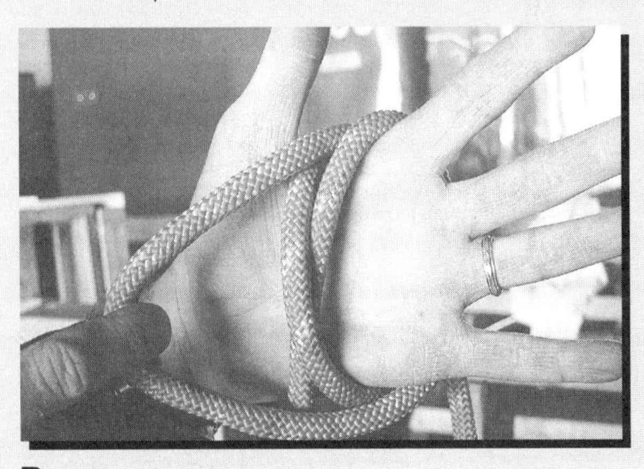

D Slack is pulled from the second loop, forming a bight.

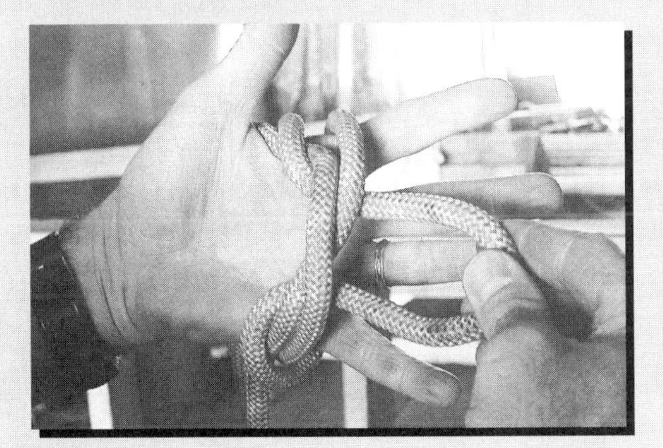

E The bight is then pulled beneath the other two remaining loops.

F The running and standing parts of the rope are pulled to set the knot.

SKILL 6-4
Tying the Handcuff Knot

A Two loops opposite of each other are formed in the rope.

B The two loops are passed to the inside of each other.

C The two loops are passed to the inside of each other.

D The loops need to be pulled through to be sized large enough to fit over the downed firefighter's forearms.

E The handcuff knot needs to be placed on the firefighter's forearms where it will cinch down on the turnout coat and avoid causing injury to the wrists.

The Handcuff Knot

The **handcuff knot** is one of the most important knots for firefighters to know. Its greatest use is to pull firefighters to safety when trapped in below-grade situations. This knot is often times applied in the wrong manner, causing further injury to a downed firefighter. The handcuff knot needs to be placed on the firefighter's forearms where it will cinch down on the turnout coat and avoid causing injury to the wrists. Advantages of the handcuff knot include its ability to allow a downed firefighter to be raised in a streamlined fashion to fit through a tight or constricted space, which other knots and harnesses may not allow. The knot can also be applied to a firefighter's ankles if necessary.

Tying the handcuff knot (Skill 6-4):

1. Two loops opposite of each other are formed in the rope **(Skill 6-4A).**
2. The two loops are passed to the inside of each other **(Skills 6-4B** and **6-4C).**
3. The loops need to be pulled through to be sized large enough to fit over the downed firefighter's forearms **(Skill 6-4D).**
4. The handcuff knot needs to be placed on the firefighter's forearms where it will cinch down on the turnout coat and avoid causing injury to the wrists **(Skill 6-4E).**

The Bowline Knot

The bowline is a versatile knot that becomes very useful as a tag line or when establishing anchors when life safety is not a factor (such as anchoring a mechanical advantage system to move a firefighter over a large area). A bowline knot may have a tendency to slip when used with some synthetic-fiber ropes. For this reason, it is recommended to use a knot from the figure eight family when dealing with life-safety loads.

Tying the bowline knot (Skill 6-5):

1. A bight and loop are formed in the rope **(Skill 6-5A).**
2. The working end of the bight is fed through the loop **(Skill 6-5B).**
3. The working end is then passed behind the standing part of the rope **(Skill 6-5C).**
4. The working end is then passed back through the loop once again **(Skill 6-5D).**
5. Tighten and dress the knot. Finish the knot with a safety **(Skills 6-5E** and **Skill 6-5F).**

The Prussik or Rescue Loop

A double fisherman's knot applied in a rope upon itself to form a loop, termed the **Prussik or rescue loop,** can be very versatile for moving a downed firefighter. With proper application, it can be used as a grab handle in pulling a firefighter, or when used in conjunction with other rescue loops it can serve as a handle to lift downed firefighters up stairs and over debris.

Tying the Prussik or rescue loop (Skill 6-6):

1. Wrap one end of the rope over the other two times. The second wrap should cross over the first.
2. Pull the working end through the wraps **(Skill 6-6A).**
3. Repeat with the opposite rope end **(Skill 6-6B).**
4. Tighten both sides down to cinch upon each other **(Skill 6-6C** and **Skill 6-6D).**
5. For RIT purposes, a piece of rubberized shrink wrapping can be applied over the knot to provide a gripping surface as well as provide protection for the knot **(Skill 6-6E).**

SKILL 6-5
Tying the Bowline Knot

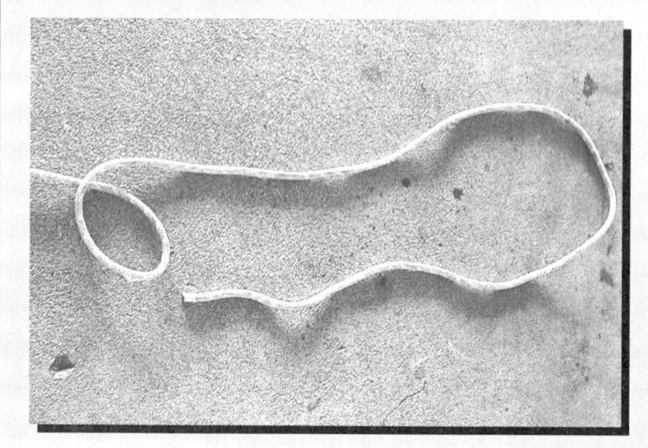

A A bight and loop are formed in the rope.

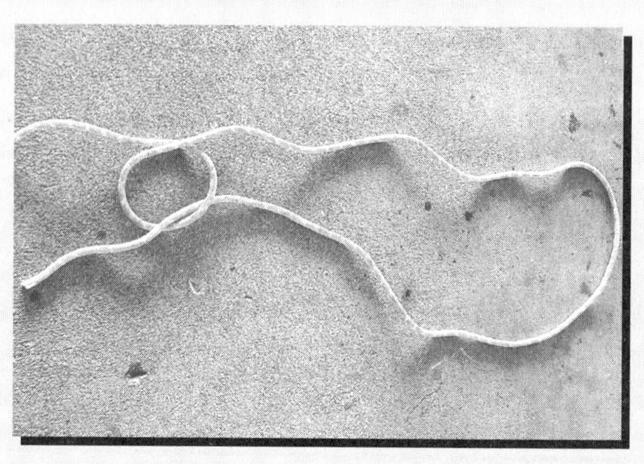

B The working end of the bight is fed through the loop.

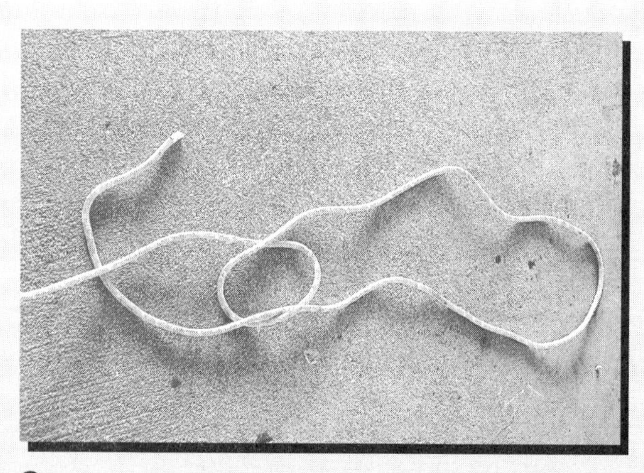

C The working end is then passed behind the standing part of the rope.

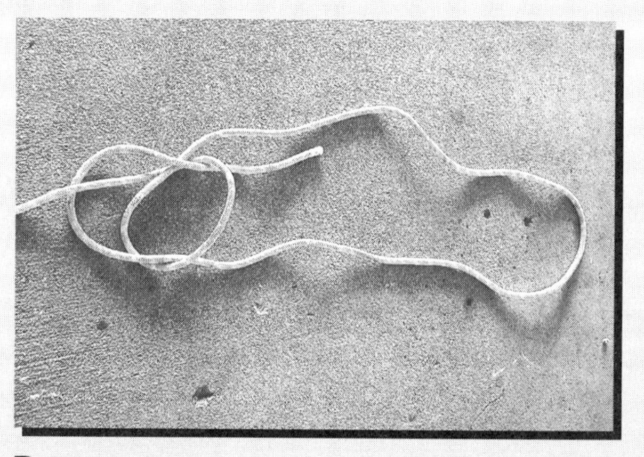

D The working end is then passed back through the loop once again.

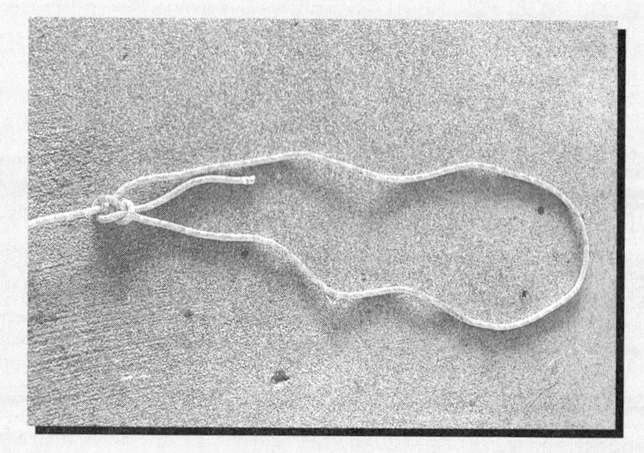

E Tighten and dress the knot. Finish the knot with a safety.

F Close up of properly tied bowline.

SKILL 6-6

Tying the Prussik or Rescue Loop

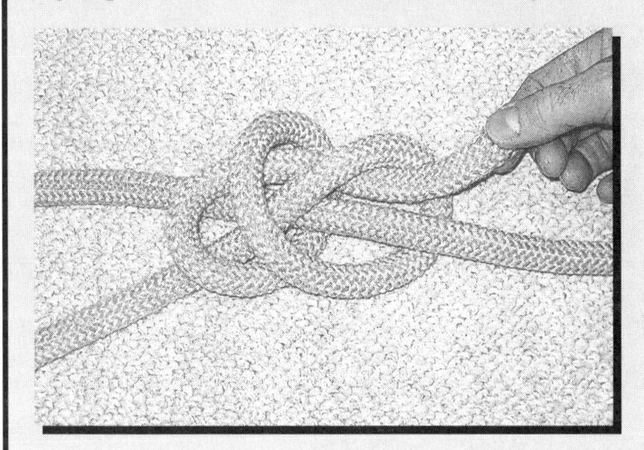

A Wrap one end of the rope over the other two times. The second wrap should cross over the first. Pull the working end through the wraps.

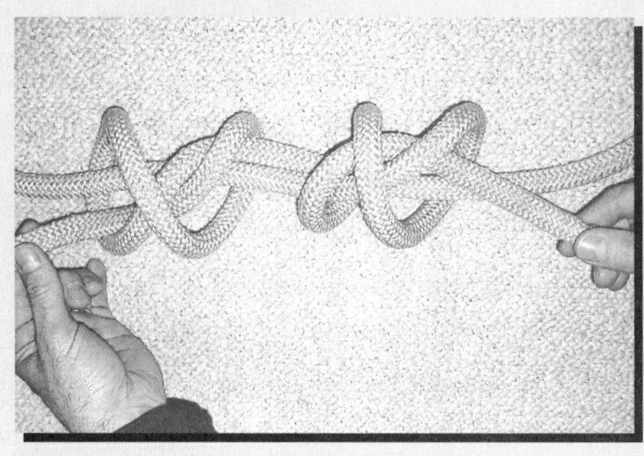

B Repeat with the opposite rope end.

C Tighten both sides down to cinch upon each other.

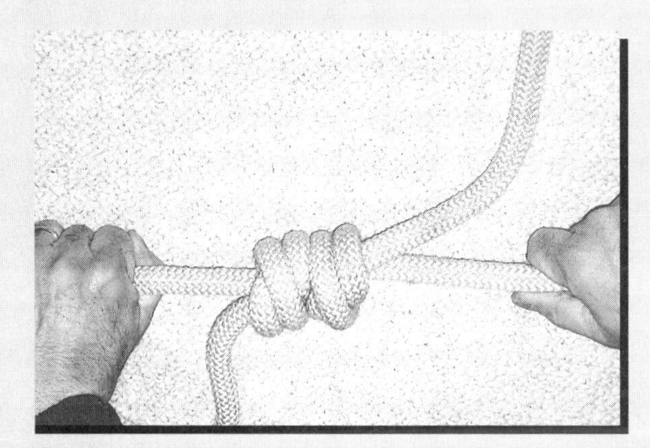

D Tighten both sides down to cinch upon each other.

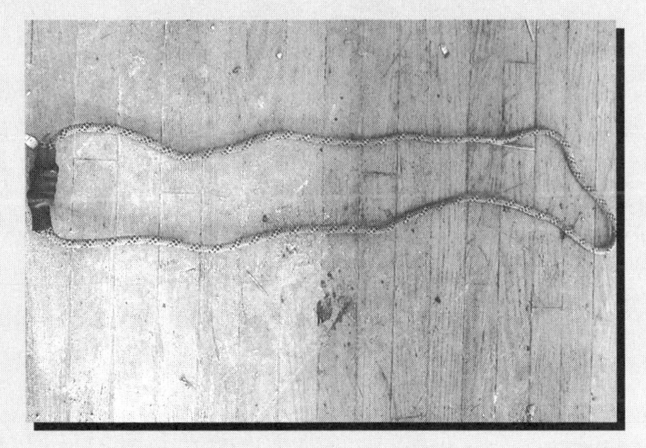

E For RIT purposes, a piece of rubberized shrink wrapping can be applied over the knot to provide a gripping surface as well as to provide protection for the knot.

SKILL 6-7

Tying the Munter Hitch (Method 1)

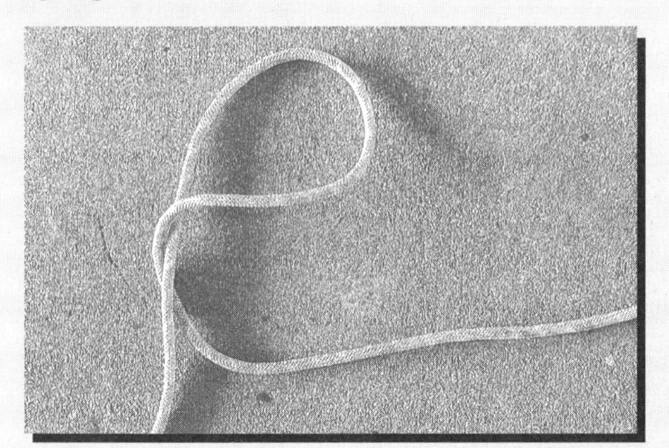

A Form a bight in the rope and twist it over itself one time.

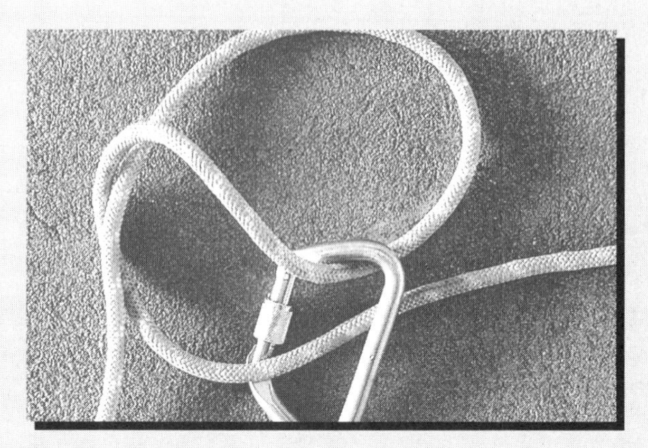

B Place a carabiner through the bight and standing part of the rope.

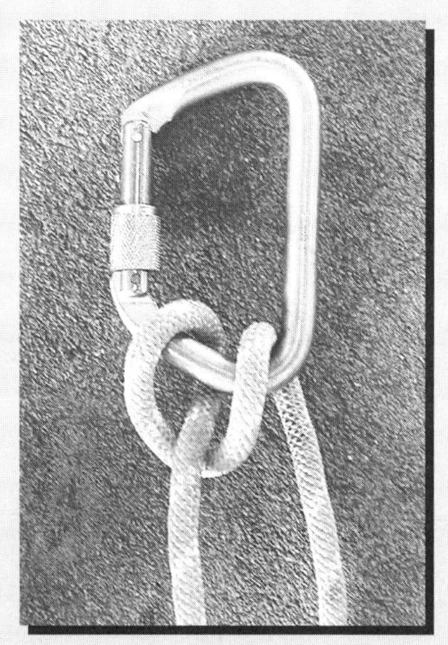

C Tighten the hitch down and test its operation. It should flip over itself when the ends are pulled.

The Munter Hitch

The Munter hitch is useful as a way to control an object's descent from an elevation difference. For example, it can be used when lowering a downed firefighter from a roof or when performing a personal rope slide for emergency escape. The use of the Munter hitch has been debated but may be a firefighter's only chance in an emergency if specialty equipment is not readily available.

Tying the Munter Hitch

Method 1 (Skill 6-7):

1. Form a bight in the rope and twist it over itself one time **(Skill 6-7A).**
2. Place a carabiner through the bight and standing part of the rope **(Skill 6-7B).**
3. Tighten the hitch down and test its operation. It should flip over itself when the ends are pulled **(Skill 6-7C).**

SKILL 6-8

Tying the Munter Hitch (Method 2)

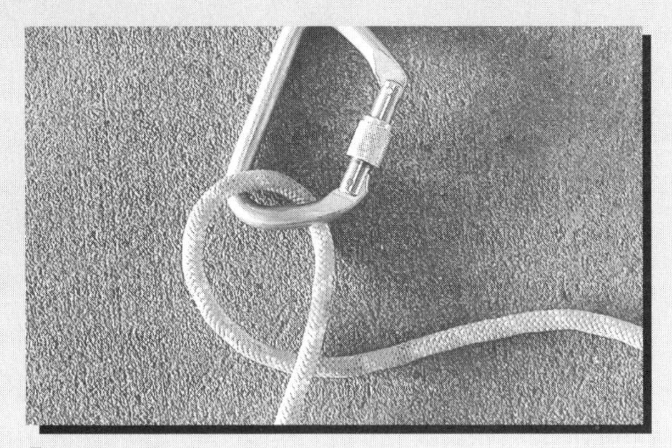

A Form a loop in the rope and place it inside of a carabiner.

B Cross the standing part of the rope over the back of the loop and place it inside the carabiner.

C Tighten the hitch down and test its operation. It should flip over itself when the ends are pulled.

Method 2 (Skill 6-8):

1. Form a loop in the rope and place it inside of a carabiner **(Skill 6-8A).**
2. Cross the standing part of the rope over the back of the loop and place it inside the carabiner **(Skill 6-8B).**
3. Tighten the hitch down and test its operation. It should flip over itself when the ends are pulled **(Skill 6-8C).**

Water Knot or Tape Knot

The **water** or **tape knot** is used to join two separate ends of webbing together to form a large loop. An overhand knot is made on one end and the other end simply traces the first overhand from the opposite side. The knot should be backed up by an overhand knot on each side to prevent it from slipping.

SKILL 6-9
Tying the Water or Tape Knot

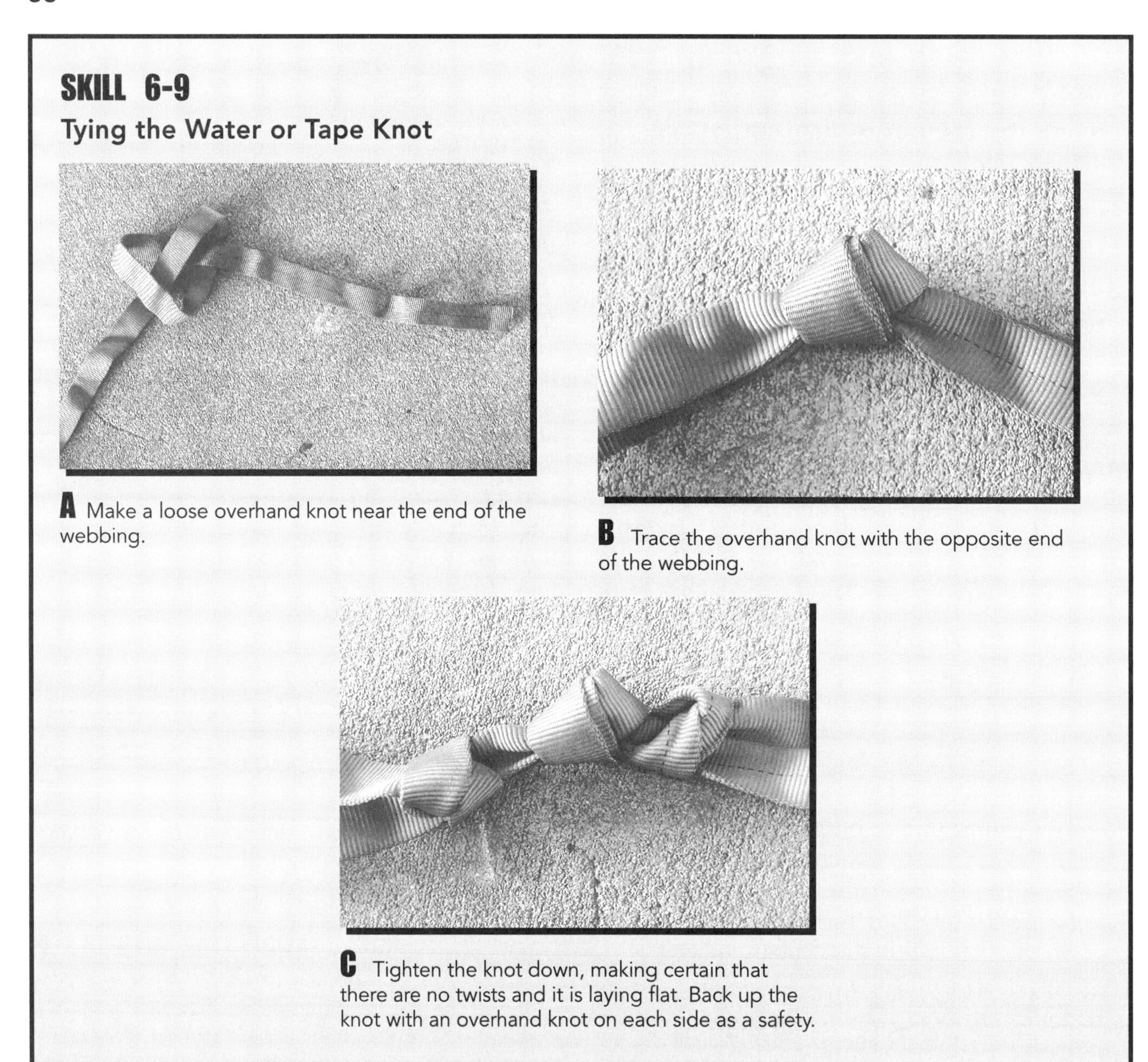

A Make a loose overhand knot near the end of the webbing.

B Trace the overhand knot with the opposite end of the webbing.

C Tighten the knot down, making certain that there are no twists and it is laying flat. Back up the knot with an overhand knot on each side as a safety.

Tying the water or tape knot (Skill 6-9):

1. Make a loose overhand knot near the end of the webbing **(Skill 6-9A).**
2. Trace the overhand knot with the opposite end of the webbing **(Skill 6-9B).**
3. Tighten the knot down, making certain that there are no twists and it is laying flat. Back up the knot with an overhand knot on each side as a safety **(Skill 6-9C).**

Modified Diaper Harness

The **modified diaper harness** is made from webbing and can function as a class 3 rescue harness if needed. It is useful for dragging firefighters and raising or lowering them.

Tying the modified diaper harness (Skill 6-10)

1. Webbing formed into a loop utilizing a water knot is placed behind the firefighter **(Skill 6-10A).**
2. A bight is pulled from the webbing located between the firefighter's legs **(Skill 6-10B).**
3. The webbing located on each side of the firefighter is pulled through the bight from the inside **(Skill 6-10C).**
4. The two newly formed loops are placed over the firefighters shoulders **(Skill 6-10D).**
5. A carabiner attached to the two loops behind the firefigh90ter's head will provide a point for lifting or lowering. Grabbing these two loops can provide a handle for dragging the firefighter **(Skill 6-10E).**

Anchors

One of the most crucial steps involved when using rope for rapid intervention or self-survival is the establishment of an adequate **anchor,** or point used to attach a rope. This can be very difficult when the distressed firefighter or RIT may be working in near-zero visibility. A firefighter who is unable to see may make a mistake when selecting what appears to be a sound anchor. The major consideration when establishing an anchor is whether it will support the weight of the load being placed on it. There are a few options when anchoring a rope to be used for an emergency maneuver.

Wall Anchor without a Tool

One of the easiest ways to secure a wall anchor without a tool is to breach a hole through a wall next to a doorway. If a doorway is not present the firefighter can simply break through the interior wall on both sides and wrap the rope around several wall studs. The rope should be wrapped around a door frame, or wall studs, and secured back onto itself with a **carabiner.** The anchor should be set by pulling it in the direction that it will be loaded **(Figure 6-1).**

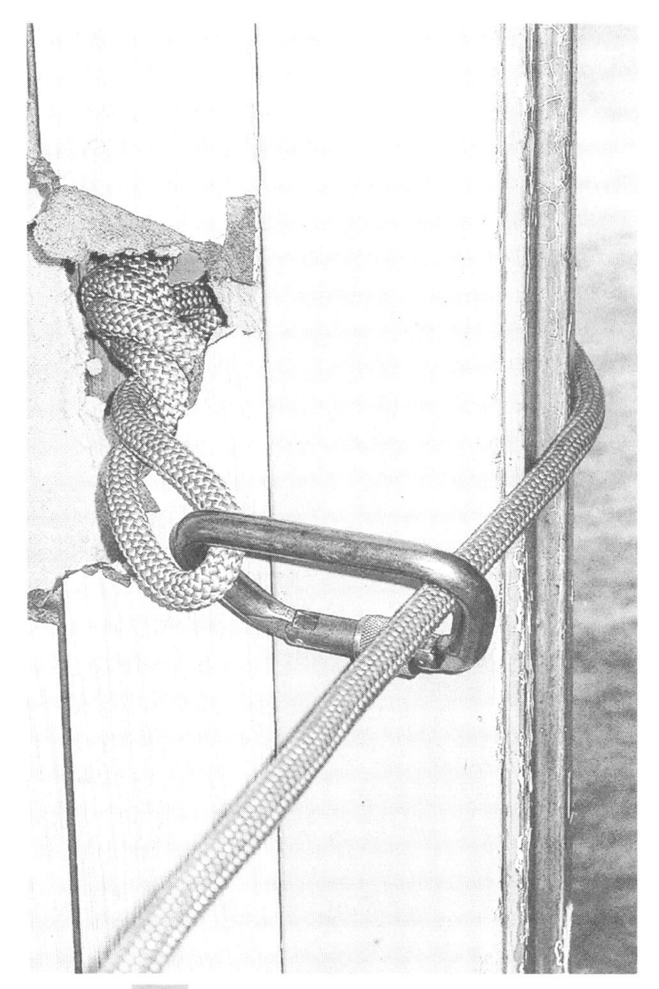

FIGURE 6-1
Wall anchor without tool.

Wall Anchor with a Tool

This technique involves attaching the rope to a tool that is secured behind two or more wall studs (the wall must be breached to set the tool in place). Attaching the rope can be done by slipping a pre-tied figure eight on a bight over the handle of the tool or wrapping the rope around the handle of the tool and securing it back onto itself with a carabiner **(Figure 6-2).** The anchor should be set by pulling in the direction that it will be loaded.

Lathe and plaster walls (Figure 6-3) will take considerably more time to penetrate than walls made of drywall. In this situation, penetrate the wall up high on one side, secure the rope to the tool, and place the tool down into the hole at an angle across the lathes of the plaster. If there are studs present, angle the tool in between them down low. Again, the anchor should be set by applying tension to it in the direction that it will be loaded.

SKILL 6-10

Tying the Modified Diaper Harness

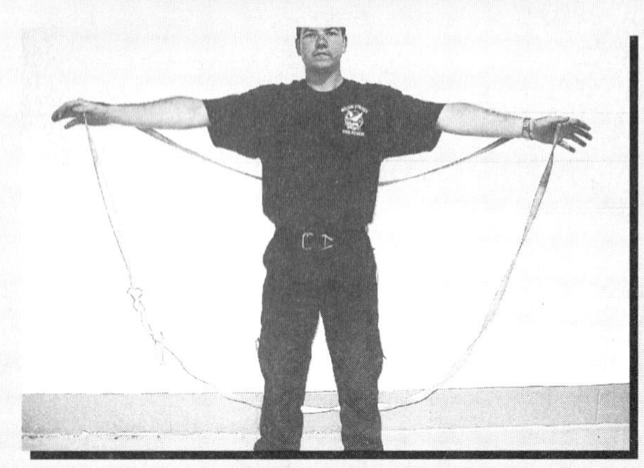

A Webbing formed into a loop utilizing a water knot is placed behind the firefighter.

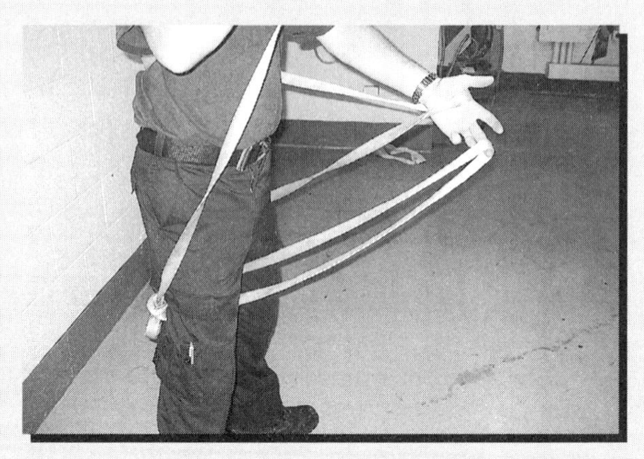

B A bight is pulled from the webbing located between the firefighter's legs.

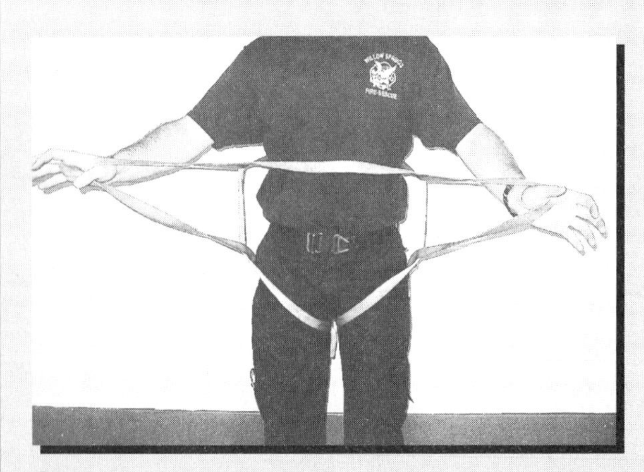

C The webbing located on each side of the firefighter is pulled through the bight from the inside.

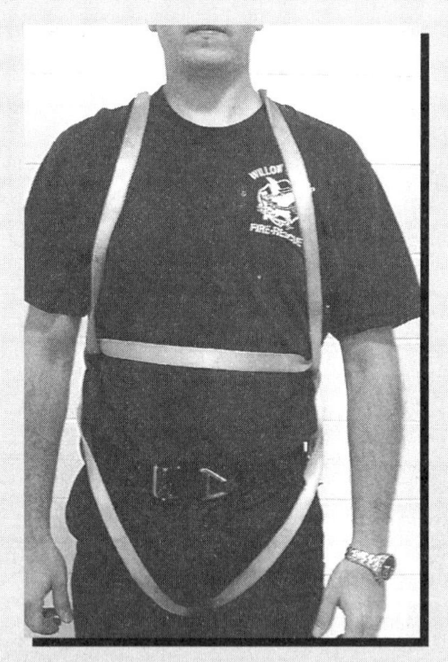

D The two newly formed loops are place over the firefighter's shoulders.

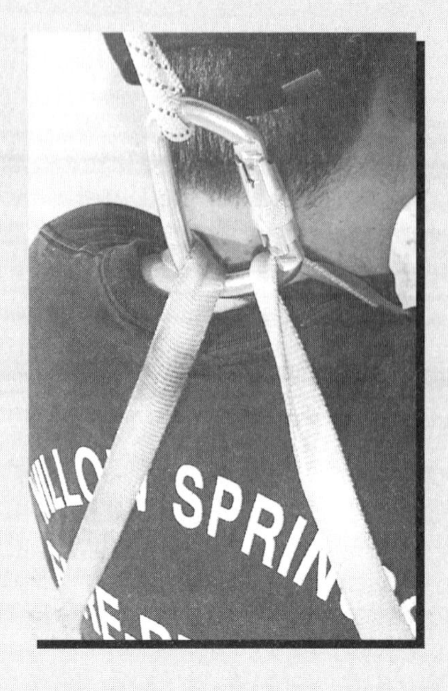

E A carabiner attached to the two loops behind the firefighter's head will provide a point for lifting or lowering. Grabbing these two loops can provide a handle for dragging the firefighter.

FIGURE 6-2

Wall anchor with a tool. (Shown with wall material removed.)

FIGURE 6-3

Lathe and plaster walls are very common in older buildings.

Corner Window Anchor

A tool can be used to create a **corner window anchor** by utilizing the corner of a window frame when the load is being applied in an opposite direction. This anchor can be applied quickly at the window but can also slip easily if tension is not maintained on the tool being used. For this reason, this anchor should be used only as a last resort.

The preferred tool when creating a corner window anchor is the Halligan bar, although an axe or steel pike pole can do the job if positioned correctly. Securing the rope is done by slipping a figure eight on a bight over the handle of the tool or wrapping the rope around the handle of the tool and securing it back onto itself with a carabiner. The tool is then placed in a diagonal position across the bottom corner of the window. The tool should be placed as low as possible in the corner. If using a tool with a pick, the pick should be driven into the wall in order to increase its stability. Once the anchor is set, constant tension needs to be maintained on the rope to prevent the anchor from moving **(Figure 6-4).**

Halligan Floor/Roof Anchor

The Halligan bar can also be driven into a floor or roof to create a **Halligan floor/roof anchor (Figure 6-5).** The tool should be driven into the surface at least 6 feet from the opening being used to prevent the bar from being lifted out of the floor or roof.

FIGURE 

Halligan bar being used as a corner window anchor.

FIGURE 6-5

A Halligan bar driven into a surface and tied off properly can act as an anchor in an extreme situation.

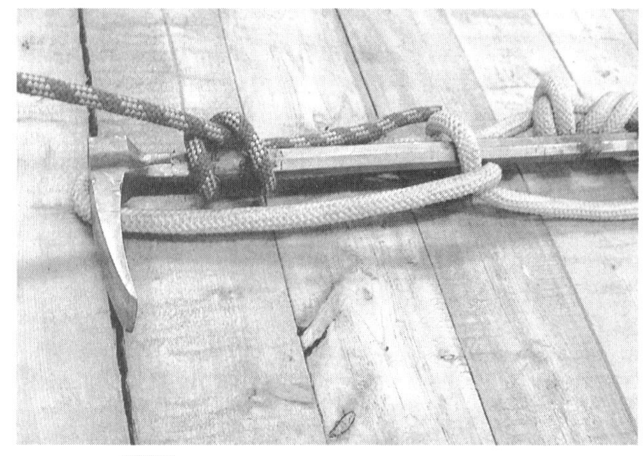

FIGURE 6-6

A Halligan bar used as an anchor should *always* be backed up with a safety to another anchor when used for training.

Safety

When establishing the Halligan bar as an anchor point, it should be no more than 6 feet from the opening to prevent it from being lifted out of the surface that it is driven into.

Tying a rope onto a Halligan bar to establish an anchor (Skill 6-11):

1. The pick of the Halligan bar is driven into the floor or roof surface **(Skill 6-11A).**
2. The rope is applied to the Halligan bar by slipping the loop of a figure eight on a bight into the fork end of the bar **(Skill 6-11B).**
3. It is then run parallel along the handle of the bar and wrapped around the adz end at the base of the pick **(Skill 6-11C).**
4. After the adz end has been wrapped, the rope is brought back down the handle of the bar to its halfway point, where a half-hitch is applied **(Skill 6-11D).**

Tension must be kept on this anchor at all times to maintain its stability. This type of anchor should only be used in an extreme circumstance. If used in training, make certain that it is backed up by a safety to keep it from being lifted out of the surface driven into **(Figure 6-6).**

Fixed and Nonfixed Object Anchors

Fixed objects such as stand pipes, columns, or beams can be used as anchors. When using these objects as anchors, it is important that the rope be securely attached and be within a distance from the opening that is not excessive **(Figure 6-7).**

Nonfixed objects may also be another option for an anchor. When considering using a nonfixed object such as a piece of furniture, consider the location of the object and its ability to support the weight of the load that it is being used for.

Safety

If used in training, make certain that the Halligan anchor is backed up by a safety securing it to another anchor to keep it from being lifted out of the surface driven into.

Body Anchor

In an extreme situation, an anchor point can be comprised of team members securing the rope to themselves in a safe area such as a doorway or outside the structure if necessary; this is called a **body anchor.** To counter the weight of the load, the firefighter or firefighters acting as the anchor must be certain that they are able to

SKILL 6-11

Tying a Rope onto a Halligan Bar to Establish an Anchor

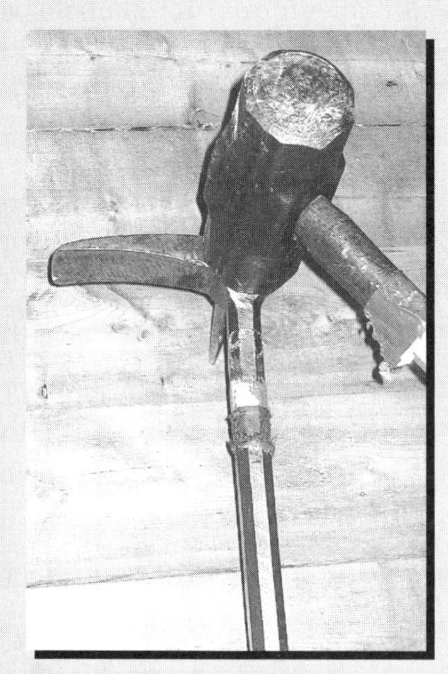

A The Halligan bar's pick is driven into the floor or roof surface.

B The rope is applied to the Hallagan bar by slipping the loop of a figure eight on a bight into the fork end of the bar.

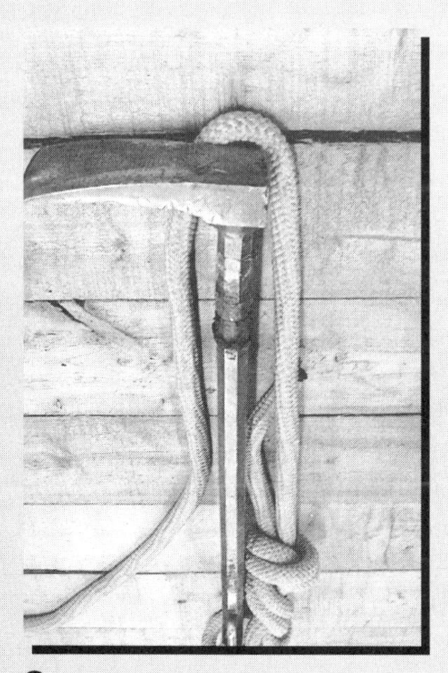

C It is then run parallel along the handle of the bar and wrapped around the adz end at the base of the pick.

D After the adz end has been wrapped, the rope is brought back down the handle of the bar to its halfway point, where a half-hitch is applied.

FIGURE 6-7

Heavy objects such as this radiator can be utilized as anchors.

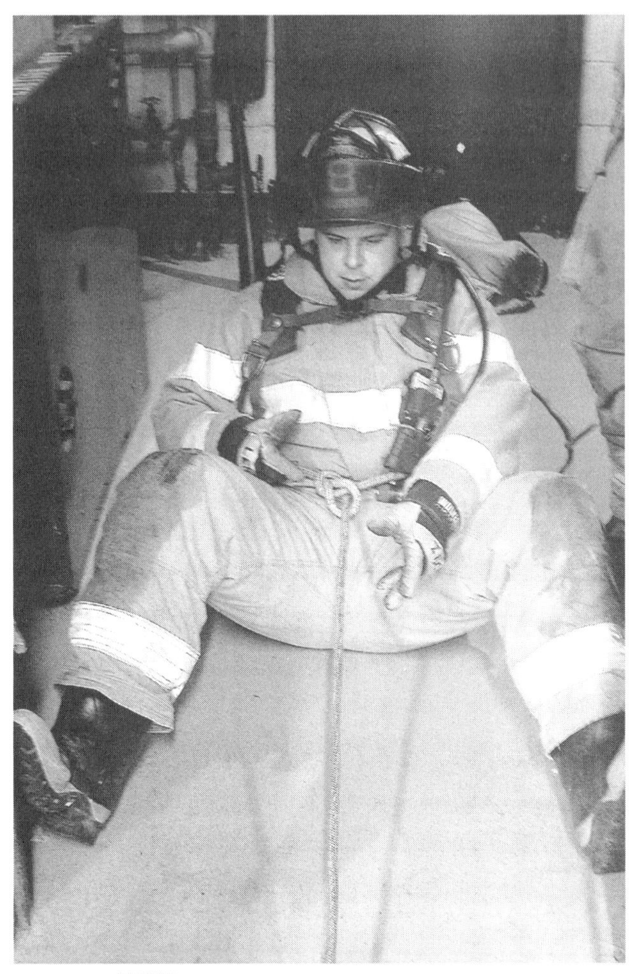

FIGURE 6-8

To counter the weight of the load, the firefighter or firefighters acting as the anchor must be certain that they are able to brace themselves behind something such as a door frame.

brace themselves behind something such as a door frame **(Figure 6-8).**

Basic Mechanical Advantage Systems

It is at this point that a lot of firefighters become intimidated by the use of ropes. These basic systems are incorporated into rapid intervention companies for the purposes of dragging, lifting, or raising downed firefighters in order to decrease the weight of a downed firefighter. They are not meant to be set up and secured in the same manner or fashion as with high-angle or vertical rescues that require additional expertise and equipment for safety and function. Again, the rescue of a downed firefighter is not meant to be a technical rescue incident; time is of the essence. The concepts of mechanical advantage are actually quite simple and easy to use. In fact, the techniques and maneuvers discussed in this text are easier and quicker to perform with a proper pre-rigged **2-to-1 mechanical advantage system** as part of the RIT's tool arsenal—even in a dark, smoke filled environment.

The 2-to-1 mechanical advantage system is the simplest to set up and the easiest to understand. For those reasons, it is the preferred system to use for rapid intervention operations. A 2-to-1 mechanical advantage system provides the RIT with one nonanchored moving **pulley** that transfers force to the rope. The weight of the load will be shared by the anchor and the firefighter or team that is pulling on the **haul line.** The one moving pulley will help to decrease the weight of the load by approximately half. A minimal amount of equipment will be required to set up a 2-to-1 system;

- a minimum of 100 feet of 1/2-inch rope in a rope bag or 60 feet of personal rope
- two large carabiners
- one pulley

The above equipment can be preassembled and deployed from a rapid intervention rope bag assigned to the RIT in order to avoid having to assemble it and waste precious time during a Mayday. An attachment point to the downed firefighter, such as a **multiple application service tool (MAST),** harness, or truck belt will also be needed and could also be included in the RIT kit. The equipment listed above can be assembled in three different ways. What is available and the purpose being used will determine which method is best. We will discuss two different ways of rigging a basic 2-to-1 mechanical advantage system.

SKILL 6-12

To Rig a 2-to-1 Mechanical Advantage System (Method 1)

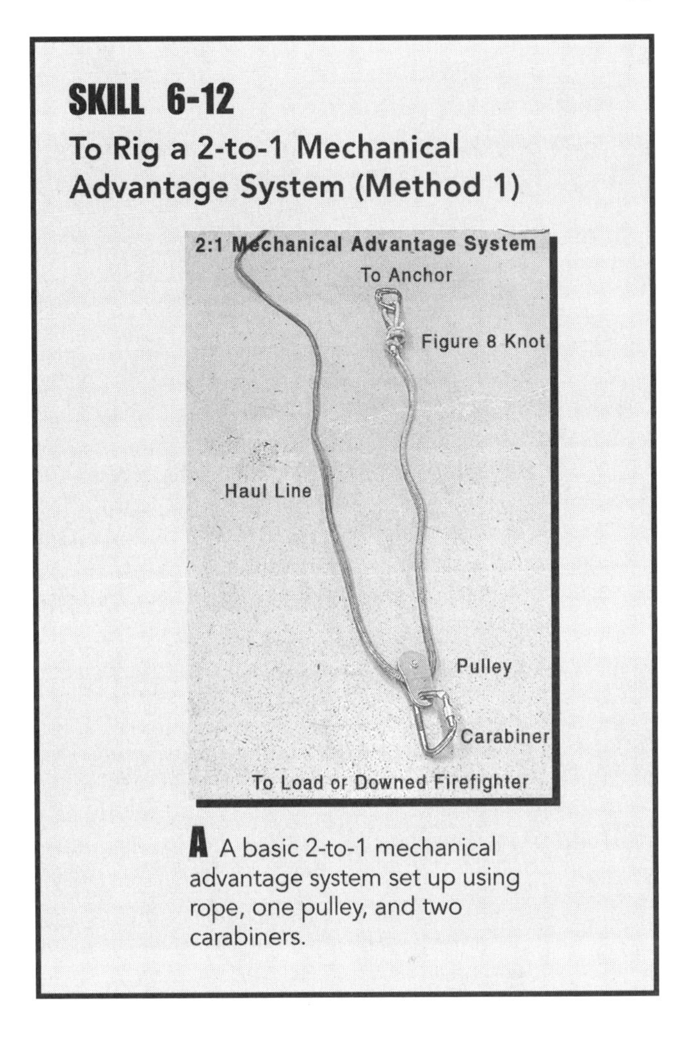

A A basic 2-to-1 mechanical advantage system set up using rope, one pulley, and two carabiners.

To Rig a 2-to-1 Mechanical Advantage System

Method 1 (Skill 6-12)

1. The first method to rig a 2-to-1 consists of taking the rope and putting a figure eight on a bight at the end of the rope and backing it up with a safety knot.
2. A large carabiner is then attached to the loop of the figure eight knot. This end of the rope will be used to establish the chosen anchor point **(Skill 6-12A).**
3. Several feet down the rope, a pulley with a large carabiner should be placed onto the rope. This carabiner will be attached to the downed firefighter or load. The remaining rope will remain inside the bag **(Skill 6-12A).**
4. The rope leading into the bag will be used as the **haul line,** which drags the load towards the anchor point **(Skill 6-12A).**

In a pinch, this system can be set up without the pulley. The rope will simply pass through the carabiner. The carabiner acts the same as a pulley, creating a mechanical advantage. Added friction will decrease the mechanical advantage from being a true 2-to-1, but it is still very advantageous if a pulley is not available.

The second method of rigging the 2-to-1 can be utilized when using a high-point anchor or wall studs with a tool as an anchor.

Method 2 (Skill 6-13)

1. A figure eight on a bight is tied at one end of the rope while backing it up with a safety knot.
2. As before, a large carabiner is then applied into the loop of the figure eight knot. This will be used again at the chosen anchor point. This time, however, the firefighter will wrap this end of the rope four times around the anchor or object, leaving the figure eight with the carabiner in its loop dangling free, forming a tensionless hitch **(Skill 6-13A).**

SKILL 6-13

To Rig a 2-to-1 Mechanical Advantage System (Method 2)

A A large carabiner is applied into the loop of the figure eight knot. This will be used again at the chosen anchor point. This time, however, the firefighter will wrap this end of the rope four times around the anchor or object, leaving the figure eight with the carabiner in its loop dangling free, forming a tensionless hitch.

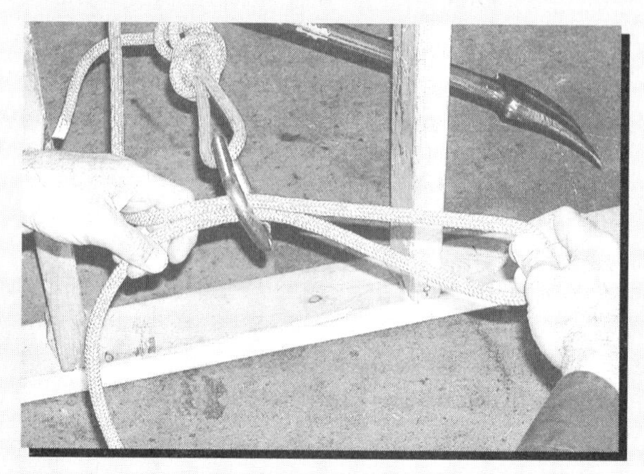

B The running part of the rope that is going into the rope bag is then formed into a bite and placed through the dangling carabiner.

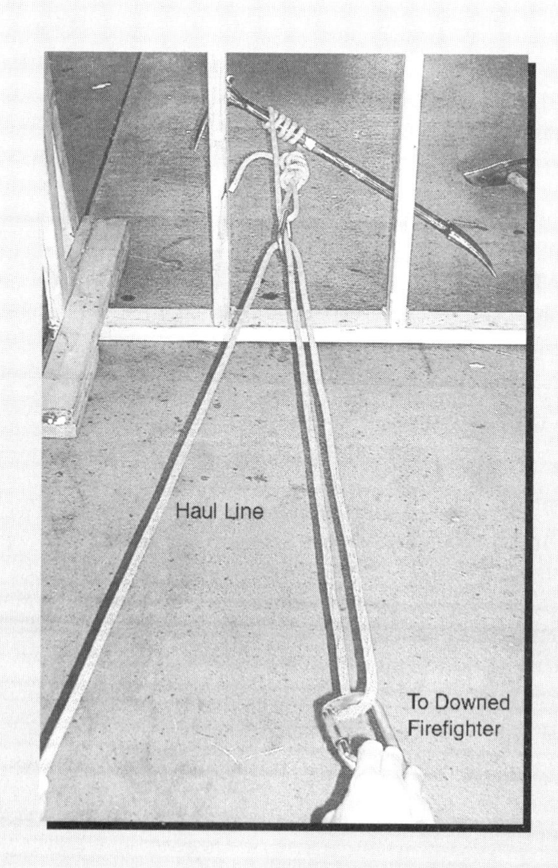

C The rope is then run through the carabiner and brought to the downed firefighter or load where an additional carabiner is placed onto the rope. This carabiner is then attached to the downed firefighter.

3. The running part of the rope that is going into the rope bag is then formed into a bite and placed through the dangling carabiner **(Skill 6-13B).**
4. The rope is then run through the carabiner and brought to the downed firefighter or load where an additional carabiner is placed onto the rope. This carabiner is then attached to the downed firefighter **(Skill 6-13C).**
5. The remaining rope will remain inside the bag. This end will be used as the haul line which will move the downed firefighter toward the anchor point.

Summary

Knowing the basics involved with ropes, knots, and basic mechanical advantage can mean the difference between life and death when attempting to rescue one of our own. Firefighters should train repeatedly in the use of the knots and anchor points discussed in order to determine what they are comfortable with and what type works best for a given emergency situation. The fireground is not the place to attempt to learn a new skill. The use of ropes, knots, and mechanical advantage is a definite plus for firefighters on the RIT. It is important that these concepts be kept as simple as possible to provide for their successful use on the fireground.

■ KEY TERMS

Anchor
Body Anchor
Butterfly Knot
Carabiner
Corner Window Anchor
Figure Eight on a Bight
Halligan Floor / Roof Anchor
Handcuff Knot
Haul Line
Lathe and Plaster Walls
2-to-1 Mechanical Advantage Systems
Multiple Application Service Tool (MAST)
Modified Diaper Harness
NFPA 1983
Prussik or Rescue Loop
Pulley
Tensionless Hitch
Water or Tape Knot

■ REVIEW QUESTIONS

1. Because ropes and mechanical advantage systems are being used, a RIT operation must be treated the same as a technical rescue incident.

   True False

2. Any rope, harnesses, or hardware utilized to support a firefighter in a life-safety situation must meet the requirements set forth by
   a. OSHA 1910.120.
   b. NFPA 472.
   c. NFPA 1983.
   d. FEMA 103.

3. The advantage of a figure eight on a bight is that it
   a. allows a loop to be placed in the rope.
   b. will not slip.
   c. can be used at any point of a rope.
   d. all of the above

4. The ___________ is used when a three-directional pull or attachment point is needed in a rope.
   a. tensionless hitch
   b. Munter hitch
   c. butterfly knot
   d. water knot

5. The advantages of the ___________ knot is that it allows a downed firefighter to be raised in a streamlined fashion.
   a. handcuff
   b. bowline
   c. butterfly
   d. Prussik
6. The handcuff knot should be placed on the downed firefighter's
   a. wrists.
   b. forearms.
   c. torso.
   d. none of the above
7. A ___________ has a tendency to slip when it is utilized with synthetic-fiber ropes.
   a. butterfly knot
   b. tensionless hitch
   c. bowline knot
   d. Munter hitch
8. A ___________ can be used to control an object's descent from an elevation difference.
   a. tensionless hitch
   b. water knot
   c. Munter hitch
   d. heavy anchor
9. When establishing a corner window anchor, the ___________ is the recommended tool of choice.
10. A 2-to-1 mechanical advantage system will decrease the weight of a 350-lb. load to approximately _____ lbs.
    a. 75
    b. 200
    c. 300
    d. 175

ADDITIONAL RESOURCES

Brown, M., *Engineering Practical Rope Rescue Systems.* Clifton Park, NY: Thomson Delmar Learning, 2004.

Firefighter's Handbook (2nd ed.). Clifton Park, NY: Thomson Delmar Learning, 2004.

CHAPTER

7 Firefighter Emergency Escape Maneuvers

Learning Objectives

Upon completion of this chapter, you should be able to:

- explain the decisions that need to be considered before and during wall-breaching maneuvers.
- describe the techniques and tools needed when deciding to breach walls of various construction types.
- discuss the various techniques and maneuvers that are involved when deciding to use a window for the purpose of emergency egress.
- demonstrate the straddle-and-hang position as a self-survival technique when utilizing a window for emergency egress.
- explain considerations and procedures when deciding to use ladders for emergency egress.
- identify two types of ladder bailout techniques and their principles.
- explain and describe the procedures involved when utilizing and performing an anchored rope slide for the purpose of emergency egress.
- discuss how webbing can be utilized during an emergency egress escape through a window.
- describe aspects of how a hoseline can be used to perform an emergency egress from an upper-floor window.
- list the steps required to establish a drywall ladder.

CASE STUDY

"On May 9, 2001, a 40-year-old male career firefighter (the victim) died after he became trapped in a third-floor apartment while searching above the fire for occupants. The victim and Firefighter #1, from Truck 2, were assigned to conduct a primary search for a mother and her children who were reported as being trapped in a second-floor apartment. The victim and Firefighter #1 conducted a primary search of three of the four apartments on the second floor while two lieutenants (from Engine 2 and Engine 3) and a captain (from Truck 2) were attacking the fire with a 1 3/4-inch handline in the fourth apartment. No civilians were found on the second floor. The victim and Firefighter #1 proceeded up the stairwell toward the third floor where they encountered heavy smoke and high heat. The victim and Firefighter #1 then descended the stairwell to the second-floor landing. Firefighter #1 told the victim to stay on the hoseline and to help the lieutenants while he went to get some box lights from the truck. Firefighter #1 had just returned to the second-floor landing when the lieutenant from Engine 3 informed him that the victim had called over the radio that he was trapped in a third-floor rear apartment. The lieutenant from Engine 3 had attempted to stretch the handline up the stairwell to the third floor but found that the line was too short to reach down the hall toward the rear apartments. The firefighter assist and search team (FAST) made several attempts to locate the victim but were unsuccessful due to the fire spread and deteriorating conditions of the building. The victim was found in an apartment bedroom on the third floor. He was unresponsive and not breathing. Two paramedics responded to the third floor, assessed the victim's condition, and found no heart activity while using a heart monitor. The victim was pronounced dead at the scene."

—Case Study taken from NIOSH Firefighter Fatality Report # 2001-18, **"Career Firefighter Dies after Becoming Trapped by Fire in Apartment Building—NJ,"** ***full report available online at http://www.cdc.gov/niosh/face200118.html.***

Introduction

Knowing your job and possessing a strong background in firefighting basics is the first and most important step that all firefighters should take when it comes to self-survival skills. Firefighters should realize that learning self-survival skills in addition to firefighting basics is an investment in themselves, their families, and their department. Firefighters should be proficient in the skills of self-survival and should continuously train to perfect those skills. All RIT members should understand that the knowledge of self-survival skills will help them in the event they get into trouble. It will also give them an understanding of what actions a firefighter in distress might take when dealing with an emergency. This knowledge may give a sense of predictability to the actions of a distressed firefighter and thus help the RIT during a rescue.

Self-Survival Techniques

Firefighters must realize that they may become disoriented, lost, or cutoff for whatever reason and may have to resort to using self-survival techniques to get out. This may necessitate retreating into another room and closing a door, breaching walls leading into a safe area, or rapidly exiting out of an opening to stay alive. In turn, RITs should be prepared to assist firefighters that have to resort to emergency egress by placing ground ladders and, if necessary, breaching exterior walls.

Everyone working on the fireground should be aware of the conditions that can create the need for emergency escape.

RITs must recognize rapidly deteriorating conditions on the fireground and be ready to anticipate and react to Mayday communications from interior firefighters. RITs responding to a Mayday must use and develop a rescue plan based on the fact that there are only so many ways that an interior firefighter will be able to exit the structure.

Safety

Any training involving self-survival maneuvers should follow strict safety measures to prevent injury. All training that consists of elevation differences should utilize appropriate safety lines.

By keeping an open mentality and training progressively and safety as the number one priority, firefighters can acquire the skills and confidence needed to react to an emergency on the fireground.

FIGURE 

In an extreme circumstance, a firefighter may have to breach a wall to escape to safety.

Wall Breaching

Firefighters in distress will have to realize that when they are lost or cutoff from their crew, they may have to resort to finding a safe area to protect themselves. **Wall breaching** is used by an interior firefighter when he is in immediate danger and unable to find a window or door for the purposes of exiting quickly to a safe area **(Figure 7-1).** It is quite evident that breaching an interior wall will be less difficult than trying to break through an exterior wall. Either way, the choice of performing such a maneuver should only be undertaken when trying to escape from immediate danger. Going through any interior wall under normal conditions can be tiring. Performed under fire conditions with an SCBA on, it becomes an extreme task that will diminish your air supply very quickly if not managed in the proper manner. All firefighters should have a tool with them when they enter a fire building. There are no acceptable excuses for not having one.

In most residential structures the interior walls are usually made up of drywall and 2×4 wood framing. Steel framing is also being utilized in newer construction and remodels. Walls will also have **electrical conduit,** HVAC duct work, and plumbing running through them **(Figure 7-2).** In older communities the firefighter may encounter walls made of lathe and plaster, which can be very difficult to breach if the proper methods are not used **(Figure 7-3).** The 2×4 framing used in older structures will also be true dimensional wood, unlike lumber used for framing in modern construction, making it considerably tougher to breach **(Figure 7-4).** In commercial structures, firefighters may have to deal

FIGURE 

Walls may also have electrical conduit, HVAC duct work, and plumbing running through them, which can make getting through difficult for a firefighter.

FIGURE 7-3

Lathe and plaster walls will be common in older communities.

FIGURE 7-4

The 2×4 framing used in older structures (right) will also be true dimensional wood unlike lumber used for framing in modern construction, making it considerably tougher to breach.

with interior walls made of concrete or masonry construction.

Firefighters must realize that breaching a wall into a safe area is only a temporary fix for the problem encountered. Once the firefighter is through the wall, advancing conditions can follow as well. It is important to continue to search for a door or window and get out.

Breaching a frame wall (Skill 7-1):

1. The first step a firefighter must undertake when deciding to utilize a wall breach is to feel the surface area of the wall chosen to penetrate. Running a hand over the wall surface will give an indication of the obstacles that may be encountered inside the wall space. A spot that is clear of electrical outlets or HVAC vents should be chosen.
2. After a location is chosen, the firefighter begins by using a tool to break through completely to the other side **(Skill 7-1A).** It is not necessary to make a large opening at first; only a hole adequate to put the tool or an arm through to check for obstructions is necessary **(Skill 7-1B).** If an obstruction is present, the firefighter can move over and try another inspection hole.
3. Once the decision is made that the other side is clear of obstructions and hostile conditions, the firefighter should proceed to make a hole big enough to get through **(Skill 7-1C).** Most interior walls encountered will have 2 × 4 wood or metal studs on 16-inch centers, so it may be necessary to remove one of the studs out of the way. (This may be especially true if a RIT is trying to remove a downed firefighter; **Figure 7-5**). This can be done by knocking the base of the stud off of the sole plate with a tool. Be cognizant that the nails from the sole plate will now be exposed and can cause possible injury.
4. After the interior of the wall is exposed, the firefighter needs to make certain that any other obstacles are removed. Material on the opposite side of the wall should be pushed away, while material on the side of the firefighter should be pulled **(Skill 7-1D).**

SKILL 7-1
Breaching a Frame Wall

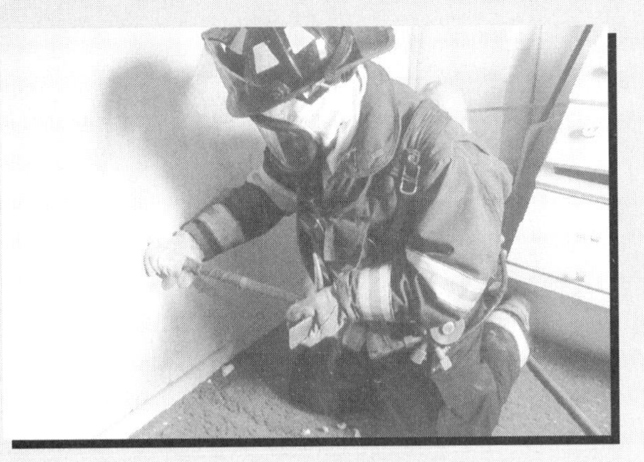

A After a location is chosen, the firefighter begins by using a tool to break through completely to the other side.

B Tools should penetrate completely through the wall to ensure that nothing will stand in the way of the firefighter trying to escape.

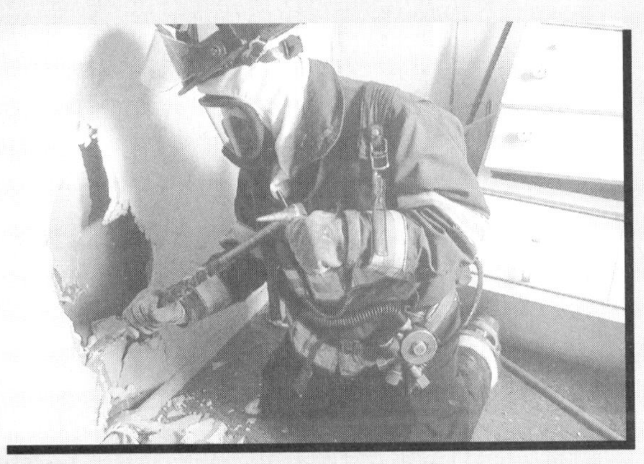

C Once the decision is made that the other side is clear of obstructions and hostile conditions, the firefighter should proceed to make a hole big enough to get through.

D Material on the opposite side of the wall should be pushed away, while material on the side of the firefighter should be pulled.

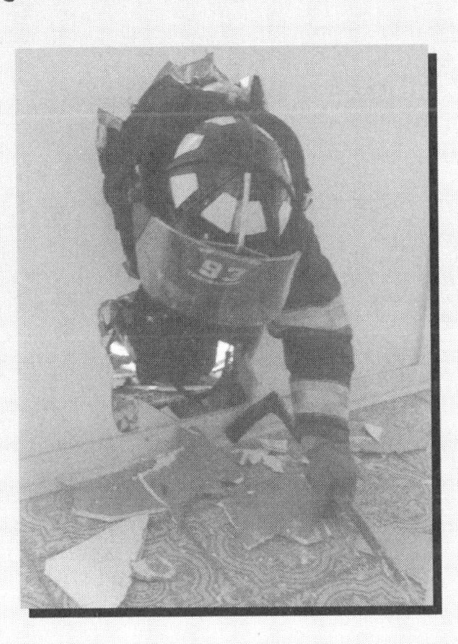

E Proceed through the wall with the upper extremities and head first in order to allow the arms to assist pulling through the wall.

FIGURE 7-5

It may be necessary to remove one of the wall studs out of the way if a RIT is trying to remove a downed firefighter through a wall breach.

5. The firefighter should be positive that he checks the integrity of the space on the other side of the wall. This means checking for fire conditions along with making sure that a solid floor exists on the other side. The firefighter may be required to adjust his SCBA into a reduced profile maneuver in order to get through and between the studs (refer to Chapter 5). It is also a good idea to proceed through the wall with the upper extremities and head first in order to allow the arms to assist pulling through the wall **(Skill 7-1E).**
6. Once through the wall, again continue to look for a means of escape.

A firefighter may encounter lathe and plaster walls in buildings within older parts of the community. Lathe and plaster walls can be very challenging to penetrate, especially when the plaster is backed by wire lathe.

To breach a lathe and plaster wall (Skill 7-2):

1. The firefighter will size up the wall, making certain to be clear of electrical outlets and HVAC vents.
2. The wall should be penetrated forcefully with the tool to check the opposite side for obstructions.
3. The tool should then be dropped behind the lathe and pulled back toward the firefighter as a fulcrum. This will break the wall apart from the framing **(Skill 7-2A).**

SKILL 7-2

To Breach a Lathe and Plaster Wall

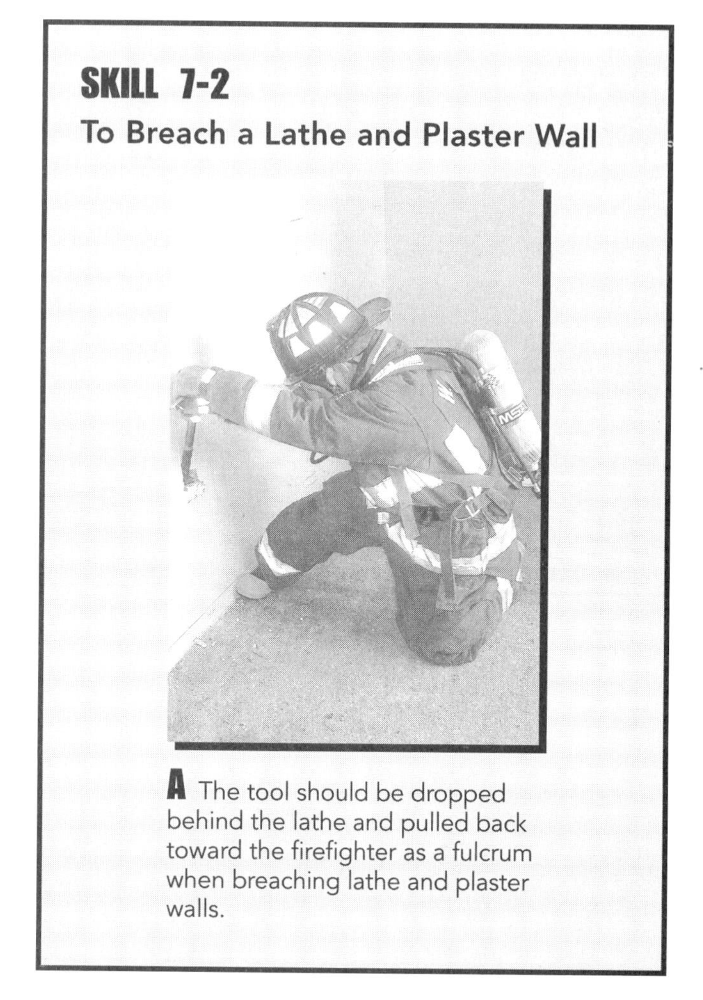

A The tool should be dropped behind the lathe and pulled back toward the firefighter as a fulcrum when breaching lathe and plaster walls.

If the firefighter finds himself in a situation where he does not have a tool, he will have to proceed by what is known as **mule kicking** the wall in order to penetrate it.

To perform the mule kick breach technique (Skill 7-3):

1. The firefighter will size up the wall, making certain to be clear of electrical outlets and HVAC vents **(Skill 7-3A).**
2. The firefighter will face away from the wall on his hands and knees **(Skill 7-3B).**
3. By lifting the right or left leg and flexing, the firefighter will thrust backward in a kicking motion into the wall. The kick should be hard enough to provide an opening all the way through in order to check the other side **(Skill 7-3C).**

SKILL 7-3

To Perform the Mule Kick Breach Technique

A The firefighter will size up the wall, making certain to be clear of electrical outlets and HVAC vents.

B The firefighter will face away from the wall on his hands and knees.

C By lifting the right or left leg and flexing, the firefighter will thrust backward in a kicking motion into the wall. If clear, the firefighter will continue to mule kick the wall until there is an opening big enough to fit through.

4. If clear, the firefighter will continue to mule kick the wall until there is an opening big enough to fit through.

Kicking in a forward motion will increase the chances of the firefighter injuring his knees when they impact the wall. It is for this reason that the backwards mule kick is recommended. As with the use of the tool when penetrating drywall, the material should be pulled off the wall on the side of the firefighter and pushed away on the opposite side.

Masonry Walls

Masonry walls or concrete block are obviously not as easily breached as wood. Masonry walls are usually exterior walls but with the many architectural designs and building rehabs that are taking place, they can occur as interior walls in residential as well as commercial structures.

To breach a masonry wall (Skill 7-4):

1. When faced with a masonry wall, the firefighter should use a tool to begin a

SKILL 7-4

To Breach a Masonry Wall

A When faced with a masonry wall, the firefighter should use a tool to begin a starting point off of an existing opening or mortar joint.

B Masonry walls should be breached in a pyramid shape to prevent collapse of the wall.

starting point off of an existing opening **(Skill 7-4A)**. This opening may have been made to accommodate heating, electrical, or plumbing systems. Mortar joints are also weak points in the masonry wall.

2. As the firefighter begins to make progress on the hole, he should remember to form it in a triangle or pyramid to maintain the structural stability of the wall. The tip of the pyramid opening will always be at the top in order to avoid brick or stone from falling down onto itself, which may create a curtain effect that could produce new entrapments for the distressed firefighter **(Skill 7-4B)**.

It is important to mention that in addition to conventional wall materials, new materials and technology are constantly being developed that could have an effect on firefighting operations. At the writing of this text, drywall backed with a Lexan substrate was beginning to make an appearance in the construction market. According to the manufacturer, it is intended for use in educational and institutional facilities. This does not mean that a firefighter will not encounter this material in other occupancies. Firefighters from the Roberts Park Fire Protection District in Illinois conducted extensive testing of the material and determined that this new material would make an emergency wall breach very difficult. The striking end of tools such as an 8-pound flathead axe or maul did not breach the material after as many as eighteen swings. The cutting end of these types of tools proved to be more successful but required as many as thirty-eight swings to establish a hole large enough for a firefighter to crawl through. This material will definitely take more time and require more work to be breached in an emergency situation.

Believe it or not, we have heard this suggestion as a remedy to replace the lack of not having a tool. It is obvious to most that this is dangerous and will compromise an already precarious air supply that may be needed for survival. If wall breaching is done, it is because a firefighter is in trouble and thus it must be performed in a quick manner to expedite escape. Be trained in and give self-survival the attention that it deserves along with the basics of our job. This combined with the proper attitude will allow the greatest chance to go home safe at the end of the shift.

> **Safety**
>
> A final word when talking about breaching walls: *never* remove the SCBA from your back to use it as a battering ram for the purposes of penetrating interior or exterior walls. Possibilities of pulling off the facepiece or damaging the operational components of the unit are too great. Once we are without air, our situation only worsens.

Emergency Egress through Windows

When a firefighter decides to use a window for an emergency escape, it is because it has been determined that the interior space is about to become untenable due to fire conditions or collapse. There are only a limited number of options available when considering a window for emergency egress. Once you break the plane of the window opening with your body, the decision is made and there is no turning back. It may not be the fire and heat that will cause the fatal injury, but what is on the other side of the window and how you exit it that will determine the outcome **(Figure 7-6)**. The following survival techniques might be considered when using a window for emergency egress:

- sitting on, or straddling, a windowsill and waiting for help
- hanging out of a window using a full-body straddle technique
- straddle, hang, and drop
- using a ladder for a ladder bailout
- using an emergency rope slide
- using an aerial device or tower ladder

FIGURE  **7-6**

Conditions or window size may not allow firefighters to exit a window in a conventional manner.

Situations that require invasive actions such as these will often not allow the firefighter in trouble to be able to communicate intended actions to the IC or RIT. This further stresses the importance of proactive measures to be performed by the RIT. The distressed firefighter should realize that the equipment available will determine the method that can be used. It is important that the firefighter have the proper personal equipment to perform the window exit technique. In preparation for the use of any of these techniques many firefighters carry escape ropes, life belts, or have harnesses sewn into their turnout gear. Emergency egress techniques discussed here do not rely on specialty equipment such as that used during technical or high-angle rescue operations. Any firefighter who is trapped and required to immediately exit through a window requires the simplest possible solution that will allow a safe exit. Having a few basic tools (hand tool, personal rope, ground ladder), either brought by the firefighter or placed by the RIT, will allow an emergency window exit to take place.

Before any emergency window egress can occur, the window must be cleared of glass and obstructions. If the troubled firefighter is in a room with a door, the door should be closed to provide protection and bide time by temporarily obstructing any advancing conditions.

Straddle and Hang Position

Straddling and hanging from a window may buy enough time for the RIT or any other members on the fireground to assist the trapped firefighter. The most important issue is that the firefighter is partially removed from the untenable environment. By reducing his profile through the window straddle, he will allow fire conditions to pass over him through the upper portions of the window while reducing his exposure. When performing the technique from the first floor, a firefighter may be able to roll out of the window onto the ground, escaping the untenable environment altogether.

To perform the straddle and hang (Skill 7-5):

1. The firefighter will sit upright to one side of the windowsill with one leg outside and the other still inside.

SKILL 7-5
To Perform the Straddle and Hang

A Top view of straddle and hang. The firefighter will sit upright to one side of the windowsill with one leg outside and the other still inside and lean forward.

B Interior view of straddle and hang. As the firefighter leans forward and down, the interior leg and arm "hook" onto the interior of the windowsill.

C Exterior view of the straddle and hang. Note how the firefighter is out and below the window opening.

2. The firefighter will then proceed to lean forward and down, allowing the interior leg and arm to "hook" onto the interior of the windowsill (**Skill 7-5A**, **7-5B**, and **7-5C**).

Hang and Drop Method

If a firefighter is hanging from a second-floor window with conditions that are very serious, the **hang and drop** method may need to be used. This obviously can result in serious injury and will only be used if the firefighter has no other alternative due to extreme conditions.

1. If this has to be done, the firefighter should already be in a low-profile position, hanging out the window as stated in the straddle and hang of **Skill 7-5**.
2. The firefighter will then have to reposition by lifting the leg that is on the interior out and over the edge of the windowsill while utilizing his arms to hang and hold his body with his feet pointed to the ground **(Figure 7-7)**. Most second-story windows will be approximately 12 feet off the ground. By extending vertically in a hanging position, the firefighter will be able to reduce the fall from the point of his feet to the ground to possibly 6 feet. The firefighter may still sustain injuries, but they may not be as severe as those from staying in the window.
3. The distressed firefighter should avoid pushing off the wall when letting go of the windowsill. The reason for this is that the SCBA will create a backward gravitational pull toward the ground, causing the firefighter to land on the SCBA, which may create serious injuries to the back and central spine.

FIGURE 

Position for the hang and drop.

Safety

This manuever should be used only as a last resort—serious injuries can be sustained.

Training on this maneuver is very difficult to perform in a safe manner; even the shortest fall can seriously injure the firefighter in this instance.

Note

This maneuver should be discussed but *not* performed in training.

Emergency Escape Ladder Bails

The head-first ladder bail is another self-survival skill that every firefighter should be familiar with. The first consideration when utilizing this maneuver is to make certain that a ladder is present outside the window and that it is positioned correctly. It should be noted that ladder bailouts have been the subject of controversy due to injuries that have occurred when training in the technique. Fire departments and fire services instructors should make sure when training in these maneuvers that proper safety techniques are used. This will include the use of a rated safety line that is attached to the firefighter and that line shall be belayed in order to prevent any accidents that may be caused due to loss of handholds on the ladder or falling.

Safety

A rated safety line with a belay shall be attached to the firefighter and the ladder shall be secured from movement at all times when training on ladder bail techniques.

FIGURE 

When placing ground ladders for emergency escape purposes, the position angle will need to be less than the recommended 75 degrees or one-fourth of the distance from the ground to the window that is traditionally recommended for ladder placement.

RITs should be placing ladders to upper-floor windows around a structure to provide an alternate means of egress for interior companies. The preferred method is to exit onto a ladder in a safe, conventional manner when feasible. When interior conditions deteriorate and a firefighter is in need of an emergency escape, he may need to exit head first to avoid the severe conditions in order to reach safety. Conditions that are rapidly deteriorating, high heat, or a small window may require the firefighter to exit head first onto the ladder.

When placing ground ladders for emergency escape purposes, the position angle will need to be less than the recommended 75 degrees or one-fourth of the distance from the ground to the window that is traditionally recommended for ladder placement **(Figure 7-8).** If the ladders are placed at 75 degrees or greater, a distressed firefighter would have an increased chance of sliding down the ladder too quickly, or even worse, missing it completely because of momentum. A distressed firefighter with a SCBA on his back has a gravitational force affecting his center of balance that can pull his "head over heels" if the angle is too steep. Placing the ladder at less than 75 degrees will ensure a more controlled technique by the firefighter exiting the window.

The tip of the ladder also needs to be placed just below the windowsill **(Figure 7-9).** If the ladder is placed at the window with the rails above

FIGURE 

The tip of the ladder needs to be placed just below the windowsill to enable firefighters the ability to exit the window quickly.

the windowsill, it will be very difficult for a firefighter to exit the window. The firefighter will be forced to bring himself up into the deteriorating conditions to get onto the ladder, and window opening size will be decreased.

There are several ladder bailout techniques taught to the fire service. This text will address two techniques pertaining to ladder bailouts involving quick emergency escapes: the **butterfly ladder bailout** and the **hook-and-go ladder bailout.** The techniques shown are chosen based on the firefighter's ability to maintain control of his movements and descent.

To perform the butterfly or the extended-reach technique ladder bail (Skill 7-6):

1. The firefighter will exit the window, staying low to the sill and grasping the second rung of the ladder palm up, and place the opposite hand on the ladder rail.

Note

The following techniques have been the subject of much controversy in the fire service community. However, they are techniques that every firefighter should be proficient in performing. Conditions in a structure fire can change very rapidly. These conditions can prevent a firefighter from being able to stand up and exit a window in a conventional manner. A firefighter's profile with an SCBA may not allow exit in a conventional manner. As with all techniques introduced in training, safety measures should be properly in place before attempting.

SKILL 7-6

To Perform the Butterfly or the Extended-Reach Technique Ladder Bail

A The firefighter will exit the window, staying low to the sill and grasping the second rung of the ladder palm up, and place the opposite hand on the ladder rail. The hand on the rail is kept in an open grip position to allow it to slide.

B The hand on the rail is slid downward, bringing the body of the firefighter forward out of the window.

C As the firefighter clears the windowsill, his momentum will cause his body to pivot, at which time the hand on the rail should grasp the next closest rung across the body.

(continued)

SKILL 7-6 (CONTINUED)

To Perform the Butterfly or the Extended-Reach Technique Ladder Bail

D The pivot should take place in a controllable manner, with the firefighter's feet ending in a position to allow exit from the ladder in a safe and conventional manner.

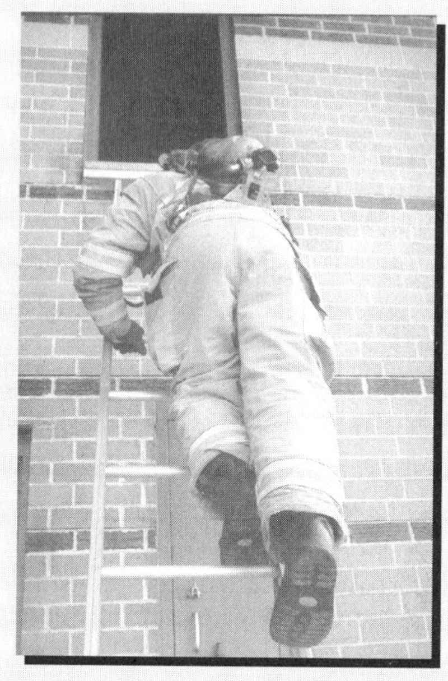

E The pivot should take place in a controllable manner, with the firefighter's feet ending in a position to allow exit from the ladder in a safe and conventional manner.

The hand on the rail is kept in an open grip position to allow it to slide **(Skill 7-6A).**

2. The hand on the rail is slid downward, bringing the body of the firefighter forward out of the window **(Skill 7-6B).**
3. As the firefighter clears the windowsill, momentum will cause his body to pivot, at which time the hand on the rail should grasp the next closest rung across the body **(Skill 7-6C).**
4. The pivot should take place in a controllable manner with the firefighter's feet ending in a position to allow exit from the ladder in a safe and conventional manner **(Skills 7-6 D** and **7-6E).**

To perform the hook-and-go ladder bail (Skill 7-7):

1. The firefighter places first arm under the second rung down, palm up, "hooking" the rung with the elbow. The opposite hand should be slid down the beam of the ladder to the fourth rung down **(Skill 7-7A).**
2. The firefighter reaches out with the hand that is on the beam and grasps the fourth rung while sliding his body forward **(Skill 7-7B).**
3. The momentum of this movement will cause the body to pivot, ending in an upright position ready to descend the ladder **(Skills 7-7C** and **7-7D).**

Sliding the Ladder

If exiting onto a ladder when other firefighters are behind and must get out of the building immediately, the firefighters can slide down the ladder to get out of each other's way quickly. The **ladder slide** is only to be used in an extreme circumstance—serious injuries can result from fire-

SKILL 7-7

To Perform the Hook-and-Go Ladder Bail

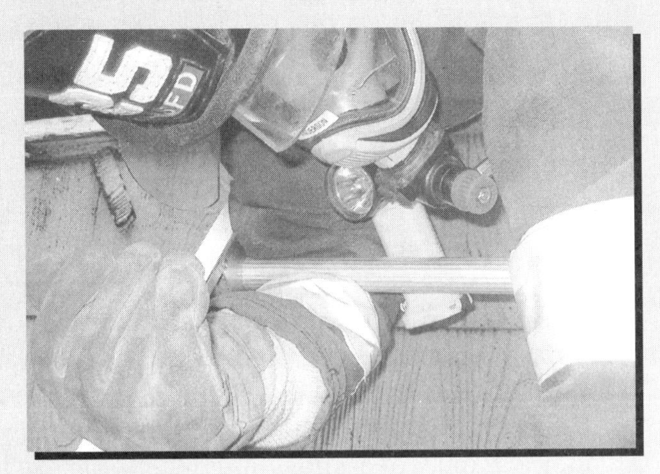

A The firefighter places first arm under the second rung down, palm up, "hooking" the rung with the elbow. The opposite hand should be slid down the beam of the ladder to the fourth rung down.

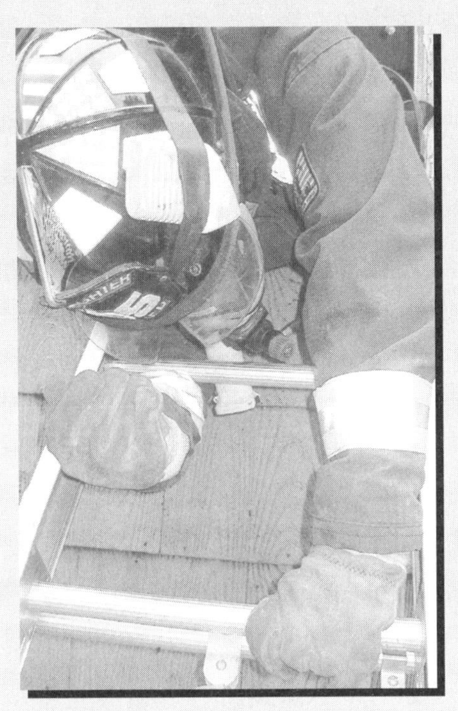

B Firefighter reaches out with the hand that is on the beam and grasps the fourth rung while sliding the body forward.

C The momentum of this movement will cause the body to pivot, ending in an upright position ready to descend the ladder.

D The firefighter should exit off the ladder quickly and safely.

fighters hitting the ground too hard with their feet if they slide down too quickly or lose control. Like the ladder bail, firefighters can easily lose control of their descent if the angle of the ladder is too steep.

To perform the ladder slide (Skill 7-8):

1. The firefighter will place his feet and knees to the outside of the ladder beams while holding onto the beams with his hands. His body will be in a crouched position, which will place his weight in a position that will cause him to slide down **(Skill 7-8A).**
2. The firefighter will tighten his grip with his hands and knees to slow down and control himself.

SKILL 7-8

To Perform the Ladder Slide

A The firefighter will place both feet and knees to the outside of the ladder beams while holding onto the beams with his hands. His body will be in a crouched position, which will place his weight in a position that will cause him to slide down. The firefighter will tighten his grip with his hands and knees to slow down and control himself.

3. Once down on the ground, the firefighter should get out of the way so that other firefighters can also exit quickly.

Again, the safest and preferred way to exit is to climb down the ladder in a conventional manner if conditions allow.

Emergency Rope Slide

The **emergency rope slide** maneuver should be reserved for the most extreme situations when exiting an untenable environment from an upper floor. When an exit from an upper-floor window requires this maneuver, there is no time to wait for a ladder or set up of specialized rope rescue equipment. Just as when using ladders for emergency escape, rope slides are a high-risk technique that should be used only if absolutely necessary.

Note

Rope slides are a high-risk technique that should be used only if absolutely necessary.

For firefighters to even consider using this maneuver, they must have the required necessities (a personal rope approximately 30 to 50 feet long and the ability to establish an adequate anchor). The most important point when dealing with the emergency rope slide is keeping it simple.

The type of rope slide technique discussed here is best described as a body wrap with a hand-controlled descent. All safety precautions should be taken when practicing this technique (safety lines, harnesses, and belay systems).

One of the most crucial steps involved when using a rope slide is the establishment of an adequate anchor. This can be very difficult when the distressed firefighter may be working in adverse conditions. The major consideration when establishing an anchor is whether it will support the weight of the firefighter during the escape. There are several options when anchoring a rope to be used for an emergency rope slide and these can be referenced in Chapter 6. Chapter 6 needs to be thoroughly reviewed so that anchoring aspects are understood before attempting the emergency rope slide.

To perform the emergency rope slide (Skill 7-9):

1. The firefighter will need to establish an anchor that will be suitable enough to support his weight throughout the maneuver (**Skill 7-9A**; see Chapter 6).
2. The rope should be wrapped behind the back, around the SCBA as high under the arms as possible. Both ends of the rope should be in front of the body (**Skill 7-9B**).
3. Both hands grasp the ends of the rope and are used to control the slide. The firefighter should make certain to grasp the rope at a point that will allow his hands to clear the windowsill and avoid being pinched (**Skill 7-9C**).
4. With the anchor secured and the rope grasped in front of the body in the proper wrapped position, the firefighter straddles the windowsill and begins to roll out of the window.
5. As the firefighter rolls out of the window, he transfers his weight from the windowsill to the rope system while maintaining control with his interior leg (**Skill 7-9D**).

SKILL 7-9
To Perform the Emergency Rope Slide

A The firefighter will need to establish an anchor that will be suitable enough to support his weight throughout the maneuver.

B The rope should be wrapped behind the back, around the SCBA as high under the arms as possible. Both ends of the rope should be in front of the body.

C Both hands grasp the ends of the rope and are used to control the slide. The firefighter should make certain to grasp the rope at a point that will allow his hands to clear the windowsill and avoid being pinched.

(continued)

SKILL 7-9 (CONTINUED)

To Perform the Emergency Rope Slide

D As the firefighter rolls out of the window, he transfers his weight from the windowsill to the rope system while maintaining control with his interior leg.

E When his weight is transferred, he controls his slide by adjusting the grip of his hands. Keeping the arms and hands close to the body increases the braking effect and moving the arms and hands away from the body decreases the braking effect.

6. When his weight is transferred, he controls his slide by adjusting the grip of his hands **(Skill 7-9E)**. The friction that is created by the rope being wrapped around the upper body and SCBA creates a **friction device** that controls the descent. Keeping the arms and hands close to the body increases the braking effect and moving the arms and hands away from the body decreases the braking effect. The firefighter should control his descent in a slow and deliberate manner and should avoid any shock loading to the rope system.

The firefighter should know the length of the rope and how far it will allow him to slide. If the rope will not reach the ground it may be used to reach a lower rooftop or window that can be entered on a lower level.

There are many descent devices and harnesses that can be used to perform this type of exit. Each system will require its own proper rigging in order for it to be safe and effective. Firefighters should look into these systems as they may provide a safe and effective egress system. However, remember that if this maneuver needs to be used on the actual fireground, time is not a luxury.

Webbing and Emergency Egress

The use of webbing for emergency escape maneuvers is obviously limited to its length, which most of the time will not be as long as a personal escape rope. Nonetheless it cannot be overlooked. Every firefighter should carry at least 20 feet of webbing with a water or tape knot tied into it to create one large loop. This loop can quickly be secured to a tool and placed in the corner of a window to allow the firefighter to egress quickly and hang from a window long enough for someone to raise a ladder for assistance. Products such as the multiple application service tool (MAST) provide

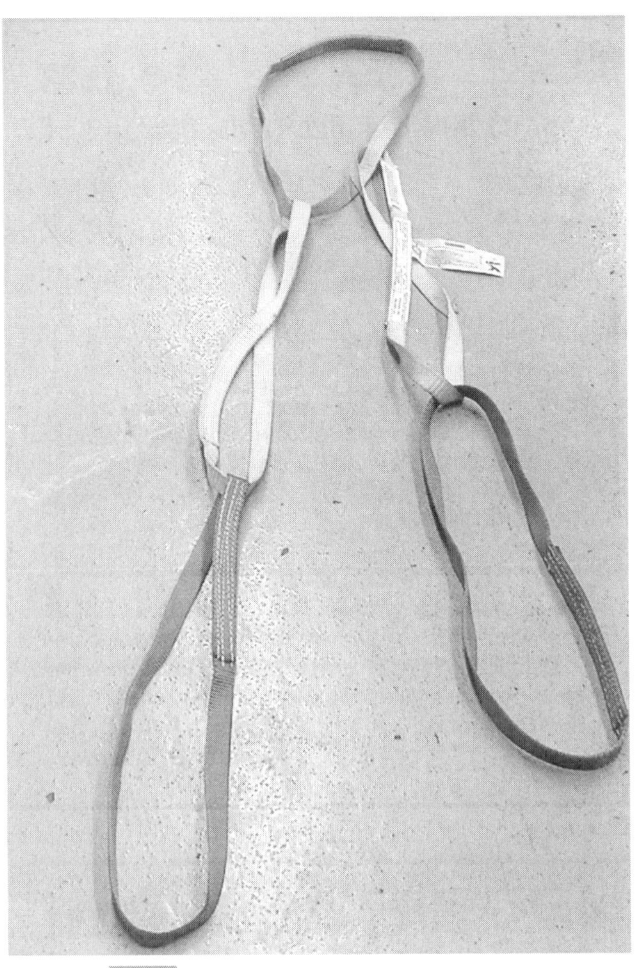

FIGURE 7-10

The multiple application service tool by Sling-Link.

FIGURE 

The multiple application service tool being used as a soft ladder.

loops linked together in a daisy-chain-type manner **(Figure 7-10).** These can be used as short, soft ladders as well as harnesses for lowering and dragging **(Figure 7-11).**

Emergency Hose Slide

Another option that exists for emergency escape from an upper floor is the use of a **emergency hose slide** to safety. Sliding along a hoseline may also be the quickest option if a firefighter goes through a floor and needs immediate water protection. Giving up the protection of a charged hoseline is generally not approved but we are referring to extreme circumstances that leave firefighters with no other alternative, such as inevitable collapse or uncontrollable fire conditions.

To perform the emergency hose slide (Skill 7-10):

1. Lower the nozzle end of the hoseline out the window to ground level **(Skill 7-10A).**
2. Anchor the hoseline to a door frame, staircase, or large object **(Skill 7-10B).** Firefighters kneeling on the line can be used as anchors when the line is being slid by rescuers to help a firefighter who has fallen through a floor.

Safety

Make certain that the line is anchored and will not move prior to sliding!

SKILL 7-10

To Perform the Emergency Hose Slide

A Lower the nozzle end of the hoseline out the window to ground level.

B Anchor the hoseline to a door frame, staircase, or large object.

C The hoseline is placed between the firefighter's legs and then around the ankle of one leg across the top of the boot. The opposite foot is crossed over the hoseline, locking it into a "figure four" between the feet.

D The firefighter's rate of descent will be controlled by the legs. The tighter the legs are kept, the slower the firefighter will descend.

E Once down, the firefighter should roll out of the way of other escaping firefighters.

3. The hoseline is placed between the firefighter's legs and then around the ankle of one leg across the top of the boot **(Skill 7-10C).**
4. The opposite foot is crossed over the hoseline, locking it into a "figure four" between the feet **(Skill 7-10C).**
5. The firefighter transfers his weight onto the hoseline by rolling over the edge of the opening. He remains in control by squeezing his legs together and grasping the *top* of the hoseline. The firefighter needs to make certain to keep his hand placement clear of the point where the hoseline changes direction in the opening. Failure to do so will cause the firefighter's hands to become pinched underneath the hose and render them unable to move.
6. The firefighter's rate of descent will be controlled by his legs. The tighter the legs are kept, the slower the firefighter will descend **(Skill 7-10D).**
7. Once down, the firefighter should roll out of the way of other escaping firefighters **(Skill 7-10E).**

Drywall Ladder Technique

Getting up to a window from an elevation difference is not generally discussed but is worth mentioning. If a firefighter cannot reach a window or an opening above to make an emergency egress, the walls can be used to make a ladder if they are made of gypsum board. This is called the **drywall ladder technique.**

To perform the drywall ladder technique (Skill 7-11):

1. Make several holes in the drywall utilizing a tool, placing them running in series parallel to each other **(Skill 7-11A).** Make certain to not place them too close to each

SKILL 7-11

To Perform the Drywall Ladder Technique

A Make several holes in the drywall utilizing a tool, placing them running in series parallel to each other.

B Use the holes to crawl upwards as a ladder to reach the window or opening for egress.

other so that the drywall breaks off into pieces.

2. Place a boot in a lower hole and use a higher one as a hand hold. Use the holes to crawl upwards as a ladder to reach the window or opening for egress **(Skill 7-11B).**

To work effectively, the firefighter must make certain that the pressure from his weight is being placed in a downward motion as opposed to pulling back on the drywall itself.

Summary

The necessity of having to utilize an emergency escape maneuver is one of life and death for a firefighter. Prevention of the Mayday is the best solution for a fireground emergency. Basic firefighting skills and training need to provide the firefighter with this knowledge. Due to the precarious nature of the fireground, there can never be a true guarantee of not getting into trouble. Emergency escape maneuvers—such as wall breaching, mule kicking, ladder bailouts, and slide technique—should be part of a firefighter's continual training. These maneuvers will need to be executed under some of the most adverse conditions on the fireground without the luxury of safety lines, complex specialized equipment, and time. It is important that these procedures be kept as quick and simple as possible to allow a firefighter in trouble the greatest chance for survival. Training is the one and only time that all variables can be controlled; *never* compromise safety when practicing self-survival escape procedures.

KEY TERMS

Butterfly ladder bailout
Drywall ladder technique
Electrical conduit
Emergency hose slide
Emergency rope slide
Friction device
Hang and drop
Hook-and-go ladder bailout
Ladder slide
Mule kicking
Wall breaching

REVIEW QUESTIONS

1. When considering using an emergency escape maneuver such as wall breaching, it should only be used
   a. as a method of choice.
   b. as a last resort to reaching a safe area.
   c. to draw attention to your position.
   d. to look for a downed firefighter.

2. When breaching a wood-frame wall the firefighter should make a
   a. very large opening to allow himself to pass through it.
   b. small opening at first in order to check the other side for obstructions, fire, and the presence of floors.
   c. triangle opening to avoid the fall-down of materials.
   d. none of the above

3. Most interior walls are made up of 2×4 wall studs on _____ centers.
   a. 24-inch
   b. 12-inch
   c. 16-inch
   d. 18-inch

4. When no tool is available to the firefighter in distress, what other means can be engaged in order to breach an interior wall?
   a. punching the wall with a gloved hand
   b. using the SCBA bottle as a battering ram
   c. mule kicking while thrusting backwards with the leg flexed
   d. using a hand light

5. Masonry walls should be breached in a __________ shape.

6. All of the following are forms of emergency escape maneuvers through windows, *except:*
   a. straddling and hanging from a windowsill.
   b. hanging and dropping.
   c. ladder bailouts and rope slides.
   d. dive method on a rope.

7. One of the first considerations when deciding to attempt an emergency escape out onto a ladder is knowing that the ladder is present outside the window of choice.

   True False

8. When conducting training on emergency escape ladder bailout, it is critical that
   a. a safety net is provided at the bottom of the ladder.
   b. a safety line is attached to the firefighter and is properly belayed by an attendant.
   c. live fire is kept away from the area of the escape maneuver.
   d. the ladder be positioned above the windowsill.

9. When positioning a ladder for an emergency bailout maneuver, the ladder should be placed
   a. above the windowsill with three rungs into the window.
   b. above the entire window opening.
   c. with the tip of the ladder just below the window sill.
   d. with the tip of the ladder above the window to the right or left of the window opening.

10. Where should the rope be positioned in relation to the firefighter's body when performing an emergency rope slide?

11. When sliding the hoseline, it is the firefighter's hand grip that controls the rate of descent.

    True False

12. List the steps to utilize drywall as a ladder to make an emergency egress.

ADDITIONAL RESOURCES

Firefighter's Handbook (2nd ed.). Clifton Park, NY: Thomson Delmar Learning, 2004.

Hoff, Robert, and Kolomay, R., *Firefighter Rescue and Survival.* Tulsa, OK: PennWell Publishing, 2003.

Lasky, R., Pressler, R., and Salka, J., "The Head First Ladder Slide: Three Methods." *Fire Engineering,* (2000, October) 59–61.

McCormack, J., and Pressler R., *Firefighter Survival.* Indianapolis, IN: Fire Department Training Network, 2002.

Norman, J., *Fire Officers Handbook of Tactics* (2nd ed.). Saddlebrook, NJ: PennWell Publishing, 1998.

Jakubowski, G., and Morton, M., *Rapid Intervention Teams.* Stillwater, OK: Fire Protection Publications, 2001.

Rapid Intervention Teams and How to Avoid Needing Them. Technical Report Series, Report #123. Washington, D.C.: U.S. Fire Administration, 2003.

Salka, J., *Training Your Firefighters to get Out Alive.* Training Program Handout.

CHAPTER

8 Removal Preparation and Emergency Air Supply

Learning Objectives

Upon completion of this chapter, you should be able to:

- explain what an immediate assessment involves pertaining to a downed firefighter's mental and physical condition.
- give details of some of the different methods of determining the presence and condition of a distressed firefighter's air supply.
- discuss what can be done for the downed firefighter who is not breathing.
- explain what is meant by a rapid removal of a downed firefighter by a RIT.
- discuss compatibility concerns involving different types of SCBA.
- identify why it is important for the RIT to be prepared to rescue at least two firefighters at any given time.
- explain the importance of bringing an additional air supply as part of the RIT team's basic and necessary equipment for a firefighter rescue.
- explain the two major disadvantages of utilizing buddy breathing.
- demonstrate the procedures involved in changing over the air supply of a distressed firefighter.

Case Study

"On March 14, 2001, at 1654 hours, Engine 24 responded to a reported exterior cardboard fire at a supermarket. A 40-year-old male career firefighter/paramedic (the victim) died from carbon monoxide poisoning and thermal burns after running out of air and becoming disoriented while fighting a supermarket fire. Four other firefighters were injured, one critically, while fighting the fire or performing search and rescue for the victim. The fire started near a dumpster on the exterior of the structure and extended through openings in the loading dock area into the storage area, and then into the main shopping area of the supermarket. The fire progressed to five alarms and involved more than 100 personnel.

"At 1729 hours, the victim radioed a transmission to the crews asking them not to back out because he needed help. The IC radioed back asking for his location. The victim replied that he was in the rear behind something, out of air, and down on the ground sucking in smoke. The IC advised him to stay calm and told him that crews were on their way to assist him. At approximately 1730 hours, the officer from Engine 21 heard someone yelling in the vicinity of the produce storage area. As he followed the voice, he ran into the victim, who was standing near one of the initial attack lines in the main supermarket area. The officer grabbed the victim, asked him to identify himself, and attempted to place him on the handline. The victim, who had removed his regulator from his facepiece because he was out of air, was resistive and reportedly appeared disoriented. On a second attempt, the officer was able to get the victim on his knees and on the handline. Soon after, the victim stood up, turned, and quickly moved toward the rear of the supermarket as the Engine 21 officer attempted to grab him and keep him on the handline. It is believed that the victim headed back toward the meat preparation room. A firefighter from Rescue 3 who was in the main storage area, upon hearing the victim's voice, headed toward the sound and ran into the victim just outside the swinging door that led from the main storage area into the meat preparation room. The victim told the Rescue 3 firefighter that he was out of air, and the firefighter told him to stay calm and follow him. The victim turned and headed away from the firefighter toward the meat preparation room as the firefighter grabbed the victim and told him that he was going the wrong way. As the victim turned, he knocked the Rescue 3 firefighter down and the two became separated. Returning to his feet, the firefighter tried to find the victim but ran out of air and was forced to leave the building.

"The Engine 25 officer located the victim in the meat preparation room. The victim was unresponsive, and his PASS device was sounding. The Engine 25 officer, who was alone with the victim, made an emergency transmission that was not received by the IC. At 1740 hours, an officer from Engine 4 heard the transmission and radioed that a firefighter was down in the southwest corner. The Engine 25 officer, unable to move the victim, ran out of air, removed his facepiece and regulator, and attempted to crawl out of the building. Simultaneously, crews from Engine 6 and Engine 710 entered the south side of the building and proceeded toward the produce storage area. As crew members from Engine 6 neared the produce storage area, they heard a PASS device sounding from their left (meat preparation room). The Engine 6 crew encountered the officer from Engine 25 (his PASS device was not sounding) and passed him to the Engine 710 crew, who assisted him out of the building. The Engine 25 officer, the third firefighter to become injured, was transported to a local hospital and treated for smoke inhalation.

"The Engine 6 crew continued toward the sounding PASS device and found the victim lying unconscious on his back with his facepiece partially removed. They checked the victim for a pulse and could not find one.

"Numerous crews participated in the removal of the victim from the building. Firefighters were hampered in their removal efforts by the victim's size (he was 6 feet, 4 inches tall, and he weighed about 289 pounds [in addition to the weight of his gear]) and the amount of debris blocking their path through the main storage area to the south-side roll-up door. Approximately 19 minutes elapsed from the point when firefighters found the unconscious victim to the point when the victim was removed from the building. Firefighters immediately began cardiopulmonary resuscitation (CPR) and advanced life support (ALS) on the victim at the scene."

—Case Study taken from NIOSH Firefighter Fatality Report # 2001-13,* "Supermarket Fire Claims the Life of One Career Firefighter and Critically Injures Another Career Firefighter—Arizona," *full report available online at http://www.cdc.gov/niosh /face0113.html.

Introduction

One of the most important issues faced during Mayday operations is the RIT's ability to provide an adequate air supply for the distressed firefighter. This must be accomplished while protecting and monitoring their own air supply.

When the RIT is assigned on the fireground, it is understood that they will be preparing for a possible Mayday situation. This includes performing proactive actions and placing their tools in a ready state **(Figure 8-1).** When the call for a Mayday is received, the action begins with the RIT reporting to the sector involved. What the RIT does after locating the downed firefighter will have a direct effect on the outcome of the situation.

Assessment and Communication

Once the RIT has located the downed firefighter, it should begin an immediate assessment of the downed firefighter's condition. This assessment should be thorough but quick. In rapid intervention terms, the assessment, preparation, and removal are performed simultaneously.

After locating the downed firefighter, the rescuers will need to establish immediate communication with the downed member to ascertain mental capacity to follow directions as well as physical ability to move. Distressed firefighters who have been exposed to toxic products of combustion can become uncooperative and even combative. Firefighters are instructed to help the search efforts of the RIT by activating their PASS alarm when they are in trouble. Once the firefighter has been found, the RIT should silence the device to allow communications with the distressed firefighter and the RIT operations chief (or rescue branch). The RIT should also communicate to the RIT operations chief or rescue branch that the victim has been located and give a description of the victim's location and condition. With this information, an exterior RIT may be able to assist by breaching a wall or enlarging an opening to make removal easier.

The RIT should immediately determine the amount of air left in the distressed firefighters' SCBA. Sometimes the need to determine this is quite evident by the sound of the low-air alarm on the SCBA (low-air alarms will activate when there is 25 percent or less of the rated capacity of the air cylinder left). Other methods include opening the purge or bypass valve to listen for air flow, breaking the mask-to-face seal of the facepiece momentarily to listen for air flow, and checking the **SCBA remote gauge** or cylinder gauge **(Figures 8-2** and **8-3).** If the air supply is critically diminished, air must be supplied either by exchanging the SCBA, or establishing some type of auxiliary air supply. Disconnecting the mask-mounted regulator may be necessary to prevent suffocation if the downed firefighter is not getting any more air from the SCBA.

If the distressed firefighter is unconscious, the RIT will have to determine if the firefighter is breathing. This can be extremely difficult in a loud environment. The rescuers should listen for noise by the downed firefighter's **exhalation valve.** While RIT members are addressing the air supply issue, they should also be determining if there is any entrapment that will require extrication. More teams and equipment may be required to complete the rescue.

If the downed firefighter is not breathing when found, the RIT should immediately begin to remove the firefighter. Just as in EMS trauma using rapid extrication on patients that are

FIGURE  8-1

An alternate source of air for the downed firefighter must be prepared prior to entry and should be brought inside as part of the RIT's basic tools.

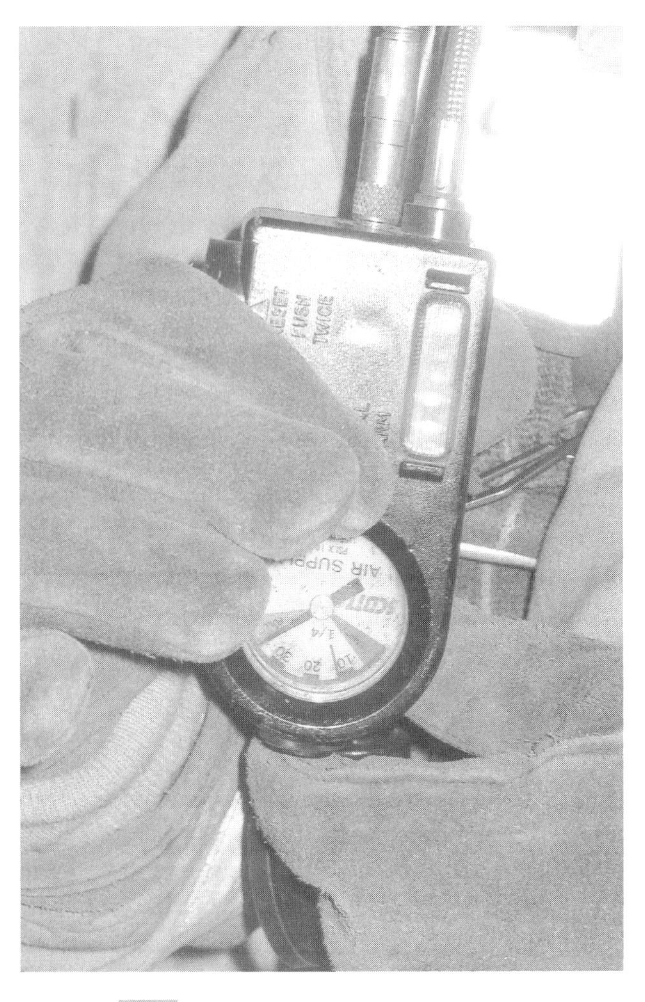

FIGURE 8-2

An example of SCBA remote gauge.

FIGURE 

Air supply can be checked by looking at the cylinder gauge. However, this may be difficult under fireground conditions.

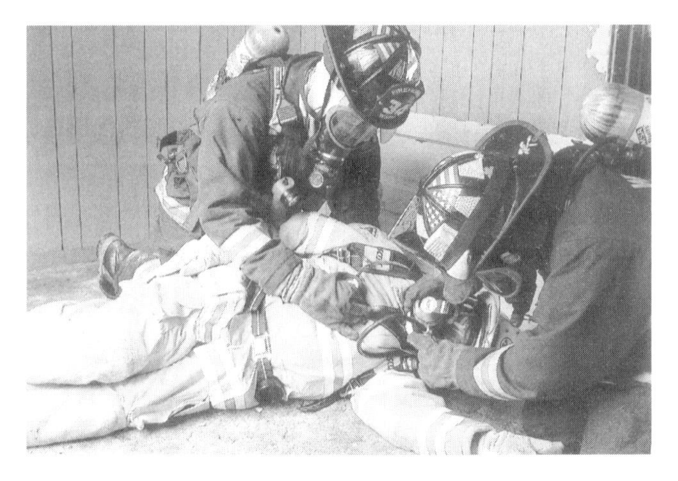

FIGURE 

Rapid assessment of the downed firefighter by the RIT. The firefighter is positioned face up as one RIT member checks the air supply by the remote gauge while the other RIT member checks to see if the firefighter is breathing.

threatened, a rapid removal process should be executed by the RIT. Once the initial RIT responds to a Mayday there should be additional RITs summoned and ready to go to work at a moment's notice **(Figure 8-4).**

Compatibility

When firefighters become lost or trapped, air supply becomes a major factor. Members assigned to a RIT should know what type of SCBA is being used on the fireground. This includes their own department and mutual aid departments. The RIT should also be aware of any compatibility issues with the different types of SCBAs. Do the SCBAs being used have quick-connect features? Are the RIT packs compatible in regard to those quick-connecting features as well as masks, regulators, and cylinders? Generally speaking, each SCBA is only compatible with components of the same model and manufacturer unless specified. Using any other components may void all warranties on the unit and may not work. There are so many different features with each manufacturer and their products that the only way to become familiar with your surrounding departments that have different brands of SCBAs is to work with them to understand the features of their SCBAs. Members of

Using components from different manufacturers of SCBAs may void warranties on the unit and may not work if not specified as acceptable.

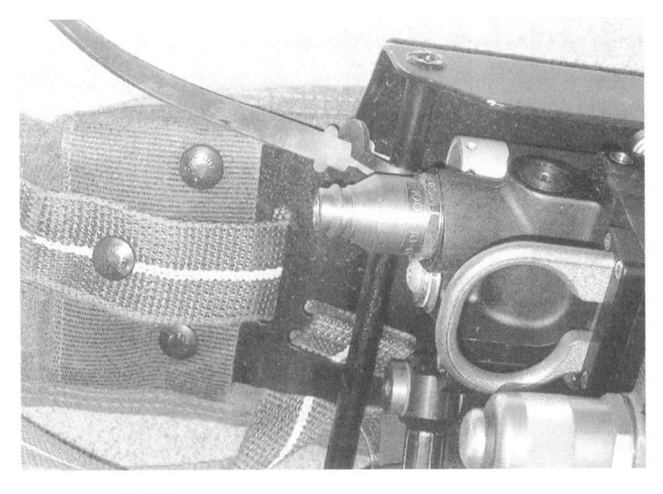

FIGURE 8-5

Rapid Intervention Crew Universal Air Coupling.

the RIT should be just as familiar with the downed firefighter's SCBA as they are with their own. An idea for promoting this knowledge is to trade one of your department's SCBAs for one from a neighboring department for a week or two for use in familiarizing and training members.

According to the 2002 edition of **NFPA 1981** (Standard on Open Circuit Self-Contained Breathing Apparatus for Fire and Emergency Services), manufacturers must produce SCBA units that contain a **Rapid Intervention Crew Universal Air Coupling (RIC UAC)** to be in compliance for firefighting **(Figure 8-5).** This connection will allow a cylinder that is low on air to be transfilled from another cylinder regardless of manufacturer. Each cylinder will then have equal amounts of air in them after the fill. This universal connection must be permanently fixed to the unit within 4 inches of the threads of the SCBA cylinder valve. These requirements only apply to newly manufactured SCBAs. Existing SCBAs are able to be retrofitted with this feature, but this is not required by the standard. Over time, this may help alleviate some of the issues with compatibility, but there will always be departments whose budgets just will not allow new purchases or retrofits to take place.

Decisions

RITs should be prepared to deal with at least two distressed firefighters at any given time. The reason for this is obvious. Firefighters are trained to work in pairs. The RIT may need to consider two RIT packs when attempting a rescue.

When the RIT responds to a Mayday, it should automatically bring an additional air supply. This will eliminate the need for members of the RIT to compromise their own precious air supplies. Before the RIT enters the building, members should realize the problems that might be encountered regarding air supply. Even when the distressed firefighter is located, the important decision of direct removal or taking the time to secure an additional air supply will have to be made. The best way to solve the distressed firefighter's low-air dilemma may be removing the victim from the harmful environment without stopping to provide a new air supply. Remember, the downed firefighter does not necessarily have to be brought out the way in which he or the RIT entered. Do not lose focus on the surroundings, such as alternate exits and windows that may be present. These decisions will have to be weighed and acted upon by the RIT.

Air Supply

SCBAs may be equipped with an emergency breathing support system (EBSS) or RIC UAC, but there is no guarantee that these will be present or even operable due to the circumstances encountered.

Buddy breathing is a term that has been with the fire service for quite some time. The term *buddy breathing* has two different variations and meanings. EBSS refers to any manufacturer-supplied SCBA connection that is connected to the SCBA cylinder and used to supply air to another SCBA. An EBSS does not require a firefighter to remove the regulator or facepiece of the SCBA, therefore maintaining a closed system and protecting the firefighter from any atmospheric contaminants. The other form of buddy breathing involves using devices such as tubing. This is a high-risk method of transferring supplied air because the seal of the SCBA facepiece must be compromised, which may expose both firefighters to toxins and contaminants. It is a last-resort measure. The major disadvantage of both of these maneuvers is that no matter which one is employed, two people are sharing an air supply that was designed for only one. It also restricts movement because both members are now connected **(Figure 8-6).**

Note

With buddy breathing or EBSS use, two people are sharing an air supply that was designed for only one and movement is restricted because both firefighters are now connected to each other.

An air supply unit set up specifically for rapid intervention use is the best option for a RIT when it comes to supplying the downed firefighter with air. These **emergency air supply units** are generally SCBAs that are equipped without the shoulder and waist harnesses to make them easier to transport during search operations **(Figure 8-7).** They can be set up with longer hoses to reach firefighters that may be trapped and may have connections to allow transfilling **(Figure 8-8).** The RIT emergency air supply unit should include a second facepiece and mask-mounted regulator (if used) in case there is a problem with either, and in case the downed firefighter is using an SCBA that is incompatible.

The RIT should also remember that they are consuming their own air supply while searching. RIT members must be consciously aware of their point of no return while performing the search. They should try to call additional companies to relieve them before they run into a low-air situation. Remember, the minimum number of RITs required to remove just one downed firefighter is be three to four teams. It should be understood that this is a realization and not an exaggeration, plan on it! The first RIT will usually only be able to address the issues of air supply and assessment of the degree of entrapment, leaving later RITs to perform the actual removal.

The successful use of an emergency air supply will depend on the RIT's knowledge of the SCBA being used by the distressed firefighter. If the RIT is familiar with the SCBA and it is com-

FIGURE  8-6

Firefighters utilizing an EBSS are restricted by the length of the line that connects them.

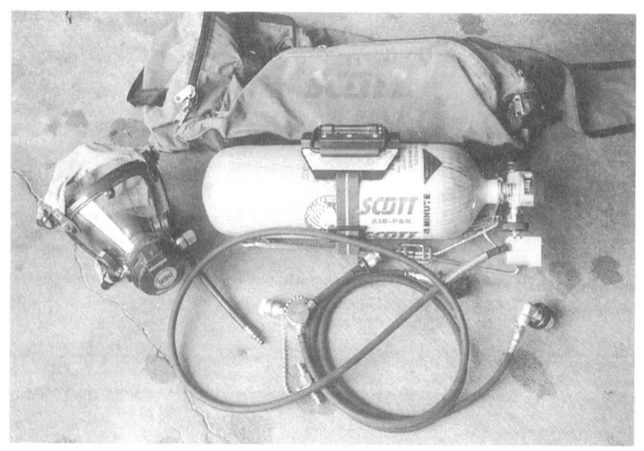

FIGURE  8-7

Rapid intervention emergency air supply unit with a second facepiece, mask-mounted regulator, and fittings.

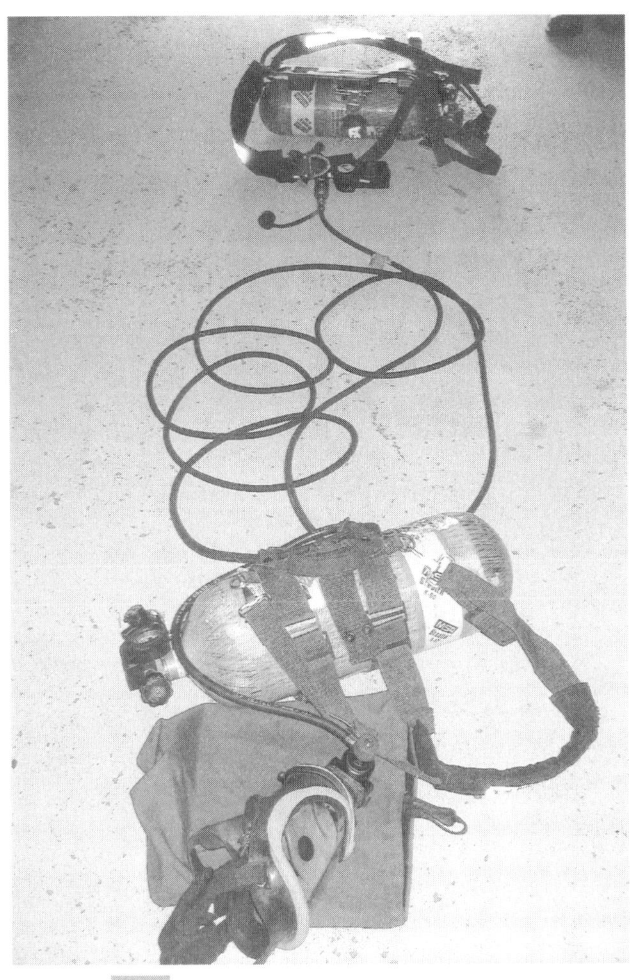

FIGURE 8-8

Some fire departments have attached longer hoses to their RIT emergency air supply units to allow hook up in circumstances where the downed firefighter may be trapped in a confined area.

patible with the emergency supply, supplying air will be a relatively simple procedure.

Note

It is the RIT's responsibility to make sure that the RIT air supply unit is prepared properly prior to the connection (full air before making entry, turned on, etc.).

To refill a downed firefighter's air supply using a RIC UAC (Skill 8-1):

1. Assess and determine need to make an air supply changeover.
2. Turn the emergency air supply unit on **(Skill 8-1A).**
3. Connect the RIC UAC, allowing the air to equalize in both cylinders **(Skill 8-1B).**
4. Disconnect the RIC UAC once the cylinders are equalized.
5. Package the downed firefighter for removal **(Skill 8-1C).**

To change out a downed firefighter's air supply when the mask-mounted regulator is inoperable (same brand SCBA) (Skill 8-2):

1. Assess and determine the need to make the air supply changeover.
2. Turn the emergency air supply unit on.
3. Disconnect regulator on the downed firefighter's facepiece **(Skill 8-2A).**
4. Connect the new regulator **(Skill 8-2B).**
5. Purge the facepiece by turning on the bypass or purge valve momentarily **(Skill 8-2C).**
6. Package the downed firefighter for removal.

When the SCBA is incompatible with the RIT emergency air supply, the RIT will have to consider securing the air supply by exchanging the facepiece on the victim with the one from the RIT emergency air supply.

To change out a downed firefighter's air supply when there is facepiece damage or incompatible units (Skill 8-3):

1. Changing over the facepiece will require removal of the downed firefighter's helmet and protective hood, which could take some time and effort and will also expose the downed firefighter to additional hazardous conditions **(Skill 8-3A).**
2. The rescuers should make sure that everything is prepared and that the facepiece of the new emergency air supply is already attached to the SCBA.
3. When changing over, facepieces should be switched as quickly as possible to limit the time that the distressed firefighter is exposed to the potentially harmful

SKILL 8-1

To Refill a Downed Firefighter's Air Supply Using a RIC UAC

A Turn the emergency air supply unit on.

B Connect the RIC UAC, allowing the air to equalize in both cylinders.

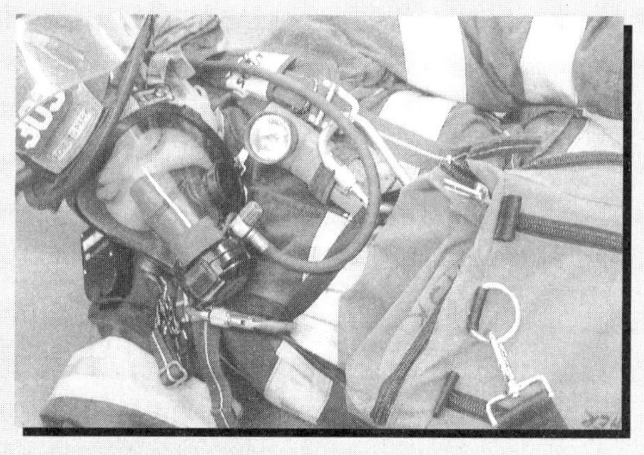

C Package the downed firefighter for removal.

atmosphere. As the change is taking place, the rescuers should open the purge or bypass valve in the new facepiece to create a positive pressure atmosphere during the change **(Skill 8-3B).**

4. Once the facepiece is secured, the bypass or purge valve should be closed.
5. The emergency air supply will have to be secured to the downed firefighter prior to removal.

Once the RIT has supplied the distressed firefighter with an alternate air supply, it will need to be secured to the firefighter to facilitate movement. One way to secure the RIT emergency air supply unit to the firefighter is to place it on top of the downed firefighter's body, securing it to the existing straps from the original SCBA **(Figure 8-9).** The downed firefighter can then be dragged by the straps of the original SCBA (refer to Chapter 9).

Once the RIT has secured the distressed firefighter's air supply, this action should be immediately communicated to the RIT operations chief or rescue branch.

The RIT will also have to determine the need for spinal mobilization and whether or not the situation will allow it. Are the interior conditions

SKILL 8-2

To Change Out a Downed Firefighter's Air Supply When the Mask-Mounted Regulator is Inoperable (same brand SCBA)

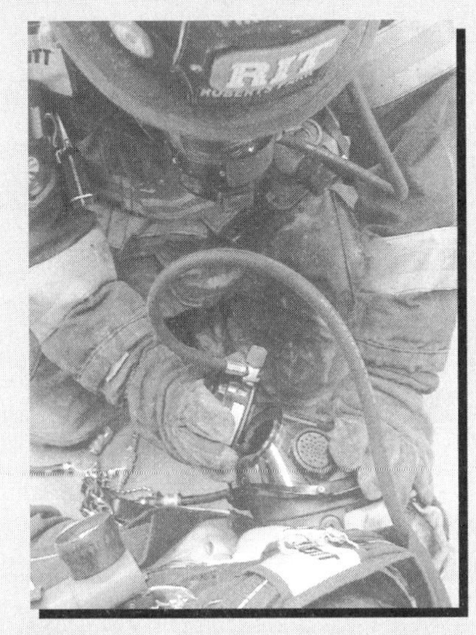

A Disconnect regulator on downed firefighter's facepiece.

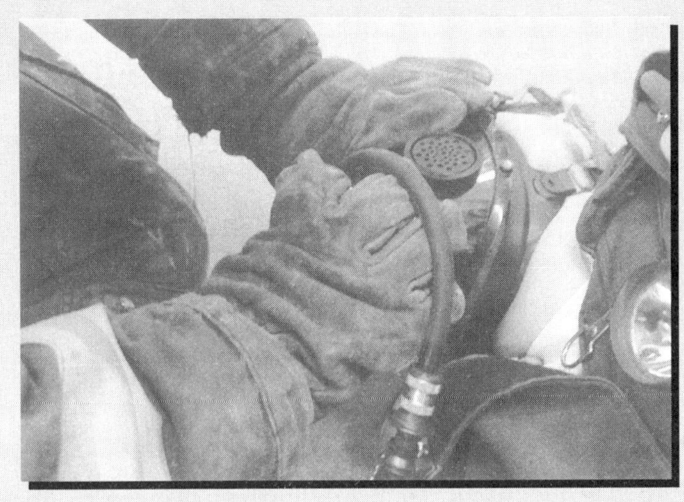

B Connect the new regulator.

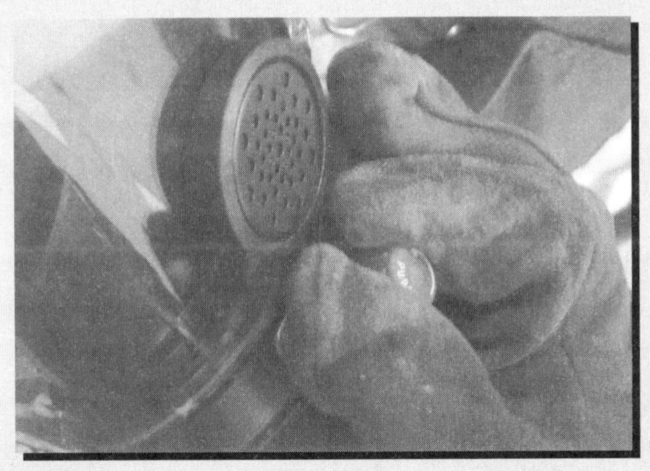

C Purge the facepiece by turning on the bypass or purge valve momentarily.

adequate to even consider these types of maneuvers? If conditions start to deteriorate the RIT will have to use whatever method is available to make a quick removal while inflicting the least amount of trauma on the downed firefighter.

One scenario that may be encountered by the RIT is the rescue of a firefighter that has suffered a heart attack or is in cardiac arrest. The RIT must be prepared to move rapidly and to make the removal as quick as possible. Supplying an emergency air supply to a conscious distressed firefighter having chest pain is one thing, trying to supply that same emergency air supply to a downed firefighter experiencing

SKILL 8-3

To Change Out A Downed Firefighter's Air Supply When There is Facepiece Damage or Incompatible Units

A Changing over the facepiece is going to require removal of the downed firefighter's helmet and protective hood, which could take some time and effort and will also expose the downed firefighter to additional hazardous conditions.

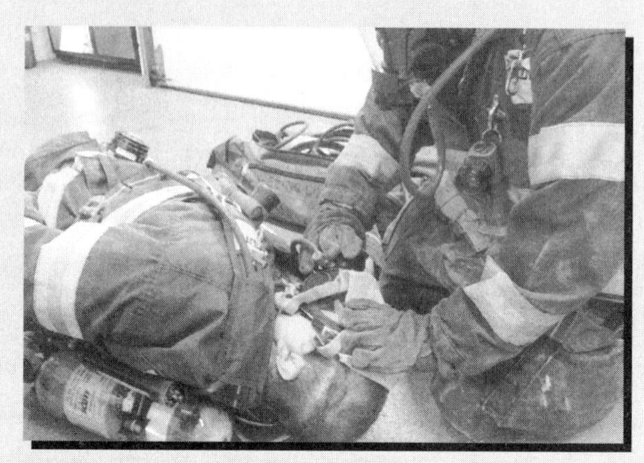

B When changing over, facepieces should be switched as quickly as possible to limit the time that the distressed firefighter is exposed to the potentially harmful atmosphere. As the change is taking place, the rescuers should open the purge or bypass valve in the new facepiece to create a positive pressure atmosphere during the change.

FIGURE  8-9

The RIT emergency air supply unit can be secured for moving the downed firefighter by attaching it to the downed firefighter's SCBA straps with a carabiner.

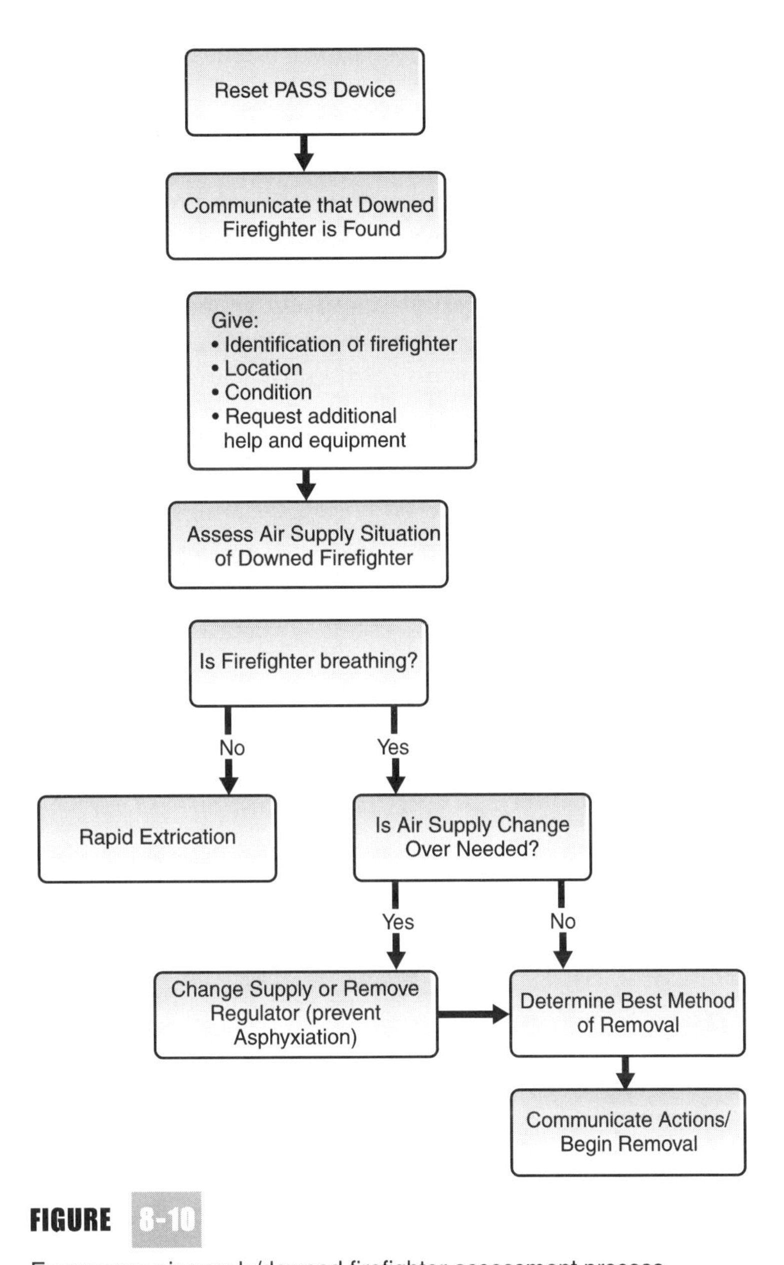

FIGURE 8-10

Emergency air supply/downed firefighter assessment process.

cardiac arrest is futile until the victim can be brought into an area where proper advanced cardiac life support can be administered.

The steps in securing a downed firefighter's air supply can be best summarized in the checklist provided in **Figure 8-10.**

Summary

The RIT may not always be required to secure the air supply with an alternate air source. An air supply can also be established by simply getting the distressed firefighter to a safe area that does not require an SCBA, such as a window. If this is the method chosen, the RIT must quickly determine the means of egress, to either the outside of the structure or a safe area and within, move the firefighter.

SCBA incompatibilities are a problem only when the RIT has not prepared prior to the emergency. The incompatibilities should be addressed prior to the incident. During an actual fireground emergency, the constraints of limited time and fire conditions will require the rescuers to begin to move the distressed firefighter immediately, sometimes even before the air supply problems have been resolved. Any additional delay caused by not being familiar with the type of SCBA will just add to the problems

The RIT members must also monitor their air supply when responding to the distressed firefighter. The RIT will be going into the same conditions that the distressed firefighter did, and with the same type of air supply. This is an important point to remember and the rescuers should remind themselves of the potential problems that could be encountered with their air supply. Training and constant improvement on techniques and skills involving rapid intervention maneuvers will help bring forward limitations of the RIT members and their air supply. All RIT members should establish a point of no return with regard to their air supply. When any member of a RIT reaches the predetermined point that has been established, the member should leave with another team member. The size of the structure will determine the amount of time that will be needed to exit based on the air supply that has been used to reach a certain point. This is a very important concept that must be understood by all RIT members. Not adhering to this principle may contribute to the problem by creating a need for the RIT to be rescued.

KEY TERMS

Emergency air supply unit
Exhalation valve
NFPA 1981
Rapid Intervention Crew Universal Air Coupling (RIC UAC)
SCBA remote gauge

REVIEW QUESTIONS

1. Which one of the following is *not* part of an appropriate rapid assessment by a RIT regarding a distressed firefighter?
   a. Once the downed firefighter is found the RIT should silence the PASS device and then communicate to Command.
   b. Determining the firefighters mental capacity.
   c. Determining the firefighter's physical ability.
   d. Assessing blood pressure and pulse.
2. What are some of the things a RIT can do into order to properly assess the distressed firefighter's air supply?
   a. Listen for the sound of the low-air alarm.
   b. Open the bypass valve to listen for air flow.
   c. Break the mask-to-face seal momentarily, listening for positive pressure flow.
   d. all of the above
3. Low-air alarms on SCBA will sound when there is _____ percent or less of the rated cylinder capacity left.
4. NFPA 1981 addresses the issue of
   a. firefighter health and safety.
   b. rapid intervention teams.
   c. staffing levels.
   d. SCBA use.
5. Disadvantages of using an EBSS include:
   a. time to connect.
   b. two people share one air source.
   c. restricted movement.
   d. both (b) and (c)
6. The minimum number of RIT teams usually needed to move a downed firefighter to safety is
   a. one.
   b. two.
   c. three.
   d. more than four.

7. When the SCBA of the distressed firefighter is different or incompatible with the RIT pack the RIT will need to
   a. exchange the facepiece on the downed firefighter with the one provided from the RIT pack.
   b. install an adapter on the regulator.
   c. do nothing—the facepiece should never be changed due to time constraints in a hazardous atmosphere.
   d. both (a) and (b)
8. One of the ways to secure the RIT emergency air supply unit to the downed firefighter is to
   a. secure and strap the pack to the downed firefighter's legs using the waist belt.
   b. remove the original SCBA from the downed firefighter and then apply the new SCBA.
   c. place the RIT emergency air supply on top of the downed firefighter while securing it to the existing straps of the original SCBA that is being worn by the downed firefighter.
   d. secure the RIT emergency air supply unit through the use of webbing tied to the waist belt of the original SCBA.
9. Getting a distressed firefighter to a window may be the easiest way to correct an air supply problem.

   True False
10. When a RIT responds to a Mayday within a structure, members should constantly be aware of their own air supply and remember to establish
    a. three ways in and three ways out.
    b. a point of no return regarding their air supply.
    c. a sector where they will be exiting from the structure.
    d. communications with Dispatch.

■ ADDITIONAL RESOURCES

Fiefighter's Handbook (2nd ed.). Clifton Park, NY: Thomson Delmar Learning, 2004.

Hoff, Robert, and Kolomay, R., *Firefighter Rescue and Survival.* Tulsa, OK: PennWell Publishing, 2003.

Jakubowski, G., and Morton, M., *Rapid Intervention Teams.* Stillwater, OK: Fire Protection Publications, 2001

McCormack, J., *Firefighter Rescue and Rapid Intervention Teams.* Indianapolis, IN: Fire Department Training Network, 2003.

CHAPTER

9 Moving the Downed Firefighter: Carries and Drags

Learning Objectives

Upon completion of this chapter, you should be able to:

- identify basic equipment that the RIT may need to move a downed firefighter.
- exhibit the procedures required to move a downed firefighter using the extremity carry, cradle carry, and blanket carry.
- discuss and demonstrate the use of an SCBA harness conversion when applying it to a downed firefighter.
- show the procedures required to move a downed firefighter using the lift-and-lead drag, the push-and-pull drag, the tool drag, and the blanket drag.
- give details on the difference of webbing and rescue loops for moving a downed firefighter.
- exhibit the procedures required to move a downed firefighter up a staircase using the multiple rescuer staircase lift, the stair raise with a tool technique, the handcuff knot, rescue loops, and a mechanical advantage system.
- demonstrate the procedures required to move a downed firefighter down a staircase.

Case Study

"On July 29, 1992 I was assigned to Station 2 driving Engine 602 for the Oak Park Fire Department. There were two other members of the department assigned to 602 that day. Ambulance 612 was also stationed there with two paramedics.

"Around 2300 hrs we got the report of a structure fire at 542 N. Humphrey. Upon arrival we found smoke coming from the basement of the structure. We laid a 1 3/4 preconnect and the other four members of the crew took it into the basement to locate and extinguish the fire, while I gained a water supply.

"From what I have been told the crew entered the side door of the house to try and locate the fire. They went down a few stairs and turned right to go through the basement and try to find the way to the rear of the basement. When they had made it to a small room, they realized that they had gone all the way across the basement. Things were getting hotter down there and they had not found the seat of the fire.

"The lieutenant told them to head back to the door they came in so they could try to get a better location of the fire. At this point, the lieutenant passed out and the man in front of him felt him fall on his ankles; he then tried to drag him to get him out of the basement, but could not and was running low on air. He had to leave the basement. I am not sure how far he got with lieutenant, but he had to leave the basement when he ran out of air. When that man got out the shift commander was informed that the lieutenant was not with them. At this time everybody else became aware of the problem.

"Many members of the shift tried to get back in to help get the lieutenant out of the basement. We believe that while trying to drag him up the stairs from the basement that some of his equipment (possibly the regulator on his SCBA) caught on the leading edge of the stairs. We could not get him up the 3 or 4 stairs leading from the basement until someone climbed over him and picked up his legs.

"When we went in later and looked the small room was a bathroom on the north side of the basement. From the best that we could figure out some of the lieutenant's equipment snagged on something in the basement, which separated him from his air supply. He crawled for a few feet before passing out. At this time a lot of the fire suppression was halted except in the area that they were trying to rescue the lieutenant from. Sometime during the operation a PPV (positive pressure ventilation) fan was placed into operation along with a couple more lines.

"Four other firefighters were injured (burns and smoke inhalation). Help was given to us from four neighboring departments. The lieutenant was pronounced dead at a local hospital at around 11:50 P.M. A short time later the shift was informed that we had lost him."

—Case study courtesy of Dennis K. Weidler, Firefighter (Ret.), Oak Park Fire Department, Oak Park, Illinois. Special thanks to Dennis Weidler for sharing his account with us.

Introduction

The process of rescuing a downed firefighter can become extensive and may involve multiple decisions and choices related to techniques that will be necessary in extraction. Often, these operations will take place under severe fireground conditions. These conditions can include the inability to stand up due to high heat, limited or zero visibility, and working in restricted or confined areas. The conditions present as well as the weight of the downed firefighter will dictate the methods utilized in moving the downed firefighter. A 180-lb firefighter may weigh well over 300 lbs with turnout gear, SCBA and water from firefighting operations absorbed into the gear. Many of the difficulties experienced in moving the downed firefighter result from the extra weight of the gear. Limited "grab points," bulkiness, and entanglement points are just a few problems that may be encountered. Moving a downed firefighter is a definite challenge.

Rescue Plan

It is important that the RIT has a rescue plan in place when it locates the downed firefighter and that each member of the RIT understands what that plan is. Communication among the RIT members is a necessity but should be kept in simple and understandable terms. Too much communication will cause confusion, waste time, use up valuable air supply, and slow the rescue process. Communication must also allow the rescuers to be synchronized in their efforts. For example, a firefighter pulling up on one side of the downed firefighter before the rescuer on the other side is ready will result in the downed firefighter not being moved and a wasted effort. Simple terms such as "Ready—Go" or "Set—Lift" should be used. A pause after the first command will give the other rescuer an opportunity to stop the operation if he is not in position or to acknowledge it and proceed. The words, "Stop" or "Ready" should be used for this acknowledgement. Remember, wasted efforts result in wasted time—make certain that members understand each other!

FIGURE 9-1

Rescuers must make certain that they keep their backs straight and use their leg muscles to move the downed firefighter to avoid injury to themselves.

Make certain that proper lifting techniques are practiced to avoid injury to rescuers.

Drastic and unconventional measures may need to be taken to remove the downed firefighter. Safety and the imagination are the only limiting factors when removing a downed firefighter in an expedient manner. The key to all of the methods discussed for moving a downed firefighter is technique. Rescuer brute strength is a great asset but is not required. Rescuers must make certain that they keep their backs straight and use their leg muscles to move the downed firefighter to avoid injury to themselves **(Figure 9-1)**.

Some suggested equipment that may help the RIT in moving a downed firefighter is as follows:

- rope bag containing 50 to 70 feet of rope with two carabiners
- 20-ft. length of webbing with a water knot pretied into the webbing to provide a large loop
- daisy chained webbing consisting of five loops or a MAST product
- Halligan bar/axe/short pike poles or closet hook
- attic ladder
- stokes basket/rescue litter
- pretied Prussik loops (approximately 3 ft long)

This basic equipment should be available to the RIT. Some of the listed equipment for RIT members can and should be personal equipment that is in their possession no matter what their role is on the fireground. The idea is to be prepared and to try to utilize the quickest and most efficient maneuvers in a given situation.

Carries

Carrying a downed firefighter will be easier than dragging if conditions will permit. Carrying the downed firefighter will allow obstacles and debris located at the floor level to be navigated successfully without slowing down the removal process. Again, conditions will dictate if this is even possible.

Extremity Carry

The **extremity carry** is a basic carry that can be performed by two rescuers. It will require the rescuers to utilize their leg muscles to lift and move the downed member.

To perform the extremity carry (Skill 9-1):

1. Locate and assess the downed firefighter, placing him on his back.
2. One rescuer will locate themselves at the head (Rescuer 1) of the downed firefighter while the second rescuer will be located at the feet (Rescuer 2).
3. The downed firefighter will be brought into a seated position with his knees bent. If unconscious, it will be necessary for Rescuer 2 to pull the downed firefighter up by the straps of his SCBA **(Skill 9-1A).**
4. Rescuer 1 will wrap his arms around the downed firefighter, grasping the downed firefighter's wrists **(Skill 9-1B).**
5. Rescuer 2 will position herself between the legs of the downed firefighter, grasping the legs underneath the knees **(Skill 9-1C).**
6. Both rescuers will utilize their legs to stand up while lifting the downed firefighter **(Skill 9-1D).**
7. Once the downed firefighter is lifted, the rescuers can proceed to a safe area. If more personnel are available, they can be used to lead the rescuers to safety.

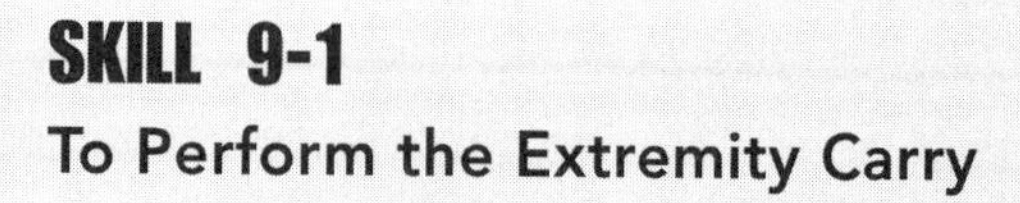

SKILL 9-1

To Perform the Extremity Carry

A The downed firefighter will be brought into a seated position with his knees bent. If unconscious, it will be necessary for Rescuer 2 to pull the downed firefighter up by the straps of his SCBA.

B Rescuer 1 will wrap his arms around the downed firefighter, grasping the downed firefighter's wrists.

(continued)

SKILL 9-1 (CONTINUED)

To Perform the Extremity Carry

C Rescuer 2 will position herself between the legs of the downed firefighter, grasping the legs underneath the knees.

D Both rescuers will utilize their legs to stand up while lifting the downed firefighter.

Cradle Carry

Another option that is available for two rescuers is the **cradle carry.** Just as in the extremity carry, conditions must allow the rescuers the ability to stand.

To perform the cradle carry (Skill 9-2):

1. Locate and assess the downed firefighter, placing the downed firefighter on her back.
2. Two rescuers will position themselves to each side of the downed firefighter.

SKILL 9-2

To Perform the Cradle Carry

A The downed firefighter's arms are placed behind the heads of the rescuers as the rescuers grasp each other's arms behind the downed firefighter's SCBA.

B Each rescuer places their free hand beneath the downed firefighter's knees while grasping the hand of the rescuer on the opposite side.

(continued)

SKILL 9-2 (CONTINUED)
To Perform the Cradle Carry

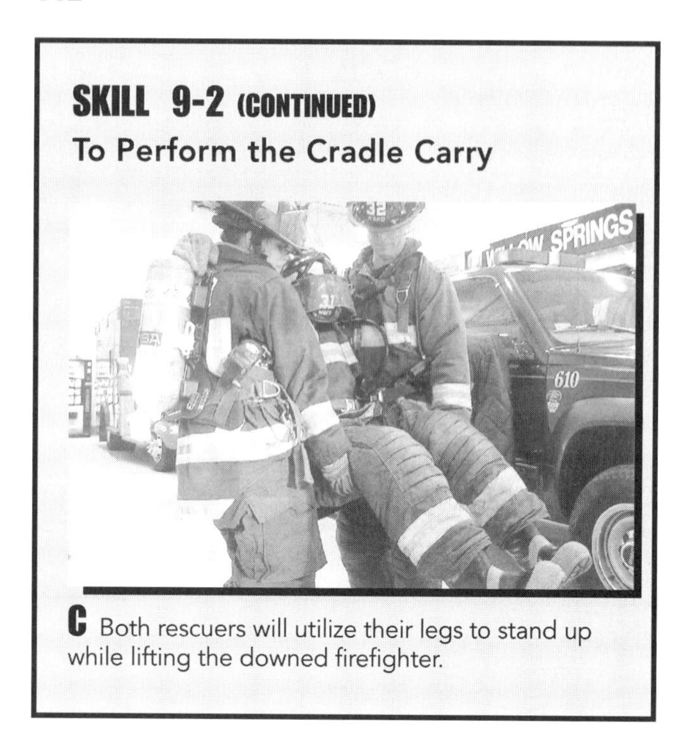

C Both rescuers will utilize their legs to stand up while lifting the downed firefighter.

3. The downed firefighter is brought into a seated position with her knees bent.
4. The downed firefighter's arms are placed behind the heads of the rescuers as the rescuers grasp each other's arms behind the downed firefighter's SCBA **(Skill 9-2A).**
5. Each rescuer places their free hand beneath the downed firefighter's knees while grasping the hand of the rescuer on the opposite side **(Skill 9-2B).**
6. Both rescuers will utilize their legs to stand up while lifting the downed firefighter **(Skill 9-2C).**
7. Once the downed firefighter is lifted, the rescuers can proceed to a safe area. If more personnel are available, they can be used to lead the rescuers to safety.

Blanket Carry

The use of a small salvage tarp will be required to utilize the **blanket carry** technique. Advantages of the blanket carry are that it provides a means for the rescuer to be able to hold onto the downed firefighter and can be used in tight or confined spaces. Some manufacturers have even produced specialty blankets that provide carrying handles as well as heat protection.

To perform the blanket carry (Skill 9-3):

1. Locate and assess the downed firefighter, placing him on his back.
2. Two rescuers will position themselves to each side of the downed firefighter.
3. The blanket or tarp is placed to one side of the downed firefighter opposite the side that the downed firefighter will initially be rolled toward.
4. The downed firefighter is rolled to one side by rescuer 1 while rescuer 2 gathers the blanket or tarp beneath the downed firefighter **(Skill 9-3A).**
5. The downed firefighter will then be rolled back toward rescuer 2, who will take control while rescuer 1 pulls the blanket or tarp from beneath the downed firefighter **(Skill 9-3B).**
6. Rescuer 1 will gather and take hold of the material on each side of the head (or handles if equipped) while rescuer 2 does the same at the feet of the downed firefighter.
7. The downed firefighter is then lifted and carried over obstacles and debris **(Skill 9-3C).**

FIGURE 9-2
Rescue litter.

SKILL 9-3

To Perform the Blanket Carry

A The downed firefighter is rolled to one side by rescuer 1 while rescuer 2 gathers the blanket or tarp beneath the downed firefighter.

B The downed firefighter will then be rolled back toward rescuer 2, who will take control while rescuer 1 pulls the blanket or tarp from beneath the downed firefighter.

C The downed firefighter is then lifted and able to be carried over obstacles and debris.

Other Alternative Carry Methods

There are numerous possibilities for carrying a downed firefighter utilizing equipment such as a rescue litter, backboard, or attic ladder. These are difficult to use in situations requiring maneuvering within tight or confined spaces. Of the three listed, the rescue litter provides the most secure measure for removal because its raised sides prevent the downed firefighter from rolling off **(Figure 9-2)**.

A rescue litter is not designed to hold a firefighter wearing an SCBA. When placing the downed firefighter with SCBA into a rescue litter, the RIT can position the victim in several different ways, depending on time available as well as the surrounding conditions that are present. Removing the victim's SCBA completely is very time consuming and should be avoided unless it is absolutely necessary. A simple solution is to loosen or disconnect the waist belt and loosen the shoulder straps of the SCBA harness in order to shift the pack to one side of the victim. This will enable the downed firefighter to be placed into the rescue litter on the left or right side depending on to which side the pack has been shifted **(Figure 9-3).** The RIT can also consider the use of a backboard with this type of packaging if time and conditions permit.

The rescue litter is placed behind the downed firefighter with the rescuers rolling the victim into the litter while maintaining the SCBA and securing it to the top of the downed firefighter's body using the litter straps or webbing **(Figure 9-4).** It is not necessary or realistic to take time to strap the victim into the basket by lacing webbing in and through the rails and bars of the litter as you would normally do for a high-angle rescue. Connecting clips, straps, or minimal webbing to accomplish the task of keeping the downed firefighter in the litter is all that is required.

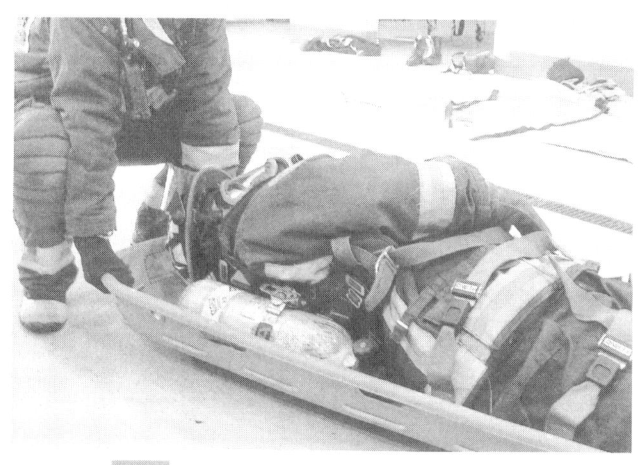

FIGURE 

The downed firefighter's SCBA will need to be shifted to the side to allow placement in the rescue litter.

FIGURE 

The rescue litter is placed behind the downed firefighter with the rescuers rolling the victim into the litter.

Dragging Downed Firefighters

Dragging a downed firefighter may be necessary when conditions dictate that rescuers remain low or when manpower is limited. Certain types of drags will require the rescuers to stand while others will allow the downed firefighter to be moved from the crawling position. Dragging a downed firefighter from the upright position will be easier than dragging from a crawl. Using leg muscles and principles of physics will make the task of dragging a downed firefighter more manageable.

The most predominant challenge when dragging a downed firefighter is attaining a solid grip on the victim. Turnout gear is especially difficult to grasp when wet. Some manufacturers of turnout gear are now outfitting their gear with handles sewn into the gear that can be easily pulled out for the purpose of rescue **(Figure 9-5).**

The SCBA can provide a place to hold onto the downed firefighter while moving him. When moving a downed firefighter a harness can become a necessity, especially in cases where firefighters must be moved up or down stairs or above/below grade. The back harness on the SCBA can be converted very easily into a body harness when time does not permit or an approved harness is not readily available. This technique of converting the SCBA into a harness, called **harness conversion,** will also prevent the SCBA from "riding up" or coming off of a downed firefighter who is being dragged.

To perform an SCBA harness conversion (Skill 9-4):

1. Unbuckle and elongate waist strap of the downed firefighter's SCBA harness.
2. Lift one leg of the downed firefighter, putting the waist strap on that side behind or underneath the raised leg and running the strap through the crotch. The shoulder straps of the

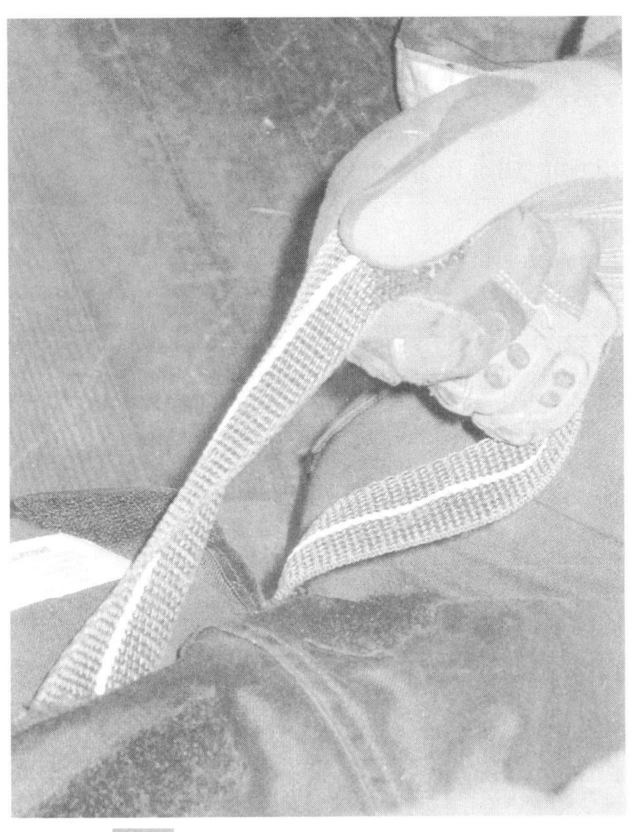

FIGURE 9-5

Some manufacturers of turnout gear are now outfitting their gear with handles sewn into the gear that can be easily pulled out for the purpose of rescue.

SCBA may have to be loosened to facilitate this step with larger-framed firefighters **(Skill 9-4A).**

3. Buckle the repositioned waist strap and tighten if possible.
4. Tighten and secure shoulder straps with half-hitch knots to prevent the harness from slipping **(Skill 9-4B).**
5. If removing the firefighter with rope, be sure to secure the rope or carabiner to the back frame assembly of the SCBA.

Not all SCBA harnesses will be able to be configured in this manner due to their design. Some manufacturer's units will not have waist

Note

No manufacturer will endorse this use of their product. Use this only as a last resort escape or life-saving technique.

SKILL 9-4

To Perform an SCBA Harness Conversion

A Lift one leg of the downed firefighter, putting the waist strap on that side behind or underneath the raised leg and running the strap through the crotch.

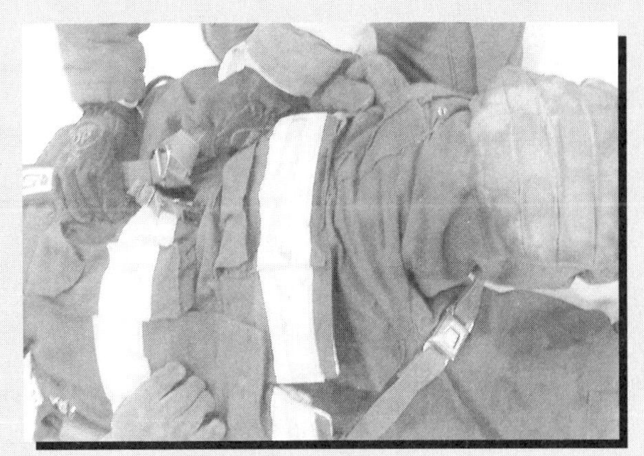

B Tighten and secure the shoulder straps with half-hitch knots to prevent the harness from slipping.

and shoulder straps that are long enough to be rebuckled when performing the harness conversion. In these cases, it is recommended to just tighten down the shoulder straps and go.

A Multiple Application Service Tool (MAST) is another quick and useful piece of equipment that can be utilized for helping to move a downed firefighter. The MAST is five large loops connected together in a daisy chain. These loops can be placed over a downed firefighter's turnout gear. The loops of the manufactured MAST are color-coded to designate which loop is placed where. The center loop of the MAST is the most important because it will provide the handle for lifting or pulling **(Figure 9-6).** If inside an environment where visibility is limited, the center loop can be easily located by counting the loops. The MAST is very quick and easy to apply.

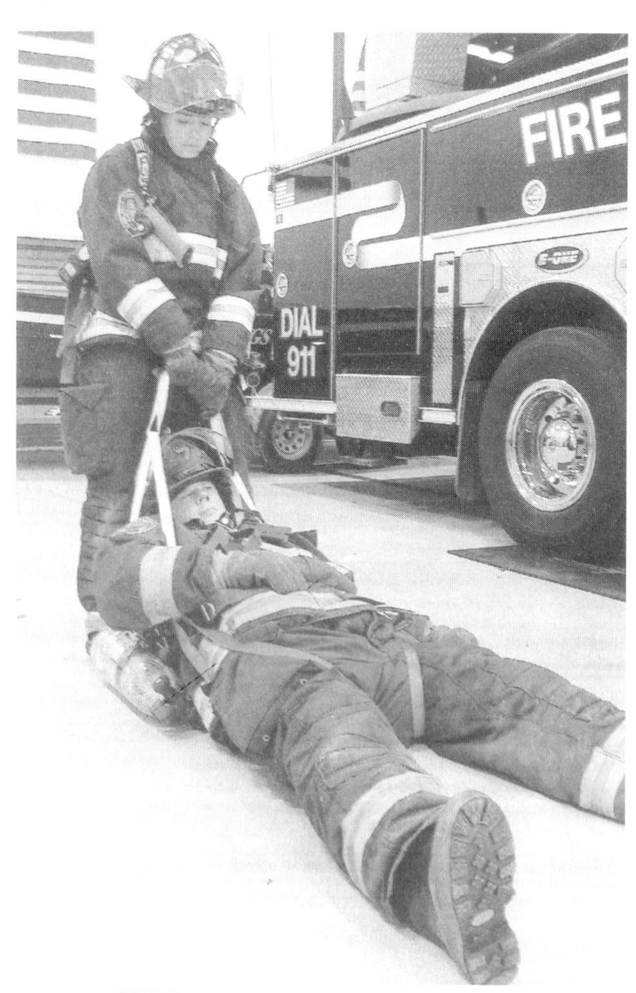

FIGURE  9-6

A MAST is a helpful piece of equipment for moving a downed firefighter.

It is the ultimate responsibility of the RIT to determine which method or equipment that will be used for dragging a downed firefighter to safety.

Side-by-Side Drag

The side-by-side drag is very basic and consists of two firefighters moving the downed firefighter by utilizing the shoulder straps of the downed firefighter's SCBA.

To perform the side-by-side drag:

1. Locate and assess the downed firefighter, placing him on his back.
2. The rescuers will locate themselves at the head of the downed firefighter on opposite sides.
3. Each rescuer will grasp a separate shoulder strap.
4. On command, the rescuers will sweep with the free hand forward while driving forward with their legs to move the downed firefighter **(Figure 9-7).**

FIGURE  9-7

Two firefighters perform the side-by-side drag.

SKILL 9-5
To Perform the Lift-and-Lead Drag

A Rescuer 1 will wrap his arms around the downed firefighter, grasping the downed firefighter's wrists.

B The rescuer will utilize his legs to stand up while lifting the downed firefighter.

C Once the downed firefighter is lifted, rescuer 2 will place a hand on rescuer 1 to guide him around obstacles to safety.

Lift-and-Lead Drag

The **lift-and-lead drag** is a basic drag that utilizes one firefighter to drag the downed member while a second rescuer provides safety by leading the way out. Conditions must allow the rescuers to stand up to use this method.

To perform the lift-and-lead drag (Skill 9-5):

1. Locate and assess the downed firefighter, placing him on his back.
2. The rescuer will locate himself at the head (rescuer 1) of the downed firefighter.
3. Rescuer 1 will wrap his arms around the downed firefighter, grasping the downed firefighter's wrists **(Skill 9-5A)**.
4. The rescuer will utilize his legs to stand up while lifting the downed firefighter **(Skill 9-5B)**.
5. Once the downed firefighter is lifted, rescuer 2 will place a hand on rescuer 1 to guide him around obstacles to safety **(Skill 9-5C)**.

Push-and-Pull Drag

The **push-and-pull drag** requires two rescuers. It is performed in the manner by which it is referred—one rescuer will be pulling with their upper body while the other will utilize their legs to push the downed firefighter. It allows the rescuers and downed firefighter to remain low and work in narrow spaces.

To perform the push-and-pull drag (Skill 9-6):

1. Locate and assess the downed firefighter, placing him on his back.
2. One rescuer will locate themselves at the head (rescuer 1) of the downed firefighter, taking hold of the downed firefighter's SCBA shoulder strap.
3. Rescuer 2 will locate himself inside the legs of the downed firefighter, lifting one of the downed firefighter's legs over his shoulder. The rescuer's head should be positioned high into the groin area with the shoulder driving into the buttocks of the downed firefighter. Correct positioning of rescuer 2 is essential for this drag to be effective **(Skill 9-6A).**
4. On command, rescuer 1 should pull the downed firefighter while rescuer 2 uses his legs to push simultaneously **(Skill 9-6B).**

Another alternative to this method that may be considered is placing rescuer 2 to the outside of the downed firefighter's legs with the downed firefighter's leg over the rescuer's inside shoulder. This may help in keeping the downed firefighter's free leg from getting in the way of the rescuers during movement.

Tool Drag

Two rescuers dragging a downed firefighter by pulling on the SCBA straps may be difficult due to the rescuers being too close to one another. A **tool drag** enables rescuers to be spaced apart while allowing a secure place to grip the downed firefighter. A tool such as a Halligan bar or closet hook works best for the tool drag. However, the tool must not be so big that it prevents extraction through tight spaces, narrow hallways, and staircases.

To perform the tool drag (Skill 9-7):

1. Locate and assess the downed firefighter, placing him on his back.
2. The rescuers will locate themselves at the head of the downed firefighter and place the downed firefighter in a seated position.
3. The tool is inserted through the shoulder straps of the SCBA, providing a handle for both rescuers to hold onto **(Skill 9-7A).** Make certain that the pick end of any tool is rotated away and facing down toward the floor to avoid injury in case the rescuer slips or falls.

Make certain that the pick end of any tool is rotated away and facing down towards the floor to avoid injury in case the rescuer slips or falls.

4. On command, the rescuers will drag the downed firefighter to safety **(Skill 9-7B).**

Blanket Drag

The use of a small salvage tarp or specialty blanket will be required to carry out the **blanket drag.** Similar to the blanket carry, advantages of the blanket drag are that it provides a means for the rescuer to be able to hold onto the downed firefighter and can be used in tight or confined spaces.

To perform the blanket drag (Skill 9-8):

1. Locate and assess the downed firefighter, placing him on his back.
2. The blanket or tarp is placed to one side of the downed firefighter, opposite the side that the downed firefighter will initially be rolled toward.
3. The downed firefighter is rolled to one side while gathering the blanket or tarp beneath the downed firefighter **(Skill 9-8A).**
4. The downed firefighter will then be rolled back. The blanket or tarp is then pulled from beneath the downed firefighter **(Skill 9-8B).**

SKILL 9-6
To Perform the Push-and-Pull Drag

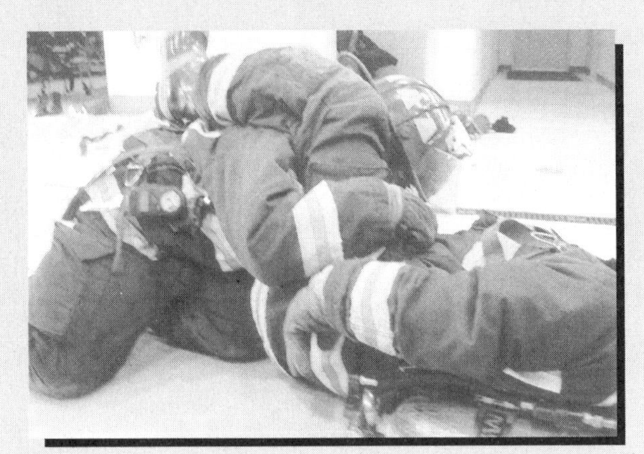

A Rescuer 2 will locate himself inside the legs of the downed firefighter, lifting one of the downed firefighter's legs over his shoulder. The rescuer's head should be positioned high into the groin area with the shoulder driving into the buttocks of the downed firefighter.

B On command, rescuer 1 should pull the downed firefighter while rescuer 2 uses his legs to push simultaneously.

SKILL 9-7
To Perform the Tool Drag

A The tool is inserted through the shoulder straps of the SCBA, providing a handle for both rescuers to hold onto. Make certain that the pick end of any tool is rotated away and facing down toward the floor to avoid injury in case the rescuer slips or falls.

B The rescuers will drag the downed firefighter to safety by using the tool as a handle.

SKILLS 9-8
To Perform the Blanket Drag

A The downed firefighter is rolled to one side while gathering the blanket or tarp beneath the downed firefighter.

B The downed firefighter will then be rolled back. The blanket or tarp is then pulled from beneath the downed firefighter.

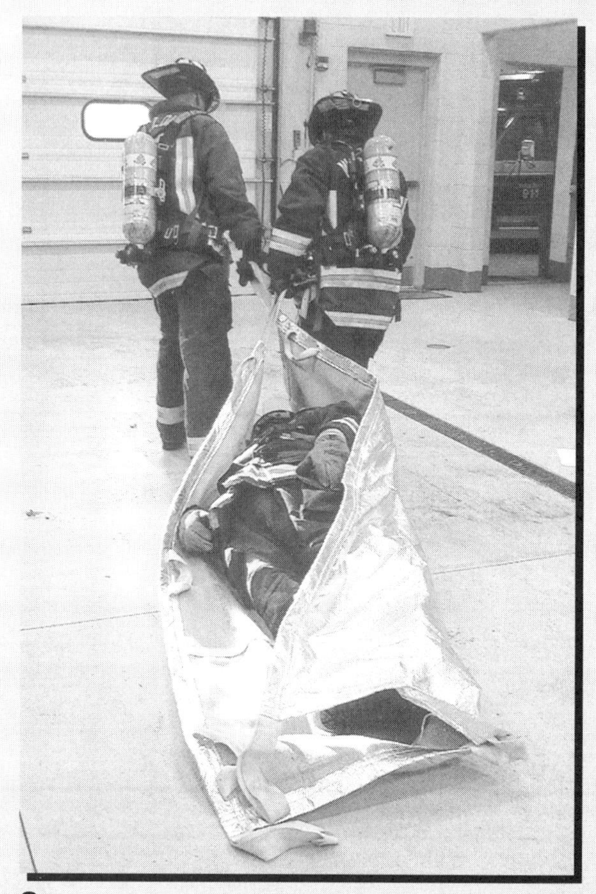

C The downed firefighter's torso is then lifted off the floor, enabling him to be dragged.

5. The rescuer will gather and take hold of material on each side of the head (or handles if equipped).
6. The downed firefighter's torso is then lifted off the floor, enabling him to be dragged **(Skill 9-8C).**

Webbing and Drags

Webbing is a versatile piece of equipment for the RIT. It can be used in establishing anchor points, tying rescue harnesses, performing emergency escape maneuvers, setting up search tethers, and can also be used to create handles and slings to assist in removing a downed firefighter. A 10- to 15-foot piece of looped webbing secured in a **girth hitch** to a downed firefighter's SCBA harness can provide a sling to pull the firefighter as a horse would pull a cart **(Figures 9-8A** and **9-8B).** Whenever webbing is used as a sling, it is important to keep in mind that the longer the distance from the downed firefighter to the rescuer,

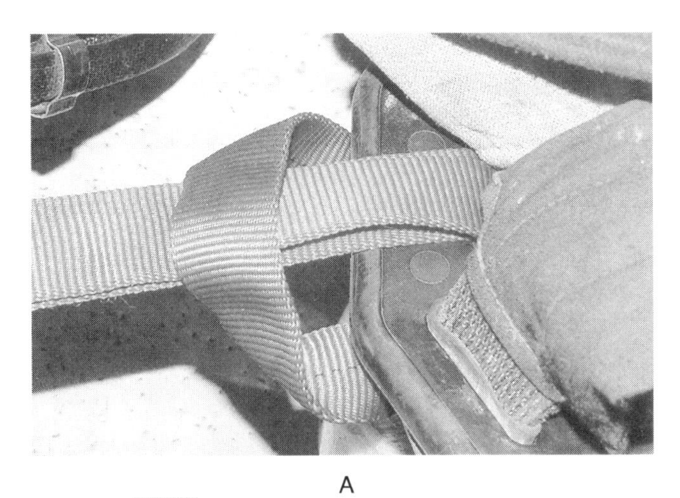
A

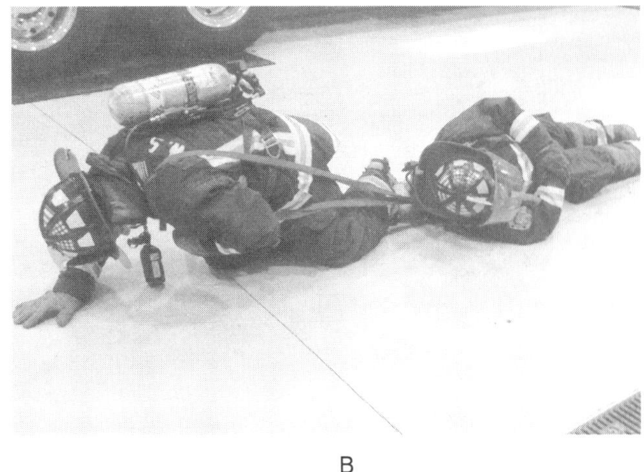
B

FIGURE 

(A) Girth hitch. (B) A 10- to 15-foot piece of looped webbing secured in a girth hitch to a downed firefighter's SCBA harness can provide a sling to pull the firefighter as a horse would pull a cart.

the more difficult controlling and dragging the downed firefighter will be. Webbing tied into a harness or formed into a girth hitch around the chest is a very effective option when an SCBA is not present or able to be used on the downed firefighter for any reason **(Figure 9-9)**.

Webbing as well as rope can be used to tie a handcuff knot on the downed firefighter for dragging. When used in this manner, the downed firefighter's arms are raised above the head, lowering his profile and thus allowing him to fit through a tight opening such as wall studs or obstacles in a collapse area **(Figure 9-10)**. Again, when using the handcuff knot, make certain that it is cinched down on the forearms of the downed firefighter—injuries to the wrist can occur if secured improperly.

Rescue Loops

An 8-mm Prussik cord tied into loops utilizing the double fisherman's knot can be very useful in

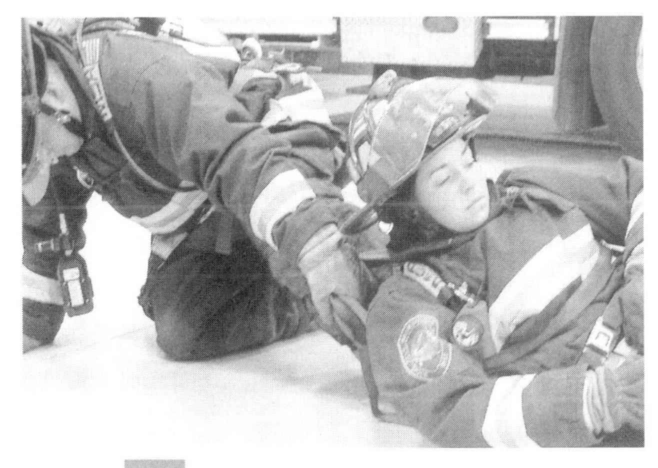

FIGURE 9-9

Webbing tied into a harness or formed into a girth hitch around the chest is a very effective option when an SCBA is not present or able to be used on the downed firefighter for any reason.

FIGURE 

Webbing as well as rope can be used to tie a handcuff knot on the downed firefighter for dragging.

FIGURE 

Placing rescue loops into a girth hitch on the extremities of the downed firefighter provides points that enable multiple rescuers to move a downed firefighter above obstacles and debris.

helping to move the downed firefighter. **Rescue loops** are only limited in use by the imagination. The Phoenix Fire Department began experimenting with the concept of utilizing these loops to establish handles and grab points to move downed firefighters. Placing the loops into a girth hitch on the extremities of the downed firefighter provides points that enable multiple rescuers to move a downed firefighter above obstacles and debris **(Figure 9-11).** The loops can also be used to form a sling for dragging the downed firefighter, as is done with webbing.

Staircases

Moving a downed firefighter up or down a set of stairs can be one of the most challenging scenarios presented to a RIT on the fireground. The strongest rescuers will be unable to move a downed firefighter up or down stairs unless proper technique is used. Teamwork and clear communication will be required to move a downed member up or down a flight of stairs. The cylinder valve of the SCBA is the most common piece of equipment that can cause difficulties when moving a downed firefighter up or down stairs. Consider what difficulties can arise when moving on staircases and make the necessary adjustments to overcome them.

Multiple Rescuer Staircase Lift

To move a downed firefighter up a staircase using the multiple rescuer staircase lift (Skill 9-9):

1. Locate and assess the downed firefighter, placing him on his back.
2. Convert the SCBA into a body harness if possible. If the SCBA harness is unusable for any reason, a looped piece of webbing can be wrapped under the downed firefighter's arms to provide a lifting point.
3. Drag the downed firefighter to the base of the staircase, positioning him facing away from the stairs on the third tread. A rescuer may have to lift the downed firefighter to accomplish this **(Skill 9-9A).**
4. Rescuer 1 will be positioned behind the downed firefighter on the stairs. This rescuer will lift the downed firefighter from the straps of the SCBA. The downed firefighter should be pulled straight up to clear the SCBA cylinder valve from being caught on the stairs.
5. Rescuer 2 will be positioned at the feet to the inside of the downed firefighter's legs with his face high into the groin area; the downed firefighter's legs will need to be positioned over the rescuer's shoulders. Do not let the downed firefighter's legs slip off the shoulders **(Skill 9-9B).**
6. Rescuer 1, who is located at the head of the downed firefighter, will give the command for the extraction. They should be kept in simple terms such as, "Ready?—Go!" with a pause in between to give the other rescuer the opportunity to stop the procedure if not ready.
7. On command, rescuer 1 will pull the downed firefighter up while rescuer 2 will push. Moving the downed firefighter will be difficult and it will be necessary to stop every few steps. Just remember that if conditions in the lower level were bad, the staircase will be even more formidable—move quickly but in a controlled manner **(Skill 9-9C).**

If the staircase is wide enough and a third rescuer is available, he can be positioned at the

SKILL 9-9

To Move a Downed Firefighter up a Staircase Using the Multiple Rescuer Staircase Lift

A Drag the downed firefighter to the base of the staircase, positioning him facing away from the stairs on the third tread.

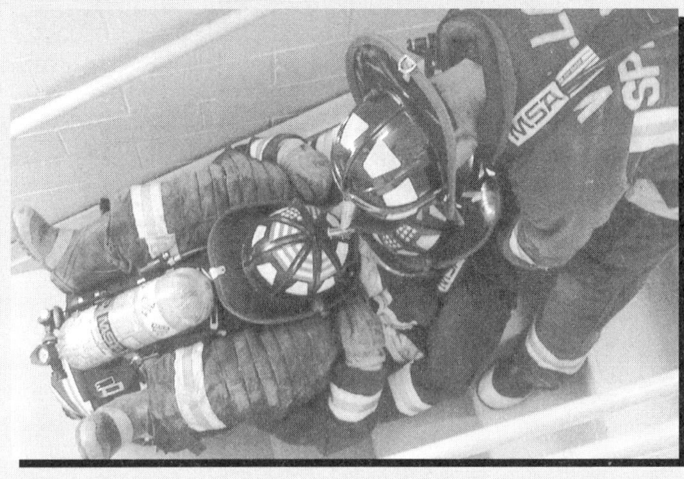

B Rescuer 2 will be positioned at the feet to the inside of the downed firefighter's legs with his face high into the groin area; the downed firefighter's legs will need to be positioned over the rescuer's shoulders.

C On command, rescuer 1 will pull the downed firefighter up while rescuer 2 will push.

head with rescuer 1. Each rescuer at the head will then have the ability to each grab a shoulder strap of the downed firefighter's SCBA harness.

Stair Raise with a Tool

If the width of the staircase will allow, a tool can be used as a handle for two firefighters to lift and carry the downed firefighter using the **stair raise with a tool technique.**

To perform the stair raise with a tool (Skill 9-10):

1. Locate and assess the downed firefighter, placing him on his back.
2. Drag the downed firefighter to the base of the staircase, positioning him facing away from the stairs.

SKILL 9–10
To Perform the Stair Raise with a Tool

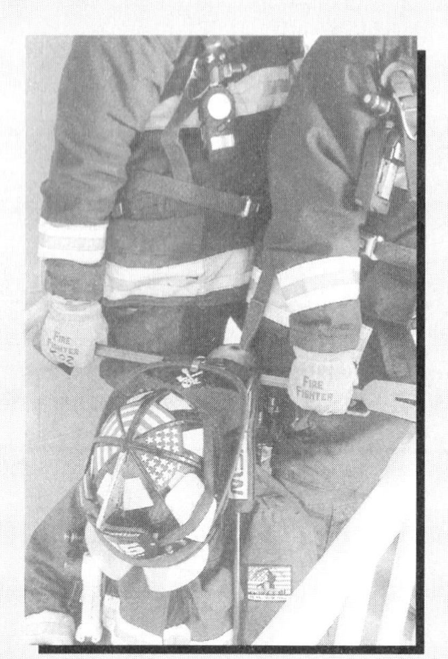

A The tool is inserted through the shoulder straps of the SCBA, providing a handle for both rescuers to hold onto.

B On command, the rescuers will pull the downed firefighter up above the stair treads with the downed firefighter's lower extremities dragging behind.

3. The rescuers will locate themselves at the head of the downed firefighter and place the downed firefighter in a seated position.
4. The tool is inserted through the shoulder straps of the SCBA, providing a handle for both rescuers to hold onto **(Skill 9-10A).** Make certain that the pick end of any tool is rotated away and facing down toward the floor to avoid injury in case the rescuer slips or falls.

Safety

Make certain that the pick end of any tool is rotated away and facing down towards the floor to avoid injury in case the rescuer slips or falls.

5. On command, the rescuers will pull the downed firefighter up above the stair treads with the downed firefighter's lower extremities dragging behind. It is important to make certain that the downed firefighter is lifted high enough to have the SCBA cylinder valve clear the stair tread **(Skill 9-10B).**

If a third rescuer is available, he can control the lower extremities by being positioned at the feet to the inside of the downed firefighter's legs with his face high into the groin area and with the downed firefighter's legs positioned over the rescuer's shoulders. He will drive the downed firefighter up, helping to clear the stair treads **(Figure 9-12).**

Stair Raise Using the Handcuff Knot

The handcuff knot can be utilized in moving a downed firefighter up stairs. It is especially useful when the staircase is narrow.

FIGURE 9-12

If a third rescuer is available, he can control the lower extremities by being positioned at the feet to the inside of the downed firefighter's legs with his face high into the groin area and with the downed firefighter's legs positioned over the rescuer's shoulders. He will drive the downed firefighter up, helping to clear the stair treads.

To perform the stair raise with the handcuff knot (Skill 9-11):

1. Locate and assess the downed firefighter, placing him on his back.
2. Drag the downed firefighter to the base of the staircase, positioning him facing away from the stairs.
3. Rescuer 1 will locate himself at the head of the downed firefighter and place the handcuff knot on the forearms of the downed firefighter. He will then pay out the rope or webbing until they are located at the landing or top of the staircase **(Skill 9-11A).** Keep in mind that conditions can be horrific at the landing or top of the staircase as heat and products of combustion will be present if the fire in the lower level has not been controlled.
4. Rescuer 2 will be positioned at the feet to the inside of the downed firefighter's legs with his face high into the groin area; the downed firefighter's legs will need to be positioned over the rescuer's shoulders. The downed firefighter should be rotated slightly to one side to allow the SCBA to slide up the stairs.
5. Rescuer 1 will call out the command and take up slack in the rope or webbing, pulling the downed firefighter up the stairs.
6. Rescuer 2 will use his legs to drive the downed firefighter up the stairs, making certain that the cylinder valve of the SCBA clears the stair treads **(Skill 9-11B).**

Using Rescue Loops to Carry a Downed Firefighter up Stairs

Rescue loops are another option to assist the RIT in moving a downed firefighter up a staircase. A minimum of two firefighters will be needed to use rescue loops.

To use rescue loops to help move a firefighter up a staircase (Skill 9-12):

1. Locate and assess the downed firefighter, placing him on his back.
2. Drag the downed firefighter to the base of the staircase, positioning him facing away from the stairs in a seated position.
3. Rescuer 1 will take position behind the downed firefighter and will grasp both shoulder straps of the downed firefighter's SCBA.
4. Rescuer 2 will take a rescue loop and place it in a girth hitch on the downed firefighter's leg as high up in the groin area as possible **(Skill 9-12A).**

SKILL 9-11

To Perform the Stair Raise with the Handcuff Knot

A Rescuer 1 will locate himself at the head of the downed firefighter and place the handcuff knot on the forearms of the downed firefighter. He will then pay out the rope or webbing until they are located at the landing or top of the staircase.

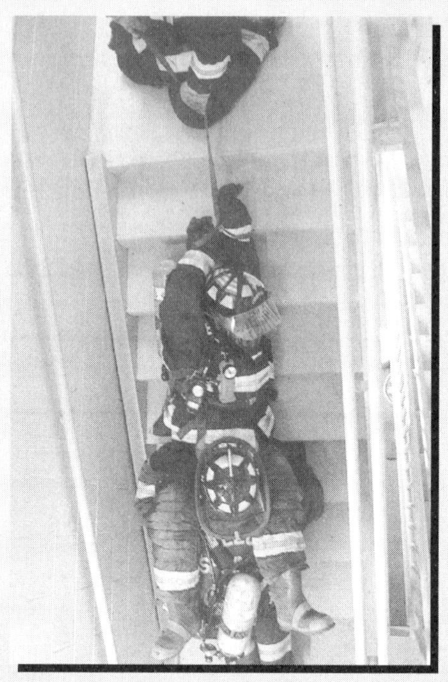

B Rescuer 2 will be positioned at the feet to the inside of the downed firefighter's legs with his face high into the groin area; the downed firefighter's legs will need to be positioned over the rescuer's shoulders.

5. This will be repeated for the second leg also.
6. Rescuer 2 will position himself inside the downed firefighter's legs, grasping a rescue loop in each hand.
7. On command, rescuer 1 will pull straight up on the SCBA shoulder straps while rescuer 2 will pull straight up on the rescue loops. At this point, the SCBA of the downed firefighter should be up high enough to clear the stair treads easily **(Skill 9-12B)**.

The rescuers should be able to navigate the stairs quite easily with proper execution of this maneuver. If needed, they can stop periodically to regroup or get a better hold. Just remember, once committed to going up the staircase, it must be performed quickly—the staircase is a ventilation outlet for any conditions on the lower level!

Using 2-to-1 Mechanical Advantage with a Rescue Litter

A rescue litter assisted with a mechanical advantage system can be used to slide a downed firefighter up a staircase as opposed to carrying. This can be beneficial when removing firefighters with neck or spinal injuries. It will require

SKILL 9-12

To Use Rescue Loops to Help Move a Firefighter up a Staircase

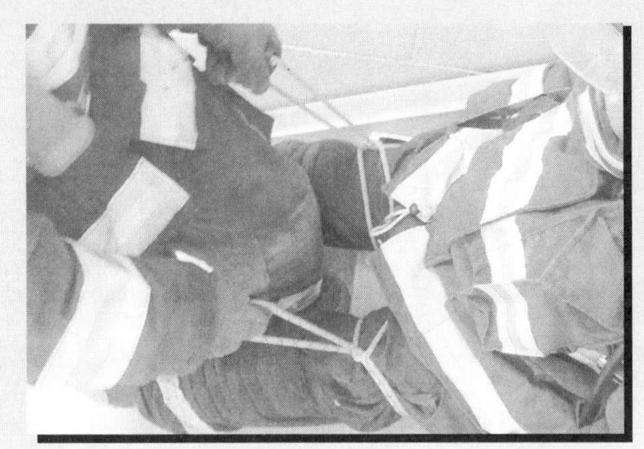

A Rescuer 2 will take a rescue loop and place it in a girth hitch on the downed firefighter's leg as high up in the groin area as possible. This will be repeated for the second leg.

B On command, rescuer 1 will pull straight up on the SCBA shoulder straps while rescuer 2 will pull straight up on the rescue loops. At this point, the SCBA of the downed firefighter should be up high enough to clear the stair treads easily.

more time than other methods but should not be discounted if conditions will allow its use.

To raise a downed firefighter up a staircase using a rescue litter and simple 2-to-1 mechanical advantage (Skill 9-13):

1. Secure the downed firefighter into the rescue litter. The litter should go up the stairs headfirst **(Skill 9-13A).**
2. Attach the 2-to-1 mechanical advantage system to an adequate anchor at the top of the stairs and the head of the rescue litter (see Chapter 6 for detailed information; **Skill 9-13B).**
3. Rescuer 1 will be positioned at the foot of the rescue litter to help guide the litter up as it is raised **(Skill 9-13C).**
4. Rescuers 2 and 3 will utilize the haul line to slide the rescue litter up the staircase **(Skill 9-13D).**

Moving a Disabled Firefighter Down a Flight of Stairs

Moving the downed firefighter down a flight of stairs is not as difficult as going up because gravity will assist to a degree, but it is still not an easy task by any means. The most important thing to consider when going down stairs with a downed firefighter is to prevent the downed member from sustaining additional injuries to the head and neck.

The simplest way of removing a downed firefighter from an upper floor using a flight of stairs is to drag the firefighter headfirst.

To drag a downed firefighter down a flight of stairs (Skill 9-14):

1. Locate and assess the downed firefighter, placing him on his back.
2. Drag the downed firefighter to the top of the staircase, positioning him face up.
3. Rescuer 1 will position himself on the stairs behind rescuer 2. He will guide rescuer 2 and the downed firefighter.
4. Rescuer 2 will roll the downed firefighter slightly to the right or left to keep the SCBA from getting caught on the stairs when dragging. Rescuer 2 will lift on the SCBA shoulder straps of the downed firefighter

SKILL 9-13

To Raise a Downed Firefighter up a Staircase Using a Rescue Litter and Simple 2-to-1 Mechanical Advantage

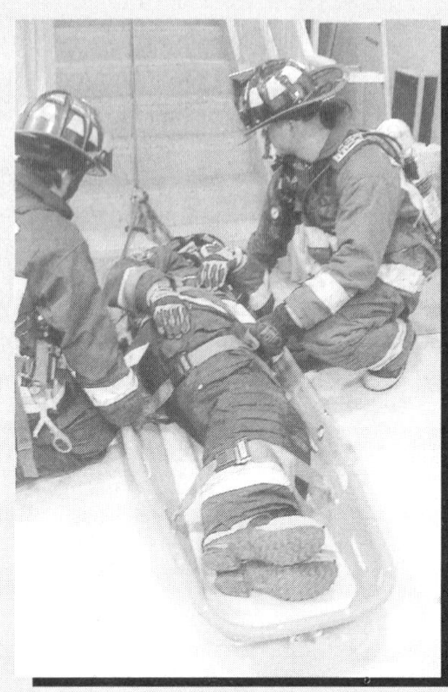

A Secure the downed firefighter into the rescue litter. The litter should go up the stairs headfirst.

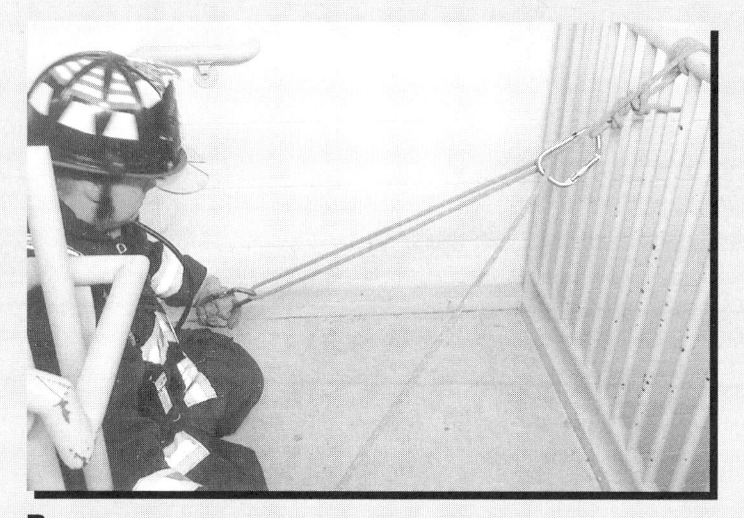

B Attach the 2-to-1 mechanical advantage system to an adequate anchor at the top of the stairs and the head of the rescue litter.

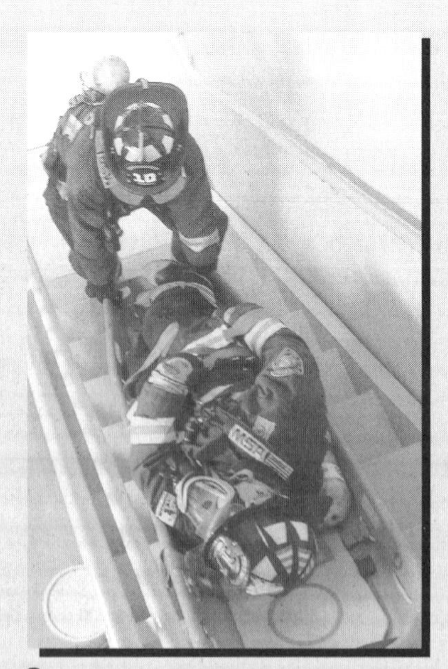

C Rescuer 1 will be positioned at the foot of the rescue litter to help guide the litter up as it is raised.

D Rescuers 2 and 3 will utilize the haul line to slide the rescue litter up the staircase.

SKILL 9-14

To Drag a Downed Firefighter Down a Flight of Stairs

A Rescuer 2 will roll the downed firefighter slightly to the right or left to keep the SCBA from getting caught on the stairs when dragging. Rescuer 2 will lift on the SCBA shoulder straps of the downed firefighter while cradling the back of the head and neck on his forearm.

B The rescuers will proceed down the stairs with the upper body of the downed firefighter supported by rescuer 2. The downed firefighter's lower body will drag down along the stairs.

while cradling the back of the head and neck on his forearm **(Skill 9-14A).**

5. The rescuers will proceed down the stairs with the upper body of the downed firefighter supported by rescuer 2. The downed firefighter's lower body will drag down along the stairs **(Skill 9-14B).**

Another option when moving the downed firefighter down a flight of stairs is to use a tool to assist in lifting the downed member over the stairs. Two rescuers can be positioned at the head of the downed firefighter while a Halligan bar or other tool is placed through the shoulder straps of the SCBA. This will allow the team members to grasp each side of the Halligan bar while raising the head and upper torso of the downed firefighter when bringing him down the stairs. If a third rescue member is available, he can help guide the rescuers down the stairs **(Figure 9-15).**

FIGURE 

Another option when moving the downed firefighter down a flight of stairs is to use a tool to assist in lifting the downed member over the stairs.

Summary

Extracting a downed firefighter from a hazard area may require the utilization of multiple techniques—such as the extremity, cradle, or blanket carry and the lift-and-lead, push-and-pull, or blanket drag—depending on the circumstances. Drastic and unconventional measures—such as the harness conversion—may need to be taken to remove the downed firefighter. Safety and the imagination are the only limiting factors when removing a downed firefighter in an expedient manner. The key to all of the methods discussed for moving a downed firefighter is technique. Practice and training is the only way to find out what works best and what adjustments can be made to make the operation more efficient.

KEY TERMS

Blanket carry
Blanket drag
Cradle carry
Extremity carry
Girth hitch
Harness conversion
Lift-and-lead drag
Multiple rescuer staircase lift
Push-and-pull drag
Rescue loops
Stair raise with the handcuff knot
Stair raise with a tool technique
Tool drag

REVIEW QUESTIONS

1. Moving the downed firefighter should only consist of one type of drag or carry throughout the rescue to avoid confusion.

   True False

2. Carrying a downed firefighter is generally considered easier than dragging because
   a. it allows the rescuers to utilize their leg muscles.
   b. it enables obstacles to be navigated.
   c. it is quicker.
   d. all of the above

3. The rescue litter is advantageous in moving the downed firefighter because
   a. it is able to be used in a tight space.
   b. it is designed to hold a firefighter wearing an SCBA.
   c. it has raised sides to keep the downed firefighter from rolling off.
   d. all of the above

4. When dragging a downed firefighter, the __________ can be made into a harness to provide a stable holding or grab point.

5. Harness conversions can be applied to all manufactured types of SCBA harnesses.

   True False

6. The MAST utilizes _____ interlinked loops into such a configuration that it can serve multiple applications when rescuing civilians and downed firefighters.
   a. two
   b. six
   c. five
   d. eight

7. The __________ is a good technique to use when working in a narrow space.
   a. push-and-pull drag
   b. tool drag
   c. cradle carry
   d. girth hitch lift

8. The key to moving a downed firefighter up a flight of stairs is
   a. a large number of rescuers
   b. strength
   c. the size of rescuers
   d. technique

9. The ________ is the most common cause of entanglement when moving a downed firefighter.
   a. helmet
   b. SCBA cylinder valve
   c. turnout coat
   d. air mask

10. __________ is the number one priority when moving a downed firefighter down a flight of stairs.

■ ADDITIONAL RESOURCES

Firefighter's Handbook, 2nd ed., Clifton Park, NY: Thomson Delmar Learning, 2004.

Hoff, Robert, and Kolomay, R., *Firefighter Rescue and Survival.* Tulsa, OK: PennWell Publishing, 2003.

Lasky, R., and Shervino, T., Saving Our Own: Moving the Downed Firefighter up a Stairwell. *Fire Engineering,* December 1997.

McCormack, J., *Firefighter Rescue and Rapid Intervention Teams.* Indianapolis, IN: Fire Department Training Network, 2003.

CHAPTER

10 Subfloor Rescues—Raising the Downed Firefighter

Learning Objectives

Upon completion of this chapter, you should be able to:

- explain the importance of proper size-up before a Mayday takes place when assigned as the RIT.
- identify the methods of lifting a downed firefighter utilizing a hoseline for the disabled, conscious, or unconscious firefighter in distress.
- demonstrate the procedure of utilizing the handcuff knot to raise a fallen firefighter.
- describe techniques involved in lowering a rescuer into a subfloor area using the handcuff knot.
- give details of the application of the handcuff cradle method as it applies to subfloor rescues of a downed firefighter.
- demonstrate the steps involved when using a four-way haul incorporating the "W" technique through the use of rope and carabiners to lift a downed firefighter.
- give explanation of the procedures involved in using a bight of rope to raise the downed firefighter through the SCBA harness.
- explain how a ladder and the use of a 2-to-1 mechanical advantage can be utilized to raise a downed firefighter through the floor.

CASE STUDY

Captain Donald Harper: "The fire was a quick three blocks from the station and as we hurried with our gear, the Battalion chief went on scene and established command. He sized up light smoke from the chimney area of a two-story frame residence. My engine was next, with the victim and I assigned as fire attack. As I looked at the residence I thought we had a small job, room and contents probably.

"The victim and I advanced a handline to side 3 as directed by the Battalion Chief for the best access. We had full turnout gear with Scott 2.2 SCBAs and made entry via a door forced by the police prior to our arrival. We encountered moderate smoke conditions and virtually no heat as we began our primary search. The Division Chief of Prevention had arrived and made a tour of the exterior, observing fire presenting in a window on the first floor of side 2. He requested a face-to face conference with me at my entry point so he could give clear directions as to the location of the fire.

"The victim and I made quick progress to what was thought to be the seat of the fire. As is my normal procedure, I was leading the way; this is something I feel strongly about as I can evaluate conditions and adjust accordingly. I entered a narrow hallway and confronted fire in a room just ahead. The victim passed the nozzle to me, and I hit the fire, knocking it down. We agreed on the need for overhaul in that room and, again due to the confines of the hall, the victim and I exchanged places so I could give a report to Command as he moved in to start overhaul.

"We exchanged places in a larger room adjacent to the narrow hallway. As we began moving back down the hallway the floor partially collapsed with no warning. I had been in that exact area just seconds before the accident and there was no indication of weakness. The victim dropped to the basement floor along with the nozzle while I went in to my waist. I was able to back fully out of the hole and I yelled to the victim, finding him to be uninjured and mobile. I instructed him to remain where he was while I requested help.

"After broadcasting emergency radio traffic requesting assistance I refocused on the victim. Still speaking to him, I was able to reach through the hole and hold his hand. The smoke and heat were increasing so the victim dropped to the floor and used the handline for protection. About this time he indicated that his air supply was getting low. I radioed a request for additional air cylinders and it seemed like an eternity—I realized something had to be done now.

"I removed my SCBA quickly; however it became entangled in debris or furniture. As I struggled to get the SCBA to the victim I was rapidly becoming overcome by smoke. At this time, the first rescue crew joined me and as one of the guys removed me, the other two continued to fight for the victim's life. I spent the remainder of the incident in the care of EMS, refusing to leave until the victim was out."

Lieutenant Brian Spini: " I was a Firefighter assigned to Quint 252 at the time of the Lamar street fire. Our crew was the third unit on the scene. We were a five-person crew, which was split into two crews a lot of the time. This day was no different. As we made entry into the first floor, we went to our left; this was uncommon—we almost always went to the right. Had we gone to the right this day, we would have fallen through the floor before the victim firefighter did. The main seat of the fire was just to the right of the front door in the basement. Very soon into the fire, this floor collapsed.

"Shortly after making entry, we were pushed out of the house by heat and smoke. The conditions in the house changed in an instant. I remember getting outside and looking back to see heavy smoke now coming from the front door; I felt we had just lost all our second-floor crews. We grabbed a line and started back towards the house to get our crews out; at that time, we saw all of the crews come falling out of the front door. Then, we heard the call from Captain Harper.

"Our crew went to the door on side three of the structure and made entry following Captain Harper's line to the point where the victim firefighter had fallen through the floor. The conditions were getting worse in the structure. We found Captain Harper starting to be overcome by smoke and we relayed him out into the yard of side three.

"Another firefighter was now at the hole trying to get the SCBA pushed through to the victim firefighter. When he went through the floor it came back up, leaving a very small opening with jagged edges. After several attempts with the SCBA, the firefighter reached through the hole and told the victim firefighter to 'grab my hand.' He replied, 'I can't reach it.' The reply was spoken in a very clear voice as if he no longer had his mask on. This was to be the last verbal contact we had with the victim.

CASE STUDY (CONTINUED)

"The fire conditions in the structure were worsening, and a basement access on side two was discovered.

"From that point, things seemed to get more chaotic. People were making entry into the basement on low air. Another firefighter actually found the victim firefighter; before he could do anything, he himself ran completely out of air and had to exit the basement. There were a growing number of people gathering at the basement door, and it appeared that no one could keep track of who was in the basement at any given time. Everyone wanted to go to the aid of his colleague, and for a while, all self-discipline was gone. Roughly twenty minutes after we lost contact with the victim firefighter, a mutual aid crew pulled him from the basement. Sadly, he was pronounced dead at a local hospital."

—Special thanks to Captain Don Harper Jr. and Lieutenant Brian Spini from Consolidated Fire District 2, Northeast Johnson County, Kansas for sharing their account for this Case Study.

Introduction

One of the highest risks to a firefighter working on the fireground is operating on a level above the fire. The situation of a firefighter becoming trapped in a subfloor situation such as a floor or stair collapse has become a common theme in case studies of firefighter fatalities.

The Rescue Plan

The RIT should always develop a two-stage plan when it is called into action. The safest, easiest, and simplest method must be taken to perform a rescue. This begins with establishing a proper size-up of the structure before a Mayday takes place. Awareness has to be obtained in relation to the construction features of the building. This is especially true when dealing with **walk-out basements** or elevation changes in which exits could offer easier access to the downed firefighter and simplify a rescue effort. The easiest method to affect a rescue in a situation such as a firefighter falling through the floor is to enter and extract the downed firefighter through an alternate entrance/exit to the space in which he had fallen. This is more than likely not going to be possible or time efficient in all circumstances. When an alternate entrance/exit is not available, it becomes obvious that the quickest method for retrieval will be the area in the floor through which the firefighter fell. Raising the downed firefighter back up through the floor will require a great amount of weight to be lifted by a minimal number of rescuers. This may sometimes have to take place in an area that may be confined or restricted. Many techniques have been introduced and perfected to rescue a firefighter in this type of situation. Whatever method is used, it is a decision that will have to be made by the RIT beforehand.

Different techniques should be trained on consistently to find out which ones work better or are more comfortable for the people performing them. During a Mayday situation is not the time to learn technique. Remember, if plan A does not work, it may be necessary for the RIT to have plan B ready to be put into battle.

Access to the Downed Firefighter

When a Mayday is called for a firefighter that has fallen through a floor, it is the RIT's priority to get someone into the area with the victim as quickly as possible and as conditions permit. Before this can take place, the RIT must find the hole that the firefighter fell through. This can be a challenging task in itself. Hot, dark, and smoky conditions may have a very profound effect on visibility. It is even possible that the hole where the firefighter fell through has closed back up to a certain degree due to the characteristics of the flooring material. Utilizing good communication procedures, following a hoseline, or responding to a PASS device may be necessary to find the hole.

Remember, the floor or stairs collapsed for a reason. The RIT needs to pay close attention to the situation and surroundings. Placing several firefighters into the same area around the hole may compound the problem, causing further collapse. Once found, the area will have to be stabilized to a degree before rescue operations can take place. Placing ladders, furniture, or even doors from inside the structure around the hole can help to distribute the weight of the firefighters performing the rescue. It is important that the weight load be distributed across an area as large as possible **(Figure 10-1).**

The hole in the floor may need to be expanded to allow firefighters to enter and pull up the downed firefighter. This may include the use of hand tools and battery-operated saws to enlarge openings in floors as well as penetrating ceilings above the hole for the purpose of exposing joists and rafters to be used as high anchors if needed. Gas-powered saws will not work due to lack of oxygen to support the combustion process needed to run the engine. The lack of visibility and the adversity of the conditions presented would also make the use of such a tool very dangerous.

Once the hole is located and the floor is safe to work on, it may be necessary to have a rescuer sent down into the hole to assist the downed firefighter. Once down, the rescuer can assess the situation and determine what actions will be necessary for the success of the retrieval. When performing a rescue of this nature it is important to descend through the hole and raise the downed firefighter when ready as quickly as possible because the hole that firefighter fell through may have become a blast furnace due to the conditions on the lower level. This also applies to the rescuers working on top around the hole. Getting a rescuer down into the hole can be accomplished by several methods.

One method is to set a ladder down into the hole and have the rescuer climb down to the lower level. If the trapped firefighter is able, this may be the only step required for self-rescue. Once placed into the hole, the ladder should not be removed. The rescuer should always have a means of egress if needed. The ladder can also be used to assist in guiding the downed firefighter up through the floor. The ladder can also be used with a simple mechanical advantage system (described later in this chapter). The type of ladder used will be dictated by the depth of the elevation difference as well as the allowable space the team has to work in. The ladder should be positioned in a corner of the hole at a 90° climbing angle to allow the maximum amount of space for the rescuers and downed firefighter to exit without having their SCBA become an encumbrance **(Figure 10-2).**

FIGURE 10-1

Placing ladders, furniture, or even doors from inside the structure around the hole can help to distribute the weight of the firefighters performing the rescue.

FIGURE 10-2

The ladder should be positioned in a corner of the hole at a 90° climbing angle to allow the maximum amount of space for the rescuers and downed firefighter to exit without having their SCBAs become an encumbrance.

Safety

Once placed into the hole, the ladder should not be removed.

It may be necessary to get a hoseline with a nozzle down to protect the downed firefighter, who may now be located in a very hostile environment. A rescuer can quickly slide a hoseline down to the lower level to protect his comrade (see Chapter 7). This may be accomplished through the use of existing hoselines or the line that the RIT brings in **(Figure 10-3).** The use of a protective hoseline cannot be over emphasized.

FIGURE 10-3

It may be necessary to get a hoseline with a nozzle down to protect the downed firefighter, who may now be located in a very hostile environment.

The use of ropes and specialized knots can also allow the rescuer to be lowered down to the victim quickly. Lowering the rescuer down on the handcuff knot or "W" will be discussed in detail later in this chapter.

However the rescuer makes it to the lower level, it is imperative that some type of tag line be in place for use in a rapid retrieval by the topside firefighters. The rescuer should always have a means of egress if needed.

Safety

The rescuer should always have a means of egress if needed.

Whatever method is chosen to raise the downed firefighter, the RIT will need to address the use of a rated harness or the downed firefighter's SCBA to help raise the firefighter. Crew members on the topside will need points to grab and lift from. The preferred method is to use a properly rated harness such as the MAST. Using the SCBA for lifting is only for extreme circumstances unless it is rated to be used as such. To do so effectively will require converting the SCBA into a body harness (see Chapter 9). Rescuers should also keep in mind that there are situations where converting the SCBA harness may not be applicable depending on the manufacturer of the SCBA or the size and height of the victim.

Whatever method is chosen to raise the downed firefighter, it is important that a concurrent attempt to access and rescue the trapped firefighter be made by an alternative means (e.g., exterior wall breach) while the rescue attempt inside is carried out.

Lifting the Downed Firefighter Utilizing a Hoseline

If a member goes down through a weakened floor and remains conscious and oriented, a rescue can be performed by simply having a bend in the hose lowered into the hole and having the victim step into the bend while holding the hose on each side of his body.

SKILL 10-1

Lifting the Conscious and Able Firefighter Using a Hoseline

A The topside crew lowers a bend in the hose down to the floor level.

B The trapped firefighter steps into the bend, holding it tightly to his torso with the bend of his elbows.

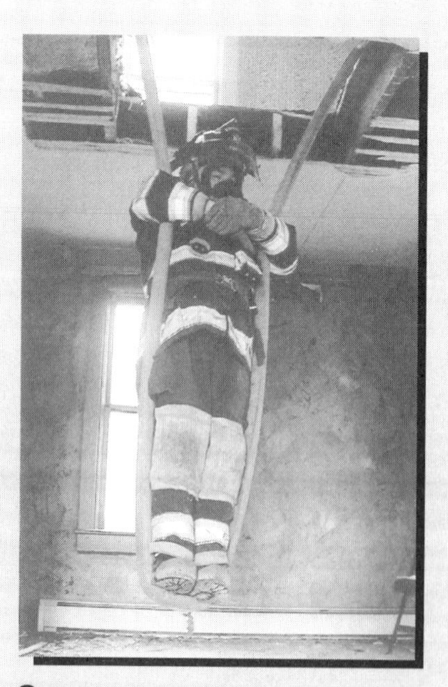

C The crew on top will then pull the hose from both sides at the same time, lifting the downed firefighter.

D Once at a high enough point, the topside firefighters will grasp the downed firefighter by the SCBA straps and pull him to safety.

Lifting the conscious and able firefighter using a hoseline (Skill 10-1)

1. The topside crew lowers a bend in the hose down to the floor level **(Skill 10-1A).**
2. The trapped firefighter steps into the bend, holding it tightly to his torso with the bend of his elbows. If the hose is only held by the hands, it will create a possibility for the firefighter to fall off the hose as it will want to straighten out as it is raised **(Skill 10-1B).**
3. The crew on top will then pull the hose from both sides at the same time. It is important that both sides pull at an equivalent pace to avoid shaking the firefighter off the line **(Skill 10-1C).**
4. Once at a high enough point, the topside firefighters will grasp the downed firefighter by the SCBA straps and pull him to safety **(Skill 10-1D).**

In cases where the downed firefighter cannot stand up because of high heat conditions or injury, he can lay over the hoseline, tucking it underneath his armpits and locking it in with his elbows, and be raised in the same manner as if he were able to stand **(Figure 10-4).**

FIGURE  10-4

In cases where the downed firefighter cannot stand up because of high heat conditions or injury, he can lay over the hoseline, tucking it underneath his armpits locking it in with his elbows, and be raised in the same manner as if he were able to stand.

Lifting the conscious but disabled firefighter using a hoseline (Skill 10-2):

1. The topside crew lowers a bend in the hose down to the floor level.
2. The trapped firefighter crawls into the bend, holding it tightly with the bend of the elbows **(Skill 10-2A).**

SKILL 10-2
Lifting the Conscious but Disabled Firefighter Using a Hoseline

A The trapped firefighter crawls into the bend, holding it tightly with the bend of the elbows.

(continued)

SKILL 10-2 (CONTINUED)

Lifting the Conscious but Disabled Firefighter Using a Hoseline

B The crew on top will then pull the hose from both sides at the same time. It is important that both sides pull at an equivalent pace to avoid shaking the firefighter off the line.

C Once at a high enough point, the topside firefighters will grasp the downed firefighter by the SCBA straps and pull the victim to safety.

3. The crew on top will then pull the hose from both sides at the same time. It is important that both sides pull at an equivalent pace to avoid shaking the firefighter off the line **(Skill 10-2B)**.
4. Once at a high enough point, the topside firefighters will grasp the downed firefighter by the SCBA straps and pull the victim to safety **(Skill 10-2C)**.

If the downed firefighter is rendered unconscious, it will be necessary to send a firefighter down to effect the rescue.

Lifting the unconscious firefighter using a hoseline (Skill 10-3)

1. The rescuer will be lowered down to the level of the downed firefighter.
2. A bend in the hose will be made, passing the nozzle back topside.
3. A twist will need to be made in the bend by the rescuer **(Skill 10-3A)**.
4. The bend of the hose will need to be placed under the downed firefighter's armpits and then crossed over behind his head above the SCBA cylinder **(Skill 10-3B)**.
5. A truckman's belt or webbing should then be used to secure the loop tightly so that the firefighter will not slip out of the hoseline **(Skill 10-3C)**.
6. The crew on top will then pull the hose from both sides at the same time. It is important that both sides pull at an equivalent pace to avoid shaking the firefighter off the line **(Skill 10-3D)**.
7. Once at a high enough point, the topside firefighters will grasp the downed firefighter by the SCBA straps and pull him to safety.
8. The downed firefighter will be released from the hoseline and the line will be sent back down to retrieve the rescuer.

Handcuff Knot Raise

The handcuff knot is another option that is available to rescuers and has many advantages:

- It is quick.
- It is easy to set up with proper training.
- It takes minimal equipment.
- It allows the firefighter to be raised in a lower-profile position.

Because of its importance, tying the handcuff knot will be reviewed once again.

SKILL 10-3

Lifting the Unconscious Firefighter Using a Hoseline

A A twist will need to be made in the bend of the hose sent down.

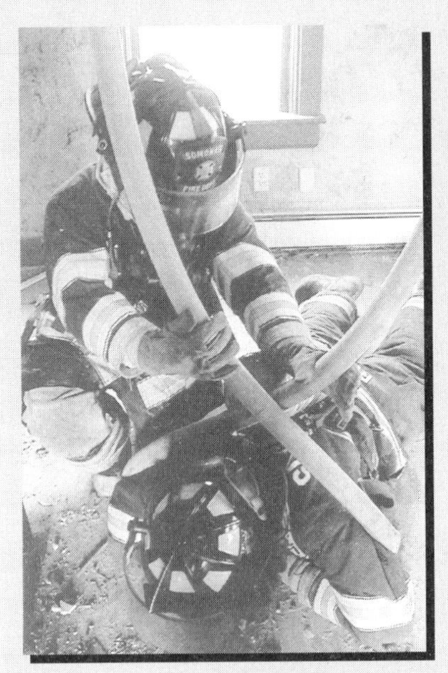

B The bend of hose will need to be placed under the downed firefighter's armpits and then crossed over behind his head above the SCBA cylinder.

C A truckman's belt or webbing should then be used to secure the loop tightly so that the firefighter will not slip out of the hoseline.

D The crew on top will then pull the hose from both sides at the same time while the rescuer assists from below.

SKILL 10-4
Tying the Handcuff Knot

A Two loops opposite of each other are formed in the rope.

B The two loops are passed to the inside of each other.

C The loops need to be pulled through to be sized large enough to fit over the downed firefighter's forearms.

D The handcuff knot needs to be placed on the firefighter's forearms where it will cinch down on the turnout coat and avoid causing injury to the wrists.

Tying the handcuff knot (Skill 10-4):

1. Two loops opposite of each other are formed in the rope **(Skill 10-4A).**
2. The two loops are passed to the inside of each other **(Skill 10-4B).**
3. The loops need to be pulled through to be sized large enough to fit over the downed firefighter's forearms **(Skill 10-4C).**
4. The handcuff knot needs to be placed on the firefighter's forearms where it will cinch down on the turnout coat and avoid causing injury to the wrists **(Skill 10-4D).**

It is important that the handcuff knot be placed far enough down on the rope to allow the two running ends enough length to reach the level where the firefighter is trapped.

The handcuff raise can be accomplished with one RIT. It will require five to six rescuers to be accomplished quickly and efficiently. If a

SKILL 10-5

Lowering a Rescuer Using the Handcuff Knot

A The rescuer to be lowered will place his feet into the loops of the handcuff knot and grasp the two running parts of the rope and position them closely in front while seated near the hole in the floor.

B The rescuer will turn onto his stomach and allow his feet to hang into the hole while being supported by RIT members holding the two running ends of the rope.

C The rescuer will apply more of his weight onto the loops of the handcuff knot by moving off the edge of the hole. Eventually he will come to a standing position while holding the two running ends of the rope.

ladder is not able to be used and the trapped firefighter is incapacitated, it will be necessary to send a rescuer down by using the rope.

Lowering a rescuer using the handcuff knot (Skill 10-5):

1. Place the handcuff knot in the rope, allowing enough slack for both running ends to reach the level of the downed firefighter.
2. For accountability and quick retrieval purposes the rescuer to be lowered should apply a bowline around his torso from another rope bag if time allows. This line should be placed to his back to avoid entanglement.
3. The rescuer to be lowered will place his feet into the loops of the handcuff knot and grasp the two running parts of the

rope and position them closely in front while seated near the hole in the floor **(Skill 10-5A)**.

4. The rescuer will turn onto his stomach and allow his feet to hang into the hole while being supported by RIT members holding the two running ends of the rope **(Skill 10-5B)**.
5. The rescuer will apply more of his weight onto the loops of the handcuff knot by moving off the edge of the hole. Eventually he will come to a standing position while holding the two running ends of the rope **(Skill 10-5C)**.
6. The RIT lowers the rescuer into the hole to the victim. If a bowline has been applied to the rescuer it should be guided down but should not carry any weight.

The hole being lowered into may present adverse conditions. For this reason, the rescuer should be lowered as rapidly and safely as possible.

Once the rescuer has located and assessed the victim, he will apply the handcuff knot to raise the downed firefighter.

Raising the downed firefighter using the handcuff knot (Skill 10-6):

1. The rescuer will bring the two arms of the downed firefighter together and place the loops of the handcuff knot over the hands and onto the forearms of the victim. The knot is then cinched tightly by pulling the two running ends from each loop. **(Skill 10-6A)**.

SKILL 10-6

Raising the Downed Firefighter Using the Handcuff Knot

A The rescuer will bring the two arms of the downed firefighter together and place the loops of the handcuff knot over the hands and onto the forearms of the victim. The knot is then cinched tightly by pulling the two running ends from each loop.

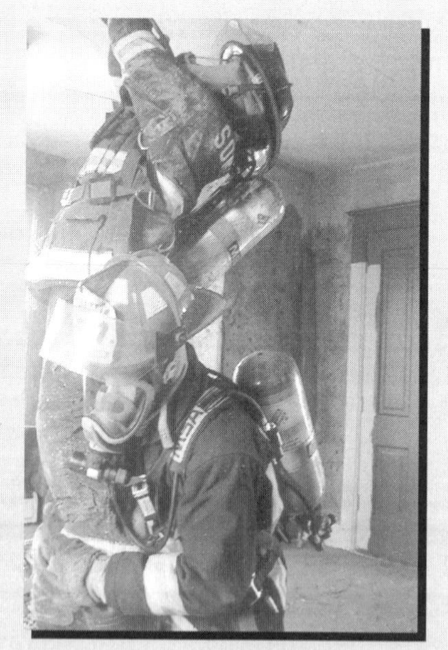

B The below-grade rescuer during the haul can assist by lifting upward under the SCBA, buttocks, and feet as the downed firefighter is being raised.

C Once at a high enough point, the topside firefighters will grasp the downed firefighter by the SCBA straps and pull the victim to safety.

The handcuff knot needs to be placed on the firefighter's forearms where it will cinch down on the turnout coat and avoid causing injury to the wrists.

2. Once the handcuff knot is applied, the rescuer will communicate to the rescuers above to begin to haul the downed firefighter upward and out of the hole. The team members above should perform the haul in a controlled manner until the downed firefighter has cleared the hole. The below-grade rescuer during the haul can assist by lifting upward under the SCBA, buttocks, and feet as the downed firefighter is being raised **(Skill 10-6B).**
3. Once at a high enough point, the topside firefighters will grasp the downed firefighter by the SCBA straps and pull the victim to safety **(Skill 10-6C).**
4. Once the downed firefighter is cleared of the hole, the rescuer can be brought back up by lowering the handcuff knot back down for him to place it on himself.

Utilizing a secondhand cuff knot will allow the RIT to have a four-line pull when raising the downed firefighter. This can give the RIT a definite needed advantage as lifting a downed firefighter will be very labor intensive. To create a four-line pull, a separate rope with a handcuff knot can be used or a handcuff knot can be tied approximately one-third the distance from each rope end if using a single rope. Just make certain that there will be enough rope on the running ends to reach the level of the downed firefighter.

Handcuff Cradle

Another variation of the handcuff knot that can be considered is the handcuff cradle. This variation utilizes two handcuff knots. One is placed on the forearms of the downed firefighter and the other is placed on the legs. Two handcuff knots will provide a **four-way haul.** When applied correctly, the downed firefighter is brought up in a seated or cradle position, which helps to distribute the weight load.

Raising the downed firefighter using the handcuff cradle (Skill 10-7):

1. The rescuer will position the downed firefighter and apply one handcuff knot to the forearms.
2. The rescuer will then apply the second handcuff knot to the downed firefighter's legs. This requires the loops of the second handcuff knot to be created larger than those for the forearms. They will be positioned into the groin area of the downed firefighter **(Skill 10-7A).**
3. After the two large loops have been placed onto each leg, they will be brought up as far as possible into the groin area with the knot portion of the rope positioned slightly above the downed firefighter's tailbone. The two running parts of the rope will then be brought up and run along each side of the SCBA cylinder, continuing up behind the downed firefighter's head and helmet **(Skill 10-7B).**
4. Once both handcuff knots are applied the rescuer will communicate to the rescuers above to begin to remove the slack and to hold tension on the four running lines coming from the handcuff knots while the rescuer positions the victim below the opening of the hole.
5. The rescuer will communicate to the team members above to haul the downed firefighter upward and out of the hole.
6. The team members above should perform the haul in a steady, controlled manner until the downed firefighter has cleared the hole. The below-grade rescuer during the haul will assist by lifting the downed firefighter under the SCBA, buttocks, and feet as the downed firefighter is being raised **(Skill 10-7C).**
7. Once the downed firefighter is cleared of the hole, the rescuer can be brought back up by lowering the handcuff knot back down for self-use.

SKILL 10-7
Raising the Downed Firefighter Using the Handcuff Cradle

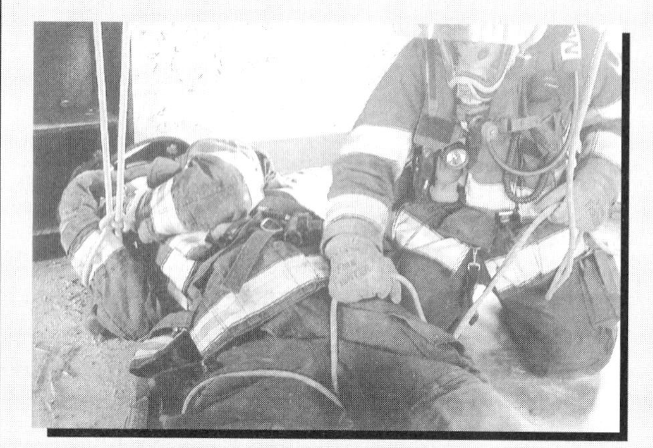

A The rescuer will apply the second handcuff knot to the downed firefighter's legs into the groin area.

B The two running parts of the rope will then be brought up and run along each side of the SCBA cylinder, continuing up behind the downed firefighter's head and helmet.

C The team members above should perform the haul in a steady, controlled manner until the downed firefighter has cleared the hole. The below-grade rescuer during the haul will assist by lifting the downed firefighter under the SCBA, buttocks, and feet as the downed firefighter is being raised.

The "W" Single Rope, Four-Way Haul Technique

The **"W" technique** is yet another alternative for the RIT when raising a downed firefighter through the floor. This technique is referred to as the "W" due to the appearance of the system when set up **(Figure 10-5).** The only equipment that is needed for the system is approximately 100 feet of rope and two large carabiners.

To assemble the "W":

1. When the system is assembled onsite, the RIT will pay out of the rope bag approximately 25 feet of rope, at which point they will place a figure eight on a bight and affix a large carabiner to its loop **(Figure 10-6).**
2. The RIT will then pay out an additional 50 feet of rope and apply another figure eight on a bight while affixing another carabiner to its loop.

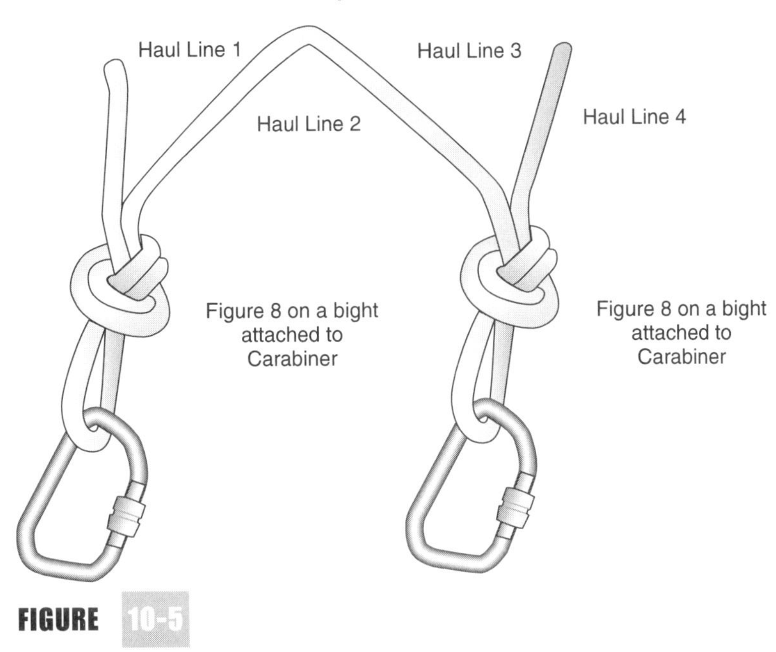

FIGURE 10-5

The "W" technique.

FIGURE 10-6

If a firefighter must measure out the length of line to place the carabiners, a quick method is to remember that the arm span of a person will be equal to their height.

3. The RIT will then pull out another 25 feet of rope from the bag.

The two points at the bottom of the "W" will have the figure eights to be attached to the downed firefighter. The outermost ends and middle point at the top of the "W" will account for the depth to the level of the downed firefighter. The "W" can be preassembled and placed into a rapid intervention rope bag or it can be assembled very quickly onsite.

Note

The 25-foot figure can change depending on the distance that the firefighter must be lifted.

In order to facilitate the use of this technique, it will be necessary to have five RIT members to perform the rescue.

Prior to utilizing the "W" to raise the downed firefighter, it will be necessary to lower a rescuer to the victim's level.

To lower a rescuer utilizing the "W" (Skill 10-8):

1. The two inside ropes with the figure eights and carabiners are placed in back alongside the SCBA cylinder (**Skill 10-8A**):
2. These ropes are then placed between the rescuer's legs and crossed over each other (**Skill 10-8B**).
3. The carabiners are brought up along the front of the chest and attached to the ropes running behind and alongside the rescuer's SCBA (**Skill 10-8C**).
4. This creates a make shift seat and shoulder harness that the below-grade rescuer can hold onto while being lowered into the

SKILL 10-8

To Lower a Rescuer Utilizing the "W"

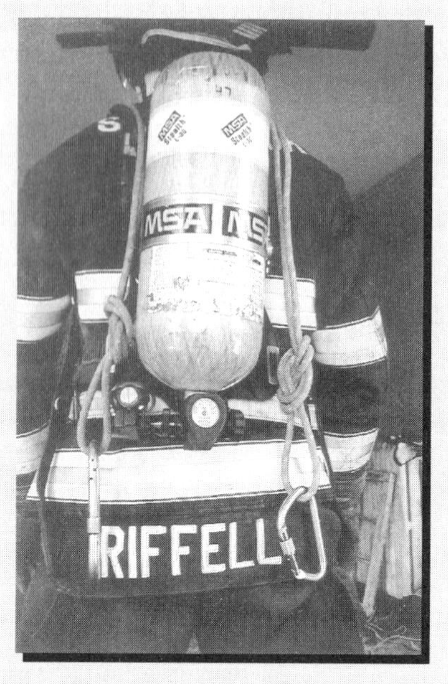

A The two inside ropes with the figure eights and carabiners are placed in back alongside the SCBA cylinder.

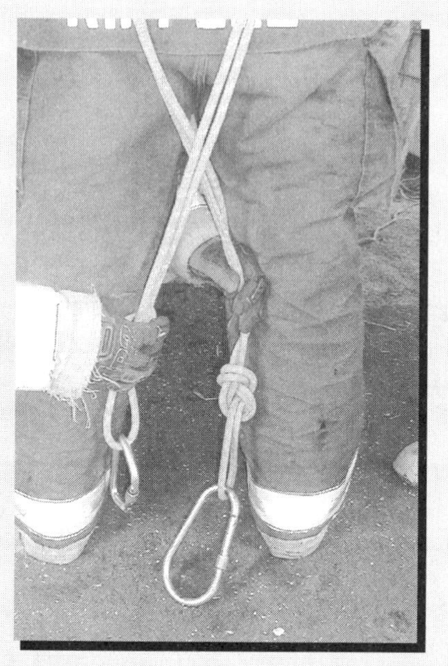

B These ropes are then placed between the rescuer's legs and crossed over each other.

C The carabiners are brought up along the front of the chest and attached to the ropes running behind and alongside the rescuer's SCBA.

D This creates a make shift seat and shoulder harness that the below-grade rescuer can hold onto while being lowered into the hole.

SKILL 10-9

Lifting the Downed Firefighter Using the "W"

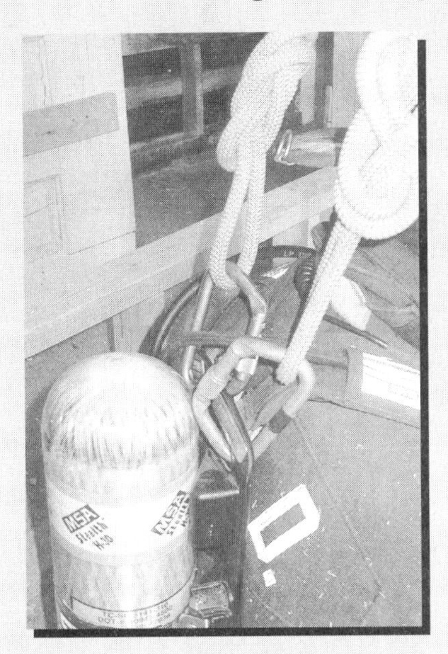

A The rescuer will then detach the "W" rescue rope from himself and attach the large carabiners to the rated harness on the downed firefighter or to the harness assembly of the downed firefighter's SCBA (if safety rated).

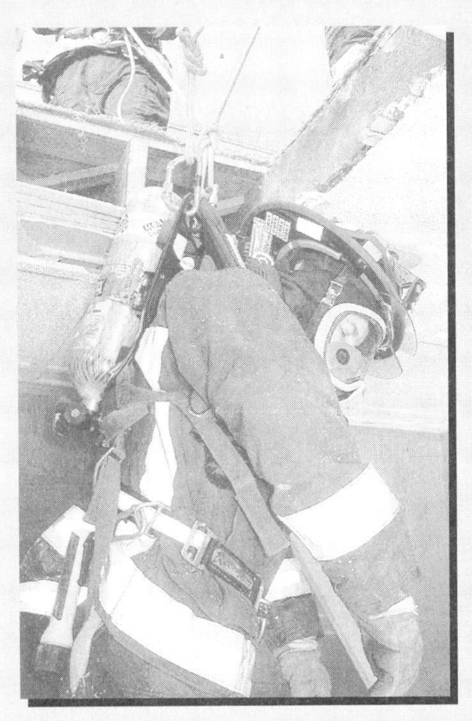

B The four lines are pulled up on to lift the downed firefighter.

hole **(Skill 10-8 D)**. The below-grade rescuer should also have a bowline tag line applied for accountability and emergency retrieval purposes.

Lifting the downed firefighter using the "W" (Skill 10-9):

1. Once the rescuer locates the downed firefighter, the rescuer will apply a rated harness such as the MAST or convert the downed firefighter's SCBA into a body harness.
2. The rescuer will then detach the "W" rescue rope from himself and attach the large carabiners to the rated harness on the downed firefighter or to the harness assembly of the downed firefighter's SCBA if safety rated **(Skill 10-9A)**.
3. The topside RIT members will position themselves around the hole by utilizing the rope's "W" pattern to create a four-line haul.
4. The rescuer will position the downed firefighter under the hole while communicating to the rescuers above to take out the slack in the lines.
5. The rescuer will then tell the RIT to haul while guiding and assisting the lift **(Skill 10-9B)**.
6. Once the downed firefighter is raised clear of the hole, the team above will lower the "W" system back down to the rescuer, who will reattach the system in the same manner used previously for lowering.

If the rescuer is unable to convert the SCBA into a body harness or a rated harness is not available, the "W" can be applied in the same manner as it was used to lower the rescuer for raising the downed firefighter.

Using a Bight of Rope to Raise a Downed Firefighter

Another very quick, "down and dirty" method of raising a firefighter can be performed with the use of rope without tying any knots. This is a useful technique when the firefighter that had fallen is still conscious and able to help himself. Some advantages to this technique are:

- The downed firefighter is removed from the hole with the upper body forward, facilitating an easier removal.
- It provides four lines for hauling.
- It is very quick.
- It is simple.
- It requires minimal equipment (only a rope).

Safety

Make certain that all straps are tight and that the waist buckle is secure.

To raise the downed firefighter using a bight of rope (Skill 10-10):

1. The first step in performing this maneuver is for the rescuer to tighten up the downed firefighter's SCBA straps after assessing the condition of the victim.
2. A rescue rope or personal line is lowered down to the rescuer. The rescuer will form a bight in the rope at the point where it reaches the floor **(Skill 10-10A).**
3. This bight is then run beneath the tightened waist strap of the firefighter's SCBA **(Skill 10-10B).**
4. The bight is pulled through and up underneath the SCBA shoulder strap to the opposite side **(Skill 10-10C).**
5. The bight is then pulled across the chest of the downed firefighter and underneath the opposite side shoulder strap **(Skill 10-10D)**
6. The topside rescuers lower the tip of a pike pole to pull the bight back up to their level. Both running ends of the rope are now with the rescuers above **(Skill 10-10E).**
7. The running ends are now pulled on to haul the downed firefighter out of the hole **(Skill 10-10F).**

SKILL 10-10

To Raise the Downed Firefighter Using a Bight of Rope

A The rescuer will form a bight in the rope at the point where it reaches the floor.

B The bight is then run beneath the tightened waist strap of the firefighter's SCBA.

(continued)

SKILL 10-10 (CONTINUED)

To Raise the Downed Firefighter Using a Bight of Rope

C The bight is pulled through and up underneath the SCBA shoulder strap to the opposite side.

D The bight is then pulled across the chest of the downed firefighter and underneath the opposite side shoulder strap.

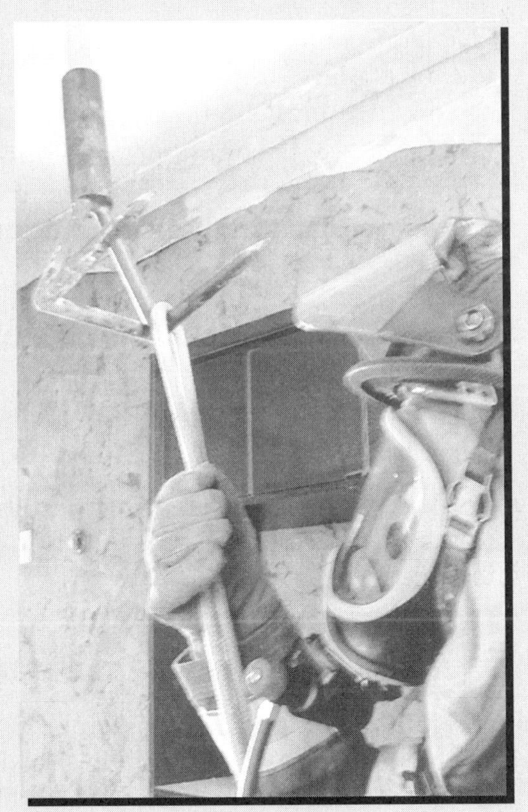

E The topside rescuers lower a tool to pull the bight back up to their level. Both running ends of the rope are now with the rescuers above.

F The running ends are now pulled on to haul the downed firefighter out of the hole.

Ladder through the Floor Using a 2-to-1 Mechanical Advantage System

This technique incorporates the use of a simple 2-to-1 mechanical advantage system to help raise the downed firefighter. A downed firefighter can be well over 300 lbs with turnout gear, SCBA, and water from firefighting operations being retained in the gear. A mechanical advantage system may be required when there are not enough personnel to lift the downed firefighter.

Note

This technique should only be utilized in an extreme circumstance.

Equipment needed for this method of extraction is:

- one ladder (14–16 ft)
- one rope bag
- two large carabiners

To raise the downed firefighter using a ladder and 2-to-1 mechanical advantage (Skill 10-11):

1. The RIT will lower the ladder into the hole at an angle that allows enough room to raise the downed firefighter and clear the opening with an SCBA. The ladder position will need to be at 75–90° due to the confines of the size of the hole. It is recommended to position the ladder on an angle in one of the corners of the hole to allow the maximum amount of working space. A 14–16-ft ladder will usually suffice; special circumstances may necessitate a larger ladder. The ladder should be positioned with four to five rungs above the hole. This will serve as a high-point anchor.
2. The RIT will take the end of the rope with a figure eight on a bight and large carabiner and wrap the second rung from the top of the ladder five times, leaving the large carabiner and figure eight hanging freely and forming a tensionless hitch **(Skill 10-11A).**
3. A bight is formed in the running part of the rope and it is passed through the hanging carabiner, bringing it down along the front of the ladder rungs to the rescuer descending into the hole **(Skill 10-11B).**
4. The rescuer will attach another large carabiner onto the bight and descend the ladder while leaving the remaining rope in the rope bag at the top of the hole. This will provide the haul line.
5. The rescuer will locate and position the downed firefighter at the base of the ladder. A rated harness will need to be placed on the downed firefighter or an SCBA body harness conversion should take place. The downed firefighter will need to be positioned with his face towards the ladder.
6. The large carabiner on the bight of rope will be attached to the harness or SCBA of the downed firefighter.
7. The rescuer will communicate to the RIT members above to take slack out of the rope and to haul the downed firefighter upwards.
8. As the downed firefighter is lifted, the rescuer will position his body beneath the downed firefighter, using his legs to help push the downed firefighter upward as he ascends the ladder back up **(Skill 10-11C).**

It is imperative that the RIT be certain that the foot of the ladder does not kick out when performing this maneuver.

2-to-1 Mechanical Advantage Systems

The possibilities of using a preassembled 2-to-1 mechanical advantage system for raising a downed firefighter through the floor are only limited by the imagination. System set-up will be determined by available anchor points. The principles introduced in Chapter 6 relating to mechanical advantage and anchors can be adapted to any presenting situation that deals with raising the downed firefighter.

SKILL 10-11

To Raise the Downed Firefighter Using a Ladder and 2-to-1 Mechanical Advantage

A The RIT will take the end of the rope with a figure eight on a bight and large carabiner and wrap the second rung from the top of the ladder five times, leaving the large carabiner and figure eight hanging freely and forming a tensionless hitch.

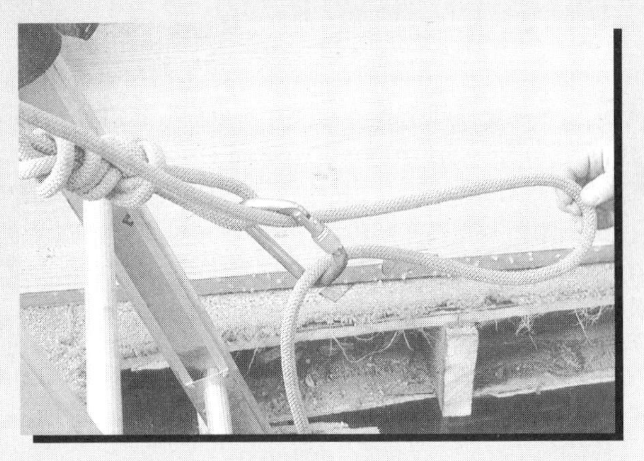

B A bight is formed in the running part of the rope and it is passed through the hanging carabiner, bringing it down along the front of the ladder rungs to the rescuer descending into the hole.

C As the downed firefighter is lifted, the rescuer will position his body beneath the downed firefighter, using his legs to help push the downed firefighter upward as he ascends the ladder back up.

Summary

The quickest and easiest method to affect a rescue in a situation where a firefighter has gone through a floor is the one that must be used. This includes making an entry through an alternate place other than the hole that the firefighter fell through. Simultaneous rescue operations should be taking place on the exterior as well as interior of the structure. Formulating a good plan of action for the RIT begins with the execution of a complete size-up on the fireground. Numerous techniques have been introduced and perfected to rescue a firefighter through the floor, such as the four-way haul and the "W" technique. Even with these, the fire service continues to lose firefighters in similar situations each year. Each technique has advantages and disadvantages. It is up to the RIT to train on and determine which techniques work for them and which ones do not. The importance of being proficient in more than one technique cannot be stressed enough.

As with all techniques, these systems should be assembled and practiced on a regular basis in order for RIT members to have the skills become second nature and instinctive.

KEY TERMS

Four-way haul
"W" technique
Walk-out basement

REVIEW QUESTIONS

1. When an alternate entrance/exit is not available, the quickest method of retrieval of a downed firefighter may be the area in the floor through which the firefighter fell.

   True False

2. When a Mayday is called for a downed firefighter that has fallen through the floor, it is a priority of the RIT to
   a. call an additional Mayday.
   b. scan the building with a thermal imaging camera.
   c. check with Command to proceed to the location.
   d. get someone into the area with the victim as quickly as possible.

3. Gas-powered saws are not recommended for enlarging the hole in which a firefighter has fallen because
   a. there may not be enough oxygen present to run a combustion engine.
   b. of the lack of visibility.
   c. they will cause excessive vibration and movement.
   d. both (a) and (b) are correct

4. Lowering a rescuer down into a hole through the floor to a fallen firefighter can be accomplished by which one of the following methods?
   a. placing a ladder into the hole
   b. the use of a handcuff knot or "W" technique
   c. utilizing a hoseline
   d. all of the above

5. When lifting the unconscious firefighter using a hoseline retrieval technique, it will be necessary to
   a. open the nozzle partially and spray the downed the firefighter to see if he will respond to stimuli.
   b. lower a rescuer down to the level of the downed firefighter.
   c. put a twist in the line to help secure the downed firefighter for raising.
   d. both (b) and (c)

6. The handcuff knot raise has many advantages, such as
   a. it is quick, easy to set up, and requires minimal equipment.

b. it can be applied to any part of the body.
c. it needs minimal manpower.
d. it has multiple loops for securing tools and equipment when lowering them into the hole.

7. The handcuff knot needs to be placed on
   a. the waist strap of the fallen firefighter.
   b. the coat and forearms of the fallen firefighter while cinching it down.
   c. the SCBA cylinder of the fallen firefighter.
   d. both (a) and (b)

8. Raising the downed firefighter using the handcuff cradle technique involves the use of
   a. two handcuff knots.
   b. one handcuff knot and a figure eight knot on a bight.
   c. one handcuff knot and the MAST.
   d. one handcuff knot and a bowline knot.

9. When utilizing the handcuff cradle maneuver,
   a. both handcuff knots will be applied to the upper arms.
   b. one handcuff knot will be applied to the forearms of the stricken firefighter while the other will be applied to the upper legs.
   c. one handcuff knot will be applied around the SCBA shoulder straps and the other on the forearms.
   d. both handcuff knots should be applied to the lower extremities.

10. The "W," single-rope maneuver has advantages, such as
    a. it requires less rope and no knots.
    b. it is quick to assemble and provides a four-way haul.
    c. it can be used to lower the rescuer.
    d. all of the above except (A)

11. The advantage of using only a bight of rope for the purpose of raising a downed firefighter is
    a. it provides four lines for hauling.
    b. it is quick and simple.
    c. it does not require the use of the SCBA.
    d. both (a) and (b)

ADDITIONAL RESOURCES

Hoff, Robert, and Kolomay, R., *Firefighter Rescue and Survival.* Tulsa, OK: PennWell Publishing, 2003.

McCormack, J., *Firefighter Rescue and Rapid Intervention Teams.* Indianapolis, IN: Fire Department Training Network, 2003.

Jakubowski, G., and Morton, M., *Rapid Intervention Teams.* Stillwater, OK: Fire Protection Publications, 2001.

CHAPTER

11 Removing and Lowering Firefighters from Windows

Learning Objectives

Upon completion of this chapter, you should be able to:

- understand and determine the appropriate time and need for removing and lowering downed firefighters from windows.
- demonstrate the procedures required to lift a downed firefighter through a window using the rescuer body ramp technique.
- display the actions required to lift a downed firefighter through a window using the two-rescuer body ramp technique.
- exhibit the measures necessary to lift a downed firefighter through a window using the seated window removal technique.
- show the actions required to lift a downed firefighter through a window by means of the knee-method window removal technique.
- demonstrate the techniques required to lift a downed firefighter through a window in a confined space such as presented in the Denver Drill.
- exhibit the measures essential to lift a downed firefighter through a window using the "piggy back" variation of the Denver technique.
- reveal the actions necessary to lift a downed firefighter through a window using the longboard and sling window removal technique.
- state the differences between a true mechanical advantage and modified mechanical advantage and how both are used in removing downed firefighters.
- communicate the procedures necessary to lift and lower a downed firefighter through a window using high-point anchors and 2-to-1 mechanical advantage systems.

Case Study

"At approximately 2:00 A.M. on Monday, September 28, 1992, Denver firefighters responded to a fire in a two-story printing office. During the fire-suppression operations one Denver firefighter died.

"Upon arrival, firefighters found that the building was full of smoke. When they entered the building, they found separate fires in several areas. The firefighters attempted to suppress the fires as they found them, and reportedly, the suppression crews felt they were beginning to control the fire.

"One firefighter was temporarily working by himself inside the fire building when a section of floor collapsed and the fire intensity suddenly increased. It was about this time that he encountered some type of difficulty. The firefighter was eventually able to reach a second-story window and shine his handlight through the window, alerting other firefighters who were outside.

"The partially collapsed floor and intense fire prevented potential rescuers from reaching the trapped firefighter through the interior of the building. Other firefighters laddered the building and entered the room containing the trapped firefighter. Over a period of approximately 55 minutes, an estimated 15 rescuers attempted to remove the victim through a window; however, they were unsuccessful due to the confinement of the space in which they were working. The fatally injured firefighter was removed through a hole that firefighters cut in a wall.

"This fire highlights the importance of firefighters in pairs during fire-suppression and related operations. This fire also reveals difficulties associated with rescue in small places."

—Reprinted with permission from the Summary Investigation Report, NFPA, Printing Office Fire, Denver, CO, September 28, 1992, copyright® NFPA 1993, all rights reserved.

Introduction

The most appropriate method for removing a downed firefighter may be to use a window in the immediate area for egress. However, it must be stressed that the safest way to remove a downed firefighter is by use of an interior staircase if feasible.

Removal via a Window

Firefighters may be forced to use windows for removal for a variety of reasons. First, the route that was taken into the structure may have been altered or changed during the course of operations by collapse or advancing fire conditions. Reduction in air supply may be another reason that will determine the need for removal via a window. To do so will be labor intensive and will require a RIT operating on the exterior as well as the interior. Communication between these companies is of highest priority. The exterior RIT will need to know the specific equipment and exact location necessary to affect the rescue. This will normally take place after the initial RIT locates the downed firefighter.

Rescuer Body Ramp Window Removal

The **body ramp** maneuver requires one or two members of a RIT. First, they must locate and move a downed firefighter into a room with a window. Second, the downed firefighter must be raised into the window where the exterior RIT will remove the firefighter onto a ladder.

To raise a downed firefighter out a window by a single rescuer using the body ramp (Skill 11-1):

1. Position the victim on his side with the SCBA still in place and his feet toward the window.
2. The downed firefighter should be positioned slightly to the side of the window. This is because the downed firefighter will be rolled toward the center of the window during the maneuver.
3. The rescuer will then remove his own SCBA, placing it on the floor next to the

victim's head so it is out of the way but still supplying the rescuer with air **(Skill 11-1A).**

4. The rescuer will then lie side by side next to the victim, reflecting the same position as the victim. The arm of the victim should be placed above his head while the rescuer grips the victim's leg and pulls the victim onto the rescuer's back. To do so requires that the rescuer pull the victim's leg as high as possible onto the rescuer's hip. The rescuer should not allow the victim's foot to become entangled between the rescuer's legs. If this happens the victim's ankle may be injured during the maneuver **(Skill 11-1B).**
5. The rescuer will then take the victim's outside arm and place it underneath his own arm as high up into the armpit as possible. It is then held in place by applying pressure from the rescuer's arm and elbow. The rescuer's other arm and elbow will be positioned underneath the rescuer's side. As the rescuer rolls over the rescuer's arm and elbow are used as a fulcrum **(Skill 11-1C).**
6. During the roll, the victim is positioned onto the rescuer's back **(Skill 11-1D).** This should place them both approximately at the center of the window. At this time, a member of the exterior RIT who is situated on a ladder outside the window will reach in and grasp the victim's feet. The first rescuer will push backwards with his hands until his hands meet his knees. The rescuer will then crawl backwards until his feet meet the wall underneath the window. The rescuer will then push the weight of the victim slightly upwards **(Skill 11-1E).** The member of the exterior RIT will pull and control the lower extremities of the victim.

Remember, the rescuer's air pack is in front of him on the floor. It will be dragged backwards

SKILL 11-1

To Raise a Downed Firefighter Out a Window by a Single Rescuer Using the Body Ramp

A The rescuer will remove his own SCBA, placing it on the floor next to the victim's head so it is out of the way but still supplying the rescuer with air.

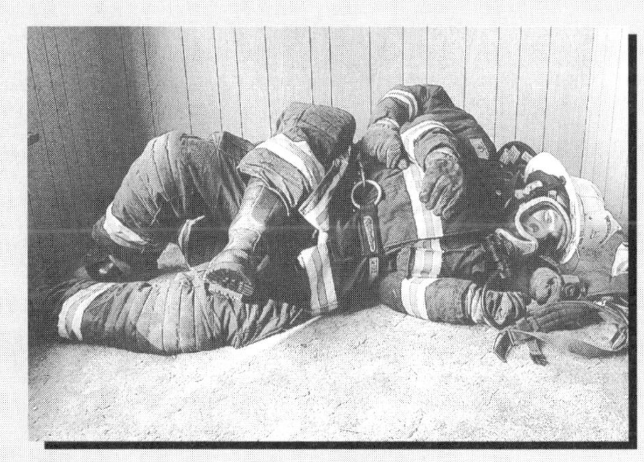

B The rescuer will then lie side by side next to the victim, reflecting the same position as the victim. The arm of the victim should be placed above his head while the rescuer grips the victim's leg and pulls the victim onto the rescuer's back.

(continued)

SKILL 11-1 (CONTINUED)

To Raise a Downed Firefighter Out a Window by a Single Rescuer Using the Body Ramp

C The rescuer will take the victim's outside arm and place it underneath his own arm as high up into the armpit as possible. It is held in place by applying pressure from the rescuer's arm and elbow. The rescuer's other arm and elbow will be positioned underneath the rescuer's side. As the rescuer rolls over the rescuer's arm and elbow are used as a fulcrum.

D During the roll, the victim is positioned onto the rescuer's back.

E Once on the rescuer's back, the rescuer can raise to his knees and crawl backwards toward the window, pushing the downed firefighter up and over the windowsill.

during the maneuver. The SCBA should be clear of all obstructions, such as the extremities of both the victim and the rescuer.

This maneuver and technique is a very efficient and a practical way of raising a downed firefighter to a windowsill when only one rescuer is available and located with the victim. Two members of a RIT can quickly ascend the ladder, gain access through the window, and begin performing this technique on the victim.

A variation of this maneuver can also be accomplished with two rescuers inside. The other rescuer will assist in placing and rolling the downed firefighter.

To perform the body ramp with two rescuers (Skill 11-2):

1. Position the victim on his side with his SCBA still in place and his feet toward the window.
2. The downed firefighter should be positioned slightly to the side of the window. This is because the downed firefighter will be rolled toward the center of the window during the maneuver.
3. Rescuer 1 will remove his own SCBA, placing it on the floor out of the way but still supplying the rescuer with air **(Skill 11-2A).**
4. Rescuer 1 will lie face down next to the victim with his feet at the window. Rescuer 2 will roll the victim to one side **(Skill 11-2B).**

SKILL 11-2

To Perform the Body Ramp with Two Rescuers

A Rescuer 1 will remove his own SCBA, placing it on the floor out of the way but still supplying the rescuer with air.

B Rescuer 1 will lie face down next to the victim with his feet at the window. Rescuer 2 will roll the victim to one side.

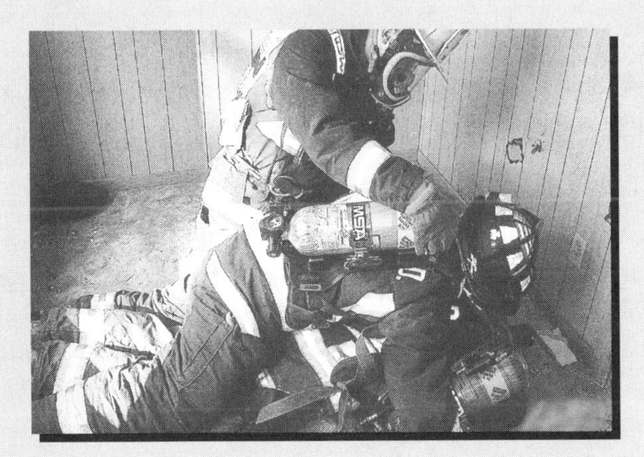

C Rescuer 2 will roll the victim onto rescuer 1's back.

D Rescuer 1 will push up onto all fours, crawling back toward the window with the victim on his back. Rescuer 2 will assist in making certain that the victim clears the windowsill.

5. Rescuer 2 will roll the victim onto rescuer 1's back **(Skill 11-2C)**.
6. Rescuer 1 will push up onto all fours, crawling back toward the window with the victim on his back. Rescuer 2 will assist in making certain that the victim clears the windowsill **(Skill 11-2D)**.

Seated Window Removal

The **seated window removal** technique will require two of the rescuers to be on the interior with the downed firefighter and one on the exterior ladder placed just beneath the windowsill. The rescuer on the ladder will be receiving the downed firefighter in a forward-sitting position. This will enable the rescuer on the ladder to descend with either the "face in the groin" method or the "knee in the groin" method. Whichever method is chosen, it should be understood that this particular delivery of the victim will require the rescuers on the interior to remove or shift the downed firefighter's SCBA at the last moment while seated in the window.

To perform the seated window removal (Skill 11-3):

1. The victim is placed directly under the windowsill with the victim's feet pointing

SKILL 11-3

To Perform the Seated Window Removal

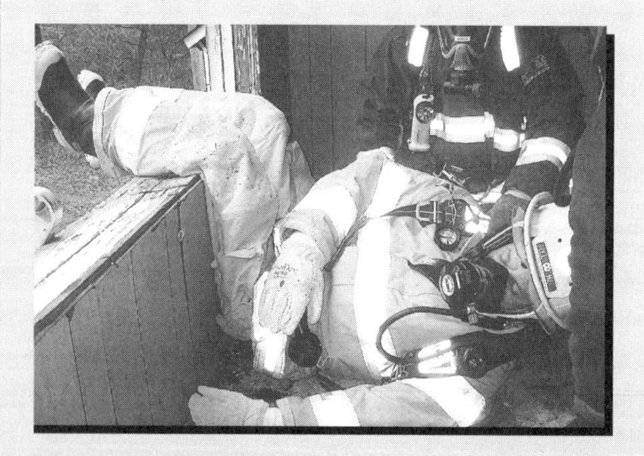

A The downed firefighter should be placed with his feet over the windowsill and buttocks resting against the wall.

B Once the rescuers have the downed firefighter in this position they can kneel or squat down on each side of the victim. They will pull up on the SCBA harness straps while reaching down and cradling the victim underneath the buttocks area or lower waist harness.

C When the downed firefighter is positioned on the windowsill and is secured by all rescuers, they can then remove and disconnect his SCBA and harness. The victim is then delivered to the rescuer on the ladder.

at the wall below the window. The victim is slid toward the window face up while sliding the SCBA cylinder along the floor.

2. As the rescuers are sliding the victim, they are positioned on each side and raise the victim's legs toward the windowsill in order to maneuver the body of the victim with the victim's buttocks coming to rest against the wall **(Skill 11-3A).**
3. If a third rescuer is available, he can help control the victim's head. This is done while grabbing the upper rear shoulder straps and guiding and pushing the victim to the wall.
4. Once the rescuers have the downed firefighter in this position, they can kneel or squat down on each side of the victim. They will pull up on the SCBA harness straps while reaching down and cradling the victim underneath the buttocks area or lower waist harness **(Skill 11-3B).**
5. The outside rescuer can also assist by reaching in while grasping and controlling the lower extremities. The outside rescuer can additionally grab the front of the harness.
6. When the downed firefighter is positioned on the windowsill and is secured by all rescuers they can then remove and disconnect his SCBA and harness. The victim is then delivered to the rescuer on the ladder **(Skill 11-3C).**

Knee-Method Window Removal

In the **knee-method window removal** the downed firefighter will be presented in the window opening in a headfirst, face-down position.

To perform the knee-method window removal (Skill 11-4):

1. Once the victim is located, the RIT members will position the downed firefighter headfirst to the bottom and center of the window in a face-down position.
2. The interior rescuers will then kneel on each side of the victim while shifting one knee and foot to the wall underneath the windowsill. This is done while grabbing each side of the downed firefighter's SCBA harness at the shoulder straps and along the waist strap **(Skill 11-4A).**
3. They will then begin to raise the victim upward by the victim's harness and place the upper chest of the victim onto his leading flexed knees, which are at the windowsill. The rescuer on the ladder will reach in to assist by grabbing the victim's shoulder straps. The downed firefighter will be brought through the window opening in this fashion **(Skill 11-4B).**
4. The two rescuers kneeling at the window will then stand. They will make certain to support and feed the waist and lower extremities of the victim out the window **(Skill 11-4C).**

One important feature of the knee method is that the downed firefighter can either be delivered through the window with or without his SCBA. If a third rescuer is present, he can assist the interior members by helping to raise and support the waist and lower extremities. This is done while the victim is passed out the window.

The Denver Drill

The 1992 fatality of Denver Fire Department Engineer Mark Langvardt has been the main catalyst for developing techniques devoted to rescuing downed firefighters within tight spaces and elevated windowsills. The techniques concerning this type of maneuver should be learned and continuously reviewed by all RITs.

Engineer Langvardt became trapped at an exterior window that was approximately 42 inches high off the floor in a corridor that was only 28 inches wide. Dimensions such as this make it very difficult to maneuver and lift a downed firefighter.

It is important for firefighters to remember that these maneuvers can be applied to other situations involving window exits and any other types of openings. These types of window removals can take place from any floor of a structure.

The **Denver Drill** and most of its variations will require a minimum of three rescuers.

SKILL 11-4

To Perform the Knee-Method Window Removal

A The interior rescuers will kneel on each side of the victim while shifting one knee and foot to the wall underneath the windowsill. This is done while grabbing each side of the downed firefighter's SCBA harness at the shoulder straps and along the waist strap.

B The rescuers will begin to raise the victim upward by the victim's harness and place the upper chest of the victim onto his leading flexed knees, which are at the windowsill.

C The two rescuers kneeling at the window will then stand. They will make certain to support and feed the waist and lower extremities of the victim out the window.

To perform the Denver Drill technique (Skill 11-5):

1. One rescuer will enter the window headfirst, crawling over the victim until he gets to the downed firefighter's feet. The rescuer will then face the victim and the window from which he entered **(Skill 11-5A).**
2. The rescuer will then grasp the downed firefighter's SCBA shoulder straps and begin to lean backwards into a sitting position, pulling the downed firefighter into a sitting position also. This allows the second rescuer to enter the window onto the floor area beneath the windowsill and have enough room to position himself with his back to the wall **(Skills 11-5B** and **11-5C).**
3. The second rescuer will then bend his knees and place his feet at the victim's buttocks and grasp the downed firefighter's SCBA cylinder while the first rescuer straddles over the victim's legs and positions the downed firefighter's arms onto the rescuer's upper legs.

SKILL 11-5

To Perform the Denver Drill Technique

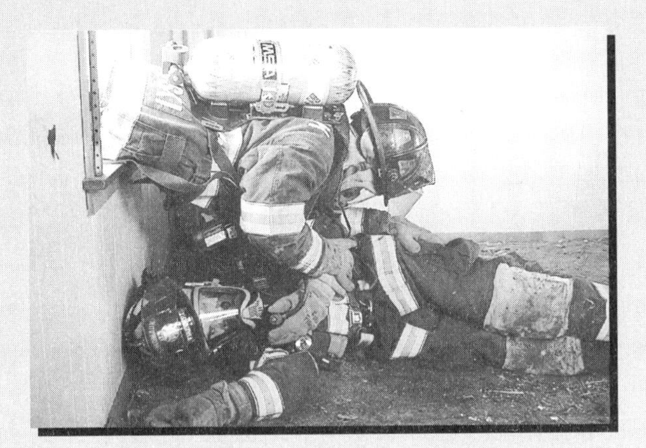

A One rescuer will enter the window headfirst, crawling over the victim until he gets to the downed firefighter's feet. The rescuer will then face the victim and the window from which he entered.

B The rescuer will then grasp the downed firefighter's SCBA shoulder straps and begin to lean backwards into a sitting position, pulling the downed firefighter into a sitting position also.

C The second rescuer will enter the window onto the floor area beneath the windowsill and position himself with his back to the wall.

D After the downed firefighter is positioned on the second rescuer's knees the first rescuer can squat down and place the victim's legs onto his shoulders. The first rescuer will place his face as deep into the groin area as possible of the downed firefighter. This is done to provide enough lift in one motion to raise the victim to the windowsill. The first rescuer will continue to lift upward while the second rescuer continues to push upward at the rear and lower portion of the SCBA cylinder.

E The rescuers on the exterior, if positioned at ground level, should assist by reaching in and grasping the downed firefighter's shoulder straps in order to help lift and clear the victim's SCBA.

4. The first rescuer will bend at the knees in order to wrap his arms around the victim and grab his SCBA harness as close to the cylinder valve as possible. The victim is then lifted onto the second rescuer's knees. The second rescuer will then grab and push upward in a bench press motion against the cylinder, assisting the lift.
5. After the downed firefighter is positioned on the second rescuer's knees the first rescuer can squat down and place the victim's legs onto his shoulders. The first rescuer will place his face as deep into the groin area as possible of the downed firefighter. This is done to provide enough lift in one motion to raise the victim to the windowsill. The first rescuer will continue to lift upward while the second rescuer continues to push upward at the rear and lower portion of the SCBA cylinder. This will help clear the victim's SCBA of the windowsill **(Skill 11-5D)**.
6. The rescuers on the exterior, if positioned at ground level, should assist by reaching in and grasping the downed firefighter's shoulder straps in order to help lift and clear the victim's SCBA **(Skill 11-5E)**.

If the maneuver is to be performed from a second- or third-story window by a single rescuer on a ladder, then that rescuer will assist in the same manner. It should be noted that this maneuver can also be assisted with two rescuers on two ladders next to each other. Another point to mention is that the downed firefighter will be presented face up while being carried out of the window. His SCBA will most likely run against the ladder. This will prove to be more difficult for the descent, but can be managed safely if two rescuers on two ladders placed side by side are used instead of one.

Piggy Back Variation

The piggy back variation is performed as follows (Skill 11-6):

1. The first rescuer will enter the window headfirst and position himself over the downed firefighter. The first rescuer grabs the shoulder straps of the victim and leans backwards, pulling the victim away from the window in order to allow the second rescuer to enter through the window **(Skill 11-6A)**.
2. The second rescuer will turn around and face the window. While on his hands and knees he will remove his own SCBA and

SKILL 11-6
The Piggy Back Variation is Performed as Follows

A The first rescuer grabs the shoulder straps of the victim and leans backwards, pulling the victim away from the window in order to allow the second rescuer to enter through the window.

B The second rescuer will turn around and face the window. While on his hands and knees he will remove his own SCBA and place it on the floor in front of or to the side of the victim.

(continued)

SKILL 11-6 (CONTINUED)

The Piggy Back Variation is Performed as Follows

C The downed firefighter is placed on rescuer 2's back while rescuer 1 positions himself with the downed firefighter's legs over his shoulders.

D As rescuer 1 drives the downed firefighter forward, rescuer 2 will raise to his feet to help lift the downed firefighter over the windowsill.

place it on the floor in front of or to the side of the victim. The second rescuer will remain connected to his air supply. The reason for this positioning and removal of the SCBA is to create a flat surface area along the second rescuer's back. The victim will be placed on top of the rescuer's back **(Skill 11-6B).**

3. The first rescuer straddles the victim's legs. The first rescuer will bend at the knees in order to wrap his arms around the victim. The rescuer should grab as far down as possible around the victim. The rescuers should reach around the victim's SCBA harness as close to the cylinder valve as possible. The victim is then positioned on the second rescuer's back. The second rescuer is unable to help the first in this process. This makes the maneuver difficult but still attainable if performed properly. Because the downed firefighter is face up, his SCBA will rest on the back of the second rescuer. This will result in the victim leaning left or right while on the second rescuer's back.
4. The first rescuer gets into position by squatting down. He places the victim's legs onto the rescuer's shoulders as far as the thigh. He will then place his face as deep into the groin area of the downed firefighter as possible to provide enough lift in one motion to raise the victim upward to the windowsill **(Skill 11-6C).** The first rescuer will continue to lift upward, creating enough lift for the victim's SCBA to clear the windowsill while driving the downed firefighter forward. The second rescuer will assist by raising himself up onto his feet **(Skill 11-6D).** This creates enough height and force to clear the victim's SCBA of the windowsill.

Longboard and Sling Window Removal

This technique is limited to ground-floor or flat-surface conditions on which the exterior rescuers are based. This maneuver is not recommended when working on ladders at upper floors because of the unstable situation that would be created. The technique is also designed to be useful in confined areas when only one rescuer may be able to enter the space around the window opening. This maneuver can be applied to any given ground-level window removal of a downed firefighter. A ladder may also be substituted for the longboard if not available.

To perform the longboard and sling window removal (Skill 11-7):

1. The first rescuer of the RIT will enter the window headfirst. If the confined area allows, a second rescuer can also enter to assist the first.
2. Once the first rescuer has positioned the downed firefighter he will apply the MAST or sling made of webbing material to the downed firefighter. Another option is to perform the SCBA harness conversion on the downed firefighter, or both can be applied if necessary.
3. The first rescuer will grab the downed firefighter's SCBA shoulder straps. By leaning backwards he will move the downed firefighter into sitting position. This will create enough space for a longboard to be used as a lever **(Skill 11-7A).** The foot end will be placed under the victim's buttocks while the head of the board is resting on the windowsill.
4. The rescuers on the interior will place the victim on the board using the sling or SCBA harness conversion with assistance from the outside rescuers. To accomplish this, the exterior rescuers will reach in and grab the sling handle found at the head of the victim above the SCBA cylinder **(Skill 11-7B).** If the rescuers are not using the sling but are instead using a SCBA harness conversion, they will grab the upper shoulder straps of the victim's SCBA. If the exterior rescuers are unable to reach the sling handle or the

SKILL 11-7
To Perform the Longboard and Sling Window Removal

A The first rescuer will grab the downed firefighter's SCBA shoulder straps. By leaning backwards he will move the downed firefighter into sitting position. This will create enough space for a longboard to be used as a lever.

B The rescuers on the interior will place the victim on the board using the sling or SCBA harness conversion with assistance from the outside rescuers. To accomplish this, the exterior rescuers will reach in and grab the sling handle found at the head of the victim above the SCBA cylinder.

(continued)

SCBA shoulder straps, an additional short piece of rope or webbing with a carabiner is used to pull the downed firefighter onto the board.

5. The victim is positioned on the board. The exterior rescuers will hold the sling or SCBA shoulder straps with tension. The interior rescuers will raise the bottom of the board until it is parallel with the floor **(Skill 11-7C).** The victim will be lying slightly on his side because of the SCBA. Rescuers will continue to maintain this position, thereby preventing the victim from sliding off the board while

SKILL 11-7 (CONTINUED)

To Perform the Longboard and Sling Window Removal

C The victim is positioned on the board. The exterior rescuers will hold the sling or SCBA shoulder straps with tension. The interior rescuers will raise the bottom of the board until it is parallel with the floor.

D The victim will be lying slightly on his side because of the SCBA. Rescuers will continue to maintain this position, thereby preventing the victim from sliding off the board while being moved through the window.

E The exterior RIT crew will grab each side of the board as it exits the window.

being moved through the window **(Skill 11-7D).**

If using a ladder in place of the longboard, the exterior rescuers can further assist by using their weight as a counterbalance and pushing down on the ladder.

6. The exterior RIT crew will grab each side of the board as it exits the window **(Skill 11-7E).** It will be lowered to the ground or directly onto a cot.

Alternative Single Rescuer Technique

The following is an alternative technique that may be utilized when only one rescuer can make entry into the window opening.

To perform the alternative single rescuer technique (Skill 11-8):

1. The rescuer enters into the window opening.
2. The downed firefighter is moved to the window with his knees bent and feet placed against the wall.
3. The interior rescuer will lift the downed firefighter from behind while an exterior rescuer will take control of the downed firefighter's hands **(Skill 11-8A).**
4. The interior rescuer will grasp beneath the downed firefighter's SCBA cylinder, lifting up while the exterior rescuer pulls the downed firefighter's arms **(Skill 11-8B).**
5. The downed firefighter should be positioned on the windowsill where he can be removed to the exterior.

SKILL 11-8

To Perform the Alternative Single Rescuer Technique

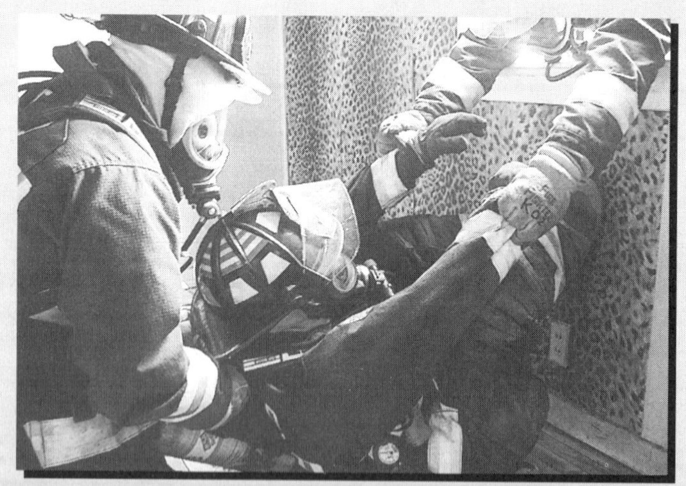

A The downed firefighter is moved to the window with his knees bent and feet placed against the wall. The interior rescuer will then lift the downed firefighter from behind while an exterior rescuer will take control of the downed firefighter's hands.

B The interior rescuer will grasp beneath the downed firefighter's SCBA cylinder, lifting up while the exterior rescuer pulls the downed firefighter's arms. The downed firefighter should be positioned on the windowsill where he can be removed to the exterior.

High Point Anchors and 2-to-1 Mechanical Advantage for Window Removals

The use of modified and **true mechanical advantages** systems provides quick and easy assembly. Additionally, the use of a 2-to-1 mechanical advantage system can remove a downed firefighter from a confined space such as that presented in the Denver Drill.

These techniques should only be used when there are no other safe or efficient means for removing a downed firefighter from a structure. It is important to note that under fire conditions these techniques can become very hazardous and are only to be considered after all other maneuvers have been dismissed.

The rescuers should be thoroughly trained in the setup of mechanical advantage systems. These should be practiced so that the use of these systems becomes second nature. Only life-safety rope that is inspected regularly should be used. The carabiners should be made of the appropriate rated material and they should be large enough to be manageable with gloved hands **(Figure 11-1)**. Rated harnesses such as the MAST will enhance the degree of safety when using these techniques.

High-Point Anchor Removal with Modified 2-to-1 Mechanical Advantage

This method requires a minimum of:

- four to five rapid intervention company members

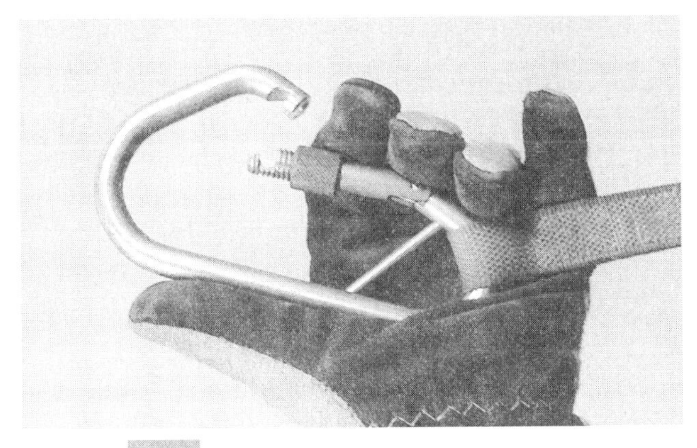

FIGURE 11-1

Carabiners used for rapid intervention should be large enough to be manageable with gloved hands.

- one rope bag containing at least 100 feet of life-safety rope
- two carabiners
- ladder of the appropriate length

This technique should be used if it is not feasible to move a firefighter through an entire building or egress is unobtainable. The simple mechanical advantage system can be applied if the downed firefighter is located away from the window and cannot be moved toward it. With the assistance of the exterior rescuers, the victim can now be moved to the window by use of this system. It will also lower him to the ground.

To utilize a high-point anchor with a 2-to-1 mechanical advantage to remove a downed firefighter from an upper-floor window (Skill 11-9):

1. After the first interior RIT locates and assesses the victim, members will determine the need for this method. They will relay information to Command or a rescue sector officer that additional RITs will be required to facilitate the rescue. The first RIT should indicate the equipment that will be needed. This may include an additional air supply for the downed firefighter. The first interior RIT team will apply a rescue harness and move the victim to a position near a window.
2. The exterior RIT will raise a ladder to the building so the tip of the ladder is approximately three to four rungs above the top of the window. Enough space must be available to fit the downed firefighter between the building and the back of the ladder **(Skill 11-9A)**.
3. A rope with a figure eight on a bight with a carabiner will be brought up to a rung as high above the window as possible. The rescuer on the ladder will wrap the end with the carabiner around the rung above the window four to five times. This will form a tensionless hitch, leaving the figure eight with a carabiner hanging freely just below the rung that was wrapped **(Skill 11-9B)**. It is important for the hanging figure eight with carabiner to be as high above the window as possible to gain the proper

SKILL 11-9

To Utilize a High-Point Anchor with a 2-to-1 Mechanical Advantage to Remove a Downed Firefighter from an Upper-Floor Window

A The exterior RIT will raise a ladder to the building so the tip of the ladder is approximately three to four rungs above the top of the window. Enough space must be available to fit the downed firefighter between the building and the back of the ladder.

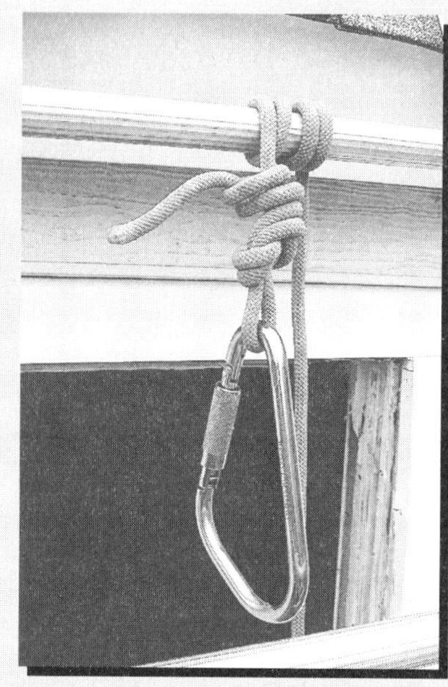

B A rope with a figure eight on a bight with a carabiner will be brought up to a rung as high above the window as possible. The rescuer on the ladder will wrap the end with the carabiner around the rung above the window four to five times. This will form a tensionless hitch, leaving the figure eight with a carabiner hanging freely just below the rung that was wrapped.

C The rescuer on the ladder will form a bight in the rope and put it through the hanging carabiner. The rescuer will apply an additional large carabiner onto the bight that has been pulled through the free-hanging carabiner. It will extend to the interior rescuers at the window.

(continued)

height and clearance for moving the downed firefighter through the window.

4. The rescuer on the ladder will form a bight in the rope and put it through the hanging carabiner. The rescuer will apply an additional large carabiner onto the bight that has been pulled through the free-hanging carabiner. It will extend to the interior rescuers at the window **(Skill 11-9C)**. The rescuer on the ladder will bring the rope bag and the remaining length of rope to the exterior rescuers on the ground.
5. The rescuers on the ground will weave the rope through the bottom three or four rungs of the ladder. A control firefighter will provide tension and braking to the line when needed during the course of lowering the victim to the ground **(Skill 11-9D)**. The interior rescuers will then attach the large carabiner onto an approved rescue harness placed on the downed firefighter.
6. The exterior rescuers on the ground will pull on the line to the inside of the ladder while the interior rescuers help guide and

SKILL 11-9 (CONTINUED)

To Utilize a High-Point Anchor with a 2-to-1 Mechanical Advantage to Remove a Downed Firefighter from an Upper-Floor Window

D The rescuers on the ground will weave the rope through the bottom three or four rungs of the ladder. A control firefighter will provide tension and braking to the line when needed during the course of lowering the victim to the ground.

E The exterior rescuers on the ground will pull on the line to the inside of the ladder to raise the victim into and out of the window opening.

F By allowing the rope to slowly pass through the rungs of the ladder, the victim is lowered in a controlled fashion once out the window.

raise the victim into the window opening **(Skill 11-9E)**. The control firefighter will remove slack in the line once the victim is clear of the window. By allowing the rope to slowly pass through the rungs of the ladder the victim is lowered in a controlled fashion **(Skill 11-9F)**.

High-Point Anchor Removal with True 2-to-1 Mechanical Advantage

Another technique available to the RIT is the use of a pre-made 2-to-1 system that incorporates the use of a pulley with large carabiners **(Figure 11-2)**. The principles of each system are identical in their purpose and are only slightly modified from

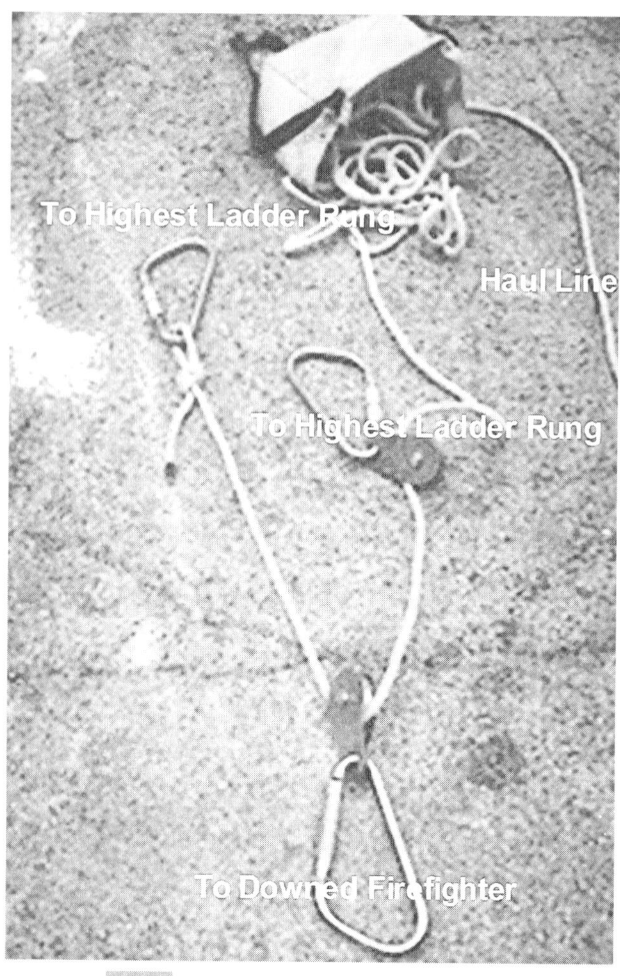

FIGURE 11-2

Pre-rigged 2-to-1 system.

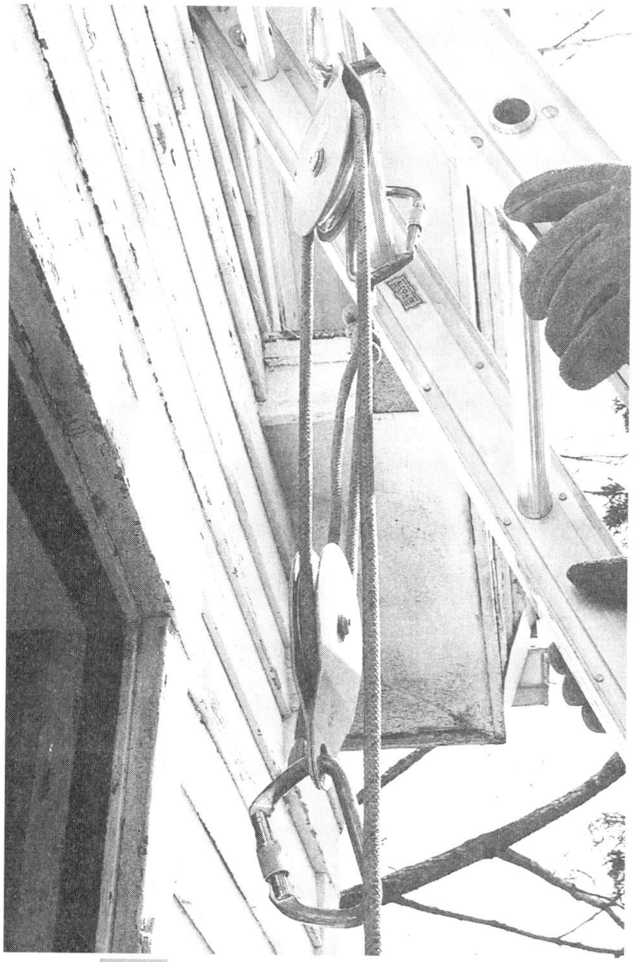

FIGURE 11-3

Two carabiners will be attached to the highest rung above the window. Between these two will be a pulley used for raising the downed firefighter.

each other. The main difference is that this system will be pre-rigged and connected to a life-safety rope in a rope bag. Two carabiners will be attached to the highest rung above the window. Between these two will be a pulley used for raising the downed firefighter **(Figure 11-3).** The victim is lowered in the same way as described in the high-point anchor modified 2-to-1 method.

Summary

Once a RIT locates a downed firefighter, it will be faced with the labor intensive task of removing the victim. Oftentimes, a window in the immediate area may be the most appropriate way for removal. Lifting a firefighter up and over a windowsill can be a very challenging task. Numerous techniques have been introduced and perfected to help overcome this challenge, such as the body ramp method and the seated and knee-method window removal techniques. Each technique has its unique advantages and disadvantages as well as applicability to the situation at hand. It is up to the RIT to train on and determine which techniques work for them and which ones do not. The importance of being proficient in more than one technique cannot be stressed enough.

The Denver Drill reveals the importance of having a method to remove a downed firefighter from a tight space and should be trained on and periodically reviewed. In addition, the benefits of using true mechanical advantage and a high-point anchor to conduct a rescue should not be overlooked. As noted, many rescue techniques are available and the RIT should be familiar with and proficient in using each.

KEY TERMS

Body ramp
Denver Drill
High-point anchor
Knee-method window removal
Seated window removal
True mechanical advantage

REVIEW QUESTIONS

1. The safest method for removing a downed firefighter from a structure is
   a. through a window.
   b. with a ladder.
   c. by use of an interior staircase.
   d. in a tower ladder.
2. RITs may be forced to use windows for the removal and rescue of a downed firefighter for a variety of reasons. Name four.
   1.
   2.
   3.
   4.
3. When utilizing the rescuer body ramp window removal, it will be necessary for the rescuer to
   a. lift his leg up and over the window while grabbing the victim's SCBA straps.
   b. remove his own SCBA, placing it on the floor next to the victim's head so it is out of the way but still supplying air.
   c. remove the victim's SCBA and harness completely and drag him over to the window by either the use of webbing or his coat.
   d. none of the above
4. The seated window removal will require
   a. one rescuer at the window and one rescuer on the interior.
   b. two rescuers on the interior and one rescuer on an exterior ladder at the windowsill.
   c. four rescuers—two on the interior and two on the exterior with a ladder at the windowsill.
   d. two ladders and one rescuer.
5. When performing the seated window removal, the interior rescuers will be presenting the downed firefighter to the rescuers on the exterior
   a. face down.
   b. feet up, legs straight.
   c. face up.
   d. feet down, legs straight.
6. When performing the knee-method window removal, rescuers will position the downed firefighter
   a. feet first to the bottom of the windowsill.
   b. headfirst to the bottom of the window, face down.
   c. feet first on the victim's side to the bottom of the windowsill.
   d. both (a) and (c) are correct

7. The longboard and sling window removal
   a. requires the technique be limited to ground-floor and/or flat-surface conditions on which exterior rescuers are based.
   b. will require one to two members of an RIT to accomplish the technique efficiently.
   c. is mainly used for second-story window removals on ladders.
   d. is used for both single- and multiple-story rescues requiring spinal immobilization of the stricken firefighter.

8. When positioning a ladder for the high-point anchor, modified 2-to-1 mechanical advantage technique, where should the ladder be placed?
   a. with the tip placed approximately three to four rungs centered above the window opening
   b. two rungs into the opening of the window
   c. just below the windowsill
   d. to the right or left of the window opening

■ ADDITIONAL RESOURCES

Hoff, Robert, and Kolomay, R., *Firefighter Rescue and Survival.* Tulsa, OK: PennWell Publishing, 2003.

McCormack, J., *Firefighter Rescue and Rapid Intervention Teams.* Indianapolis, IN: Fire Department Training Network, 2003.

NFPA, Fire Investigations, *Printing Office Fire, Denver, CO, September 28, 1992.* Quincy, MA: National Fire Protection Association, 1992.

Rapid Intervention Teams and How to Avoid Needing Them, Technical Report Series, Report #123. Washington, D.C.: U.S. Fire Administration, 2003.

C H A P T E R

12 Removing Distressed Firefighters Using Ladders and Aerial Devices

Learning Objectives

Upon completion of this chapter, you should be able to:

- explain the importance and proper placement of ground ladders for the purpose of removing a downed firefighter.
- demonstrate the proper techniques for removing a downed firefighter down a ladder using the face-in-groin method, straddle method, and horizontal or cradle carry.
- clarify the indications of when multiple ladders are needed and explain how they can be used in rescuing a downed firefighter.
- perform the one-firefighter ladder raise when there are obstructions present.
- discuss the importance of aerial devices as related to the operations of the RIT.
- give explanation of the advantages and disadvantages of the aerial ladder, tower ladder, and snorkel.
- talk about the aspects of positioning aerial devices for RIT use.
- discuss how an aerial ladder can be used to ventilate an upper-floor window.
- give details on how to carry a downed firefighter down the aerial ladder.
- tell how a rescue litter and pike poles can be used to lower a downed firefighter using an aerial ladder.
- discuss the different ways that a tower ladder can be used to remove a downed firefighter.

CASE STUDY

In the 1970s and 1980s, the South Bronx was faced with unprecedented fire duty, especially in the 27th Battalion, which was experiencing urban decay. Fire duty was so intense that the Fire Department of New York set up an interchange program so that if companies were extremely busy one night they would be rotated into a slower district the following night and slower companies would be moved into their district to provide coverage. During this time period, many schools in the South Bronx were not heavily populated and were shut down. Vagrants and vandals took advantage of these empty buildings and oftentimes would set nuisance fires in rooms inside the building. These school buildings were five stories in height without standpipes and contained closed stairwells, which required tremendous lead outs to make it to the top floors. To combat these nuisance fires on the upper floors, the tower ladder would raise the bucket to a window that was two rooms over and stretch a line from the bucket into the hallway.

On this particular night, companies were summoned to a report of a fire in a school building that was recently closed down. This school, however, was vacated and left with all of the belongings and furniture still in place. 85 Engine was the first unit on-scene, with light, hazy smoke showing from the top floor; another nuisance fire. Two firefighters proceeded up to the top floor with forcible entry tools and a can. On the top floor they encountered very light smoke conditions. Searching, they found some smoldering debris inside of a closet in one of the classrooms. As they opened the closet, the debris fell to the floor of the classroom, which was a tongue-and-groove wood floor. During this time, the other firefighter broke out one of the windows inside the classroom for ventilation. Unknown to anyone at that time, vandals had poured accelerant on the floor, which had made its way down beneath the tongue-and-groove planks. With the oxygen being introduced into the room and smoldering debris, the room lit up in flame, trapping both firefighters.

By this time, the first due truck (31 truck) was on-scene and was positioning the bucket to stretch a line to the top floor. Inside the bucket were two firefighters. With the events taking place, the firefighters were raising the bucket as quickly as possible to rescue the trapped firefighters who were now in the window. Below, surrounding the small courtyard to the school, was an ornamental iron fence that presented an additional hazard if the trapped firefighters decided to bail. One trapped firefighter who was on fire could not wait any longer for the bucket and jumped into the bucket from 25–30 feet out. He landed in the bucket on top of one of the firefighters, breaking the leg of the firefighter in the bucket. As the bucket made it closer to the window, the other trapped firefighter jumped also. The bucket was immediately lowered to the ground where a booster line was used to put out the firefighters that were still on fire.

One firefighter received burns over 37 percent of his body while another firefighter suffered a broken leg. Without the truck being placed, there may have been two firefighter fatalities. Conditions on this particular call changed from nothing serious to an inferno in a matter of seconds without any noticeable indications. The importance of the RIT being proactive and using ladders on the fireground cannot be overemphasized.

—Case Study provided from an interview with Dan Noonan, FDNY (ret.), on January 21, 2005.

Introduction

One of the most proactive behaviors the RIT can provide is the placement of ladders raised to windows in order to provide the interior firefighters with a means of emergency egress. Ground ladders as well as aerial devices in the hands of an experienced RIT can mean the difference between success and failure in some rescue situations. As mentioned, the safest way of removing a downed firefighter from a building is by way of an interior staircase if at all possible.

Ground Ladders

Part of a RIT's proactive measures on the fireground include placing ladders to the fire floor, floors above the fire, to the sides and rear of a building, and to the roof for a second means of egress for the vent crew **(Figure 12-1)**. When the RIT is presented with a rescue that will involve getting a downed firefighter out of a window onto a ladder, the position and angle of the ladder in relation to the window becomes crucial.

For the purposes of rescuing downed firefighters or egress using a ladder, the ladder tip should always be positioned just beneath or at the sill of the window **(Figure 12-2)**. There are few exceptions to this rule, one being the use of the ladder as a high-point anchor with a 2-to-1 simple mechanical advantage to lift a downed firefighter over a windowsill and lower him to safety (see Chapter 11). The tip being placed at or just below the sill maintains the entire area of the window to be available for entry or egress. This removes the task of having to lift a firefighter up onto the ladder and overcomes the hazard of not being able to fit through the window with an SCBA being worn. It also allows a firefighter to remain low and eliminates the need to stand up in the greatest area of heat concentration when attempting to egress quickly.

Ladder angle will also have an effect on the rescue. Training manuals consistently state that a ladder should be placed at a 75° angle for optimum climbing and work **(Figure 12-3)**. As we all know, this is not always possible on the fireground. Situations such as buildings being close together and overhead obstructions such as power lines or trees may dictate that a ladder be placed at a less than ideal angle. A ladder that is

FIGURE 12-1

One of the most proactive behaviors the RIT can provide is the placement of ladders and aerial devices to areas in order to provide firefighters with a means of emergency egress.

FIGURE 12-2

For the purpose of rescuing downed firefighters or egress using a ladder, the ladder tip should always be positioned just beneath or at the sill of the window.

FIGURE 12-3

Training manuals consistently state that a ladder should be placed at a 75° angle for optimum climbing and work.

placed at too steep of an angle will be more difficult to climb and will cause difficulties in being able to control a downed firefighter while descending. If the ladder is placed at an angle that is less than preferred, it will be difficult to climb and will increase the amount of stress on the ladder components. In situations where good ladder angles are not attainable, it may be necessary to use multiple ladders with multiple rescuers to help disperse the load of the downed firefighter.

It is important to remember not to overload a ladder when rescuing a downed firefighter. Ladder failures are not common but are very possible. As a rule, ladders (with the exception of folding or collapsible ones) are rated to withstand a maximum load of 750 lbs. The weight of a downed firefighter that is soaked with water can reach well over 300 lbs. Again, multiple ladders will help to eliminate this factor.

Whatever angle is used in placing ladders, proper heeling techniques will also be required. A ladder placed at a less than ideal angle will have an increased chance of **kicking out** (suddenly moving outward) and falling. Ladders used for rescue should be aggressively heeled from the side away from the building. Positioning on the side away from the building will allow the firefighters heeling the ability to focus on the rescue, as they will be able to see it, and will make it easier for them to offset the force of the ladder kicking out **(Figure 12-4).** If possible, it is also recommended that the base of a ground ladder be secured with rope or webbing to the building or other substantial object to prevent it from kicking out.

FIGURE 12-4

Ladders used for rescue should be aggressively heeled from the side away from the building.

The fly section of the ladder should always be tied off when it is raised for rescue; this provides insurance in case the ladder locks fail under stress.

Using ladders for rescuing firefighters is a two-part process. This process will involve rescuers on the interior as well as exterior. To bring a downed firefighter down a ground ladder will require a minimum of at least four to six members of a RIT, including both rescuers on the inside of the structure with the victim as well as team members on the outside. It will be the responsibility of the rescuers inside to get the downed firefighter to and over the windowsill while the exterior team will need to get the downed member down safely. It is also important for the interior rescuers preparing the downed firefighter for the exit out the window to understand the techniques and maneuvers that the exterior team rescuers are using. Whichever presentation of a downed firefighter is shown at the window will determine the technique or maneuver that will be utilized by the rescuers on the ladder. Communication between the teams will play a very important role. Each team must know what the other is intending to do.

There are three basic ways of bringing a downed firefighter down a ladder: firefighter face in groin, cradle method, and the straddle method.

Firefighter Face-in-Groin Ladder Removal Method

This method provides a simple way of controlling the weight of the downed firefighter while descending the ladder. Two drawbacks to this method, however, are that it will be necessary to remove the downed firefighter's SCBA and it must be performed in a controlled manner to prevent the downed member's head from hitting the rungs of the ladder during the descent. In regard to the downed firefighter's SCBA, it will be the interior team's responsibility to transfer the victim to the position at the windowsill with the SCBA connected as long as possible. The interior teams may decide that the SCBA should be disconnected before lifting the victim to the window, in which case the regulator may just be connected or the entire SCBA may have been removed by the time the exterior team receives the downed member on the ladder.

It is best to have the SCBA completely removed in order for the downed firefighter's back to run along the rungs of the ladder. One method that can be employed by both the interior and exterior rescuers is to loosen and remove the shoulder straps of the downed firefighter's SCBA and loosen the waist strap. When the victim is positioned in a seated position in the window the rescuers can turn the victim's SCBA to the front or side. This provides no obstructions to the ladder rungs when the rescuer and victim descend down the ladder, and allows the downed firefighter to have a constant flow of air.

To perform the firefighter face-in-groin method (Skill 12-1):

1. The interior team places the victim's legs over the shoulders of the rescuer that is on the ladder **(Skill 12-1A)**.
2. The rescuer's face should get as close as possible into the groin area of the downed firefighter. The positioning of the rescuer in relation to the downed firefighter is the key when lowering the victim down the ladder.
3. The rescuer should step down one or two rungs while letting the victim's upper legs and weight fall more and more onto his shoulders while maintaining himself as close to the ladder as possible **(Skill 12-1B)**.
4. The rescuer's arms should be placed on the rails of the ladder underneath the victim's armpits with the victim's arms hanging on the outside of the rescuer's arms and hands, which are firmly on the rails **(Skill 12-1C)**. It is acceptable to grip the rungs firmly underneath the victim's arms as well. Both feet should be placed on each rung by the rescuer while descending to avoid excessive weight shifting and slipping off the ladder. By maintaining the proper positioning when descending with the downed firefighter, the rescuer will be able to keep the back of the victim even against the ladder rails and rungs. This provides a controlled descent and allows a good portion of the weight to be distributed to the ladder. The downed firefighter's coat may rise up, causing some increased friction, but this is negligible and will not hinder the success of the maneuver; it may actually help provide stability to the head and neck.

Straddle Method Ladder Removal

In this particular maneuver, the rescuer on the ladder will receive the victim feet first coming out of the window. The downed firefighter can be facing toward the rescuer or have his back to the rescuer on the ladder. The downed firefighter can be lowered by this method with the SCBA still on his back or removed. When present, the SCBA will cause the descent to be more challenging. This can be overcome by shifting the downed firefighter's SCBA to the side to establish a reduced profile so that the rescuer on the ladder can get his arms around the victim.

SKILL 12-1
To Perform the Firefighter Face-in-Groin Method

A The interior team places the victim's legs over the shoulders of the rescuer that is on the ladder.

B The rescuer should step down one or two rungs while letting the victim's upper legs and weight fall more and more onto his shoulders while maintaining himself as close to the ladder as possible.

C The rescuer's arms should be placed on the rails of the ladder underneath the victim's armpits with the victim's arms hanging on the outside of the rescuer's arms and hands, which are firmly on the rails.

To perform the straddle method ladder removal (Skill 12-2):

1. The interior team will guide the victim's feet and legs out the window to the exterior rescuer on the ladder **(Skill 12-2A).**
2. As the inside RIT moves the victim out the window, the exterior rescuer on the ladder will guide the downed firefighter's feet and legs to the outside of the beams of the

SKILL 12-2

To Perform the Straddle Method Ladder Removal

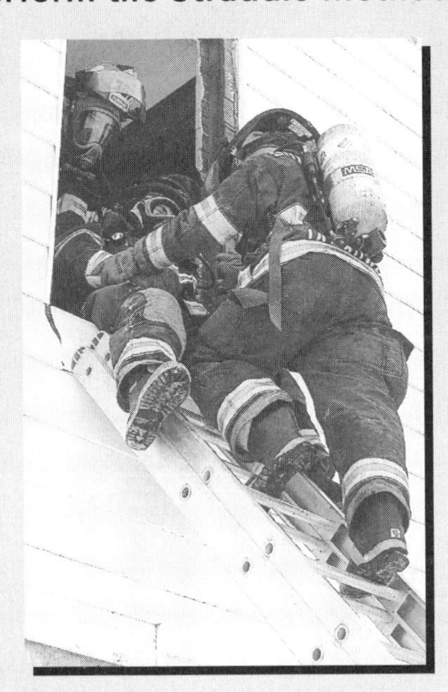

A The interior team will guide the victim's feet and legs out the window to the exterior rescuer on the ladder.

B The rescuer takes the downed firefighter's weight while holding the downed firefighter into the ladder by gripping the rails underneath the downed firefighter's armpits.

C The placement of the rescuer's arms and chest also help to hold the downed firefighter into the ladder.

D The straddle removal technique can also be performed with the firefighter facing away from the rescuer.

ladder until the downed firefighter's body comes to rest at the groin area onto one of the rescuer's knees. The rescuer takes the downed firefighter's weight while holding the downed firefighter into the ladder by gripping the rails underneath the downed firefighter's armpits **(Skill 12-2B).**

3. The downed firefighter's armpits should be resting on the rescuer's forearms while descending the ladder. As the rescuer descends the ladder, the victim's weight will rest more dominantly on the rescuer's knee in the groin area while a smaller portion of the weight will be resting on the rescuer's forearms as he shifts the weight from one knee to the other while stepping down onto each rung.
4. The placement of the rescuer's arms and chest also help to hold the downed firefighter into the ladder **(Skill 12-2C).** This additionally supports and guides the victim and the rescuer to the ground. The rescuer can also maintain the weight of the downed firefighter on the same knee throughout the descent by simply taking half-steps onto each rung and leading with the same leg and foot to each rung on the ladder. The rescuer needs to ensure that he keeps weight on the balls of his feet when descending down the rungs of the ladder. If his feet get positioned too deep on the rung, he will be unable to descend down the ladder once he has the weight of the downed firefighter. This technique can also be performed with the downed firefighter facing away from the rescuer on the ladder **(Skill 12-2D).**

Rescue Tip

The rescuer needs to ensure that he keeps weight on the balls of his feet when descending down the rungs of the ladder. If his feet get positioned too deep on the rung, he will be unable to descend down the ladder once he has the weight of the downed firefighter.

Horizontal: Cradle Carry Ladder Removal Method

In this particular situation, the interior RIT can present the downed firefighter out the window head- or feet first, facing toward or away from the rescuer. This maneuver can also be executed with or without the downed firefighter wearing an SCBA. These factors make this particular maneuver very versatile.

To perform the horizontal or cradle carry when the downed firefighter is presented headfirst (Skill 12-3):

1. The exterior rescuer on the ladder will position himself at or below the windowsill, ready to receive the victim headfirst with the victim's head off to one side or the other. This allows the remaining portion of the victim's torso and legs to come across the ladder horizontally with his back and SCBA (if wearing one) facing the rescuer on the ladder **(Skill 12-3A).** When the victim is presented in the window face-up by the interior team, the victim's SCBA will in most cases have been removed. This will allow the ladder descent to be used without the interference of the SCBA bumping against the ladder rungs. The obvious benefit with the downed firefighter being presented in the window in the face-down position is that the rescuers would be able to maintain and continue their air supply.
2. The rescuer on the ladder will position the downed firefighter by placing one forearm and hand underneath the victim's armpit in order to hold the weight of the upper torso and grasp the beam of the ladder while the other arm is placed under the top leg into the groin area of the victim. This will help to hold the weight of the victim's pelvic area and lower extremities **(Skill 12-3B).**
3. The rescuer will descend the ladder, placing both feet on each rung and grasping the rails while leaning inward to control the victim until they reach the base of the ladder and are assisted by additional rescuers in moving the downed firefighter to waiting EMS personnel **(Skill 12-3C).**

SKILL 12-3

To Perform the Horizontal or Cradle Carry When the Downed Firefighter Is Presented Headfirst

A The exterior rescuer on the ladder will position himself at or below the windowsill, ready to receive the victim headfirst with the victim's head off to one side or the other.

B The rescuer on the ladder will position the downed firefighter by placing one forearm and hand underneath the victim's armpit in order to hold the weight of the upper torso and grasp the beam of the ladder while the other arm is placed under the top leg into the groin area of the victim.

C The rescuer will descend the ladder, grasping the rails while leaning inward to control the victim until they reach the base of the ladder.

To perform the horizontal or cradle carry when the downed firefighter is presented feet first (Skill 12-4):

1. The exterior rescuer on the ladder will position himself at or below the windowsill, ready to receive the victim.
2. The rescuer on the ladder will position the downed firefighter by placing the downed firefighter's legs to the outside of the ladder beams and hooking one arm into the groin of the downed firefighter. Both hands should grasp the beams of the ladder firmly **(Skill 12-4A).**
3. The upper body of the downed firefighter should then be placed onto the rescuer's opposite arm **(Skill 12-4B).**
4. The rescuer will descend the ladder, grasping the rails while leaning inward to control the victim until they reach the base of the ladder, and will be assisted by additional rescuers in moving the downed firefighter to waiting EMS personnel **(Skill 12-4C).**

SKILL 12-4

To Perform the Horizontal or Cradle Carry when the Downed Firefighter Is Presented Feet First

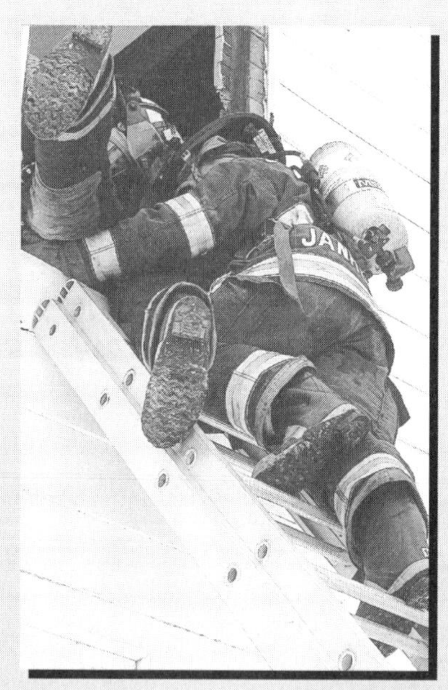

A The rescuer on the ladder will position the downed firefighter by placing the downed firefighter's legs to the outside of the ladder beams and hooking one arm into the groin of the downed firefighter. Both hands should grasp the beams of the ladder firmly.

B The upper body of the downed firefighter should then be placed onto the rescuer's opposite arm.

C The rescuer will descend the ladder, grasping the rails while leaning inward to control the victim.

Safety

If the rescuer at anytime feels as if he is losing control, he can lean into the ladder, which will stop the downed firefighter from moving.

FIGURE 12-5

Ladder removal utilizing tandem ladders.

Multiple Ladders

An important and sometimes needed addition to this technique is the added use of another rescuer and an additional ladder for removing extraordinarily heavy or taller downed firefighters. This addition can also be used to create greater stability and safety in the maneuver. This is especially true in situations where the ladder angle cannot be safely established, such as situations involving close buildings with narrow gangways, overhead power lines, or obstructions. With additional personnel, RIT teams can even incorporate the use of a third ladder and another additional rescuer if it is so desired.

Whichever method is chosen, the maneuver is accomplished by raising ladders that are placed side by side directly underneath the windowsill. The downed firefighter is delivered in the same manner by the interior RIT either face up or face down, with or without an SCBA, to rescuers on each of the ladders at the windowsill. The same arm and hand positions apply to each rescuer, with one firefighter supporting the victim's waist and legs while the second firefighter supports the victim's head and upper torso in the same manner performed by one rescuer with one ladder **(Figure 12-5).** If a third ladder is added, the weight is simply distributed among the three rescuers with the rescuer in the middle carrying the middle part of the victim's torso by placing the arms and hands underneath the side of the victim's upper thigh and just above the waist area while grasping the rails with his hands and descending the ladder. Because the victim will be centered in relation to the amount of ladders used, the rescuers may have to reposition their point of grasp on the ladder beams. Another important consideration is to remember to slightly separate the ladders to allow for hand grasps by the rescuers to the beams **(Figure 12-6).**

FIGURE 12-6

Remember to slightly separate the ladders to allow for hand grasps by the rescuers to the beams.

One-Firefighter Ladder Raise with Obstructions

A proactive action that can be taken by the RIT is to place ladders at windows and balconies in case firefighters operating on the interior need an immediate means of egress. With the limited staffing that is common, it may become necessary for a single firefighter to raise a ladder without any help. This becomes difficult when a firefighter cannot use the building to heel the ladder, such as when landscaping or obstructions are present.

To perform a one-firefighter ladder raise with obstructions (Skill 12-5):

1. Utilizing a one-firefighter carry such as the low-shoulder method, the firefighter removes the ladder from the apparatus and brings it to the position where it is needed **(Skills 12-5A and 12-5B).**

SKILL 12-5
To Perform a One-Firefighter Ladder Raise with Obstructions

A Utilizing a one-firefighter carry such as the low-shoulder method, the firefighter removes the ladder from the apparatus and brings it to the position where it is needed.

B Before placing the ladder, the firefighter must make certain that the area overhead is clear of obstructions such as power lines.

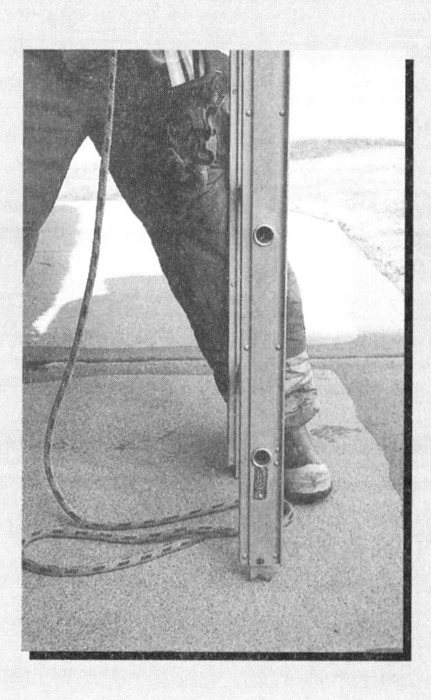

C The firefighter will place one foot to the outside of the ladder beams and place the shoulder of the same side to the inside of the ladder, forming a "V" for the ladder to rest into.

(continued)

SKILL 12-5 (CONTINUED)

To Perform a One-Firefighter Ladder Raise with Obstructions

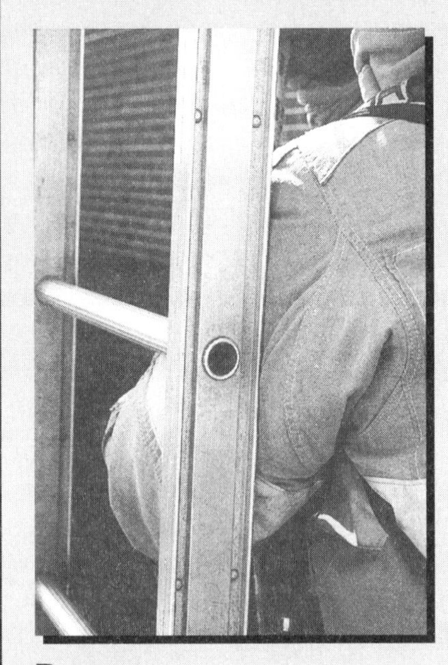

D The firefighter will place one foot to the outside of the ladder beams and place the shoulder of the same side to the inside of the ladder, forming a "V" for the ladder to rest into.

E Grasping the halyard will further stabilize the ladder.

F The firefighter is then able to begin raising the ladder to the height that is required.

Before placing the ladder, the firefighter must make certain that the area overhead is clear of obstructions such as power lines.

2. The firefighter will place his outside foot to the outside of the ladder beams and place the shoulder of the same side to the inside of the ladder, forming a "V" for the ladder to rest into. Grasping the halyard will further stabilize the ladder **(Skills 12-5C, 12-5D, and 12-5E)**.
3. The firefighter is then able to begin raising the ladder to the height that is required **(Skill 12-5F)**.
4. Once at the required height, the ladder can be laid into the building by the firefighter leaning inward toward the building and tying off the halyard. If at anytime the firefighter feels as if he is losing control of the ladder, he can allow the ladder to fall into the building.

If at anytime the firefighter feels as if he is losing control of the ladder, he can allow the ladder to fall into the building.

Aerial Devices and Rapid Intervention Team Operations

The RIT and IC should consider and be aware of the use and advantages of aerial device operations for rescuing a downed firefighter. Aerial devices not only exceed the heights of ground ladders but are in a much better position to be able to provide the ability for quick, pick-off-type rescues, and RIT members can ascend them quickly to gain entry to floors above to locate the position of a downed firefighter. They can also serve as an observation point for the RIT chief or officer to get a good look at the fireground. The best and safest way to remove a downed firefighter, however, is to utilize an interior stairwell if at all feasible.

The best and safest way to remove a downed firefighter is to utilize an interior stairwell if at all feasible.

The RIT should be aware of the availability of an aerial device that can be utilized for rescue purposes when sizing up the fireground. The closest aerial device needs to be part of the RIT's size-up. What ground ladders are available on the apparatus, what specialized equipment is present that may be needed, and whether the apparatus is in a position where it can be used effectively by the RIT need to be determined. The RIT also needs to become thoroughly proficient and familiar with set up and operation of the type of aerial devices that may be available to them. Companies routinely assigned as RIT for mutual aid departments should actively seek out training with their neighboring department's aerial devices. Incident Command along with the rescuers involved should be well versed with rescues of this nature and should have a good understanding of aerial device operations and their capabilities.

There are many considerations to take into account when utilizing aerial devices. A direct understanding that should be realized is the dividing line between a RIT rescue and the complex, technical, high-angle rescue. Some portions of the techniques used for both may be similar but there is a vast difference between lowering a downed firefighter in a rescue litter from the tip of an aerial device down to the ground and a rescue litter and rescuer coming down the rungs and beams of an aerial device. The needs and considerations of the rescue will be dictated by the incident.

The maneuvers involving aerial devices will require diligent training in order to accomplish and establish proficiency and safety. The decision to utilize these maneuvers will depend upon the ability to perform them, the consideration of time restraints, and conditions surrounding the RIT.

There are three basic aerial devices that are utilized: the **aerial ladder,** the **tower ladder,** and the **snorkel (Figures 12-7, 12-8,** and **12-9).** Each has distinct features that give the device its advantages and disadvantages for RIT operations.

An aerial ladder will consist of a power-driven ladder that is affixed to a truck. An aerial ladder on the fireground can be used to rescue a trapped firefighter, to provide quick access to up-

FIGURE 12-7

Aerial ladder.

FIGURE 12-8

Tower ladder.

FIGURE 12-9

Snorkel.

per floors for the RIT, and to ventilate upper-floor windows without putting a firefighter at risk. The disadvantage of an aerial ladder is that it does not provide abundant working room at the ladder tip and can be difficult to get a downed firefighter onto for removal.

When positioning the aerial ladder for window egress or rescue, RIT members should follow the same principles as the ground ladder: do not compromise the available window area for entrance and exit. It should be placed at or just above the windowsill. The tip of the aerial ladder should be positioned only about 2 inches away from the actual building. When weight is put onto the ladder tip, it will rest on the windowsill. This forces the ladder to be stabilized against the building when weight is applied to it.

Ventilation of windows in an expedient manner may be necessary to assist in the rescue of a downed firefighter. For whatever reason, these windows on upper levels may not be accessible by firefighters on the ground. An aerial ladder can be utilized in a pinch to accomplish this. Most chief officers will cringe at the thought of putting the department's brand new aerial ladder through a window, but we are talking about extreme measures for extreme situations—a firefighter's life is at stake. Venting a window using the tip of the aerial ladder should only be performed by extending and lowering the ladder tip through the window. Sideways pressure should never be placed on the ladder when ventilating a window with the tip. Make certain that firefighters below are clear of the area beneath the window before ventilating and that firefighters on the turntable of the aerial ladder are cognizant of glass that may slide down the rails of the ladder.

Safety

Make certain that firefighters below are clear of the area beneath the window before ventilating and that firefighters on the turntable of the aerial ladder are cognizant of glass that may slide down the rails of the ladder.

A tower ladder is the same as an aerial ladder but a basket is mounted to the end of the ladder. This basket provides a working area for firefighters at the ladder tip. The use of a tower ladder for rapid intervention is unique from the operations of other aerial devices. The basket itself offers a solid platform to work off of and railings to provide protection for the firefighters. The basket can also be utilized to establish a high-point anchor if needed **(Figure 12-10).** Working with a rescue litter can be relatively easy and is much safer with the tower ladder as opposed to descending the ladder with the downed firefighter. The rescue litter can be secured to the top of the tower basket and

FIGURE 12-10

The tower ladder basket offers a solid platform to work off of and can be utilized to establish a high-point anchor if needed.

then lowered to the ground with the aerial ladder **(Figure 12-11).** Most manufacturers of aerial devices provide special attachments for their apparatus to accomplish this maneuver **(Figure 12-12).**

Tower ladders offer a convenient means of egress to interior firefighters when positioned properly. A firefighter in trouble will be able to exit more quickly and more safely into a tower ladder basket than onto the tip of an aerial ladder. The tower ladder's advantages will be clear when there are multiple trapped firefighters or downed firefighters that are unconscious or incapacitated. The top railing of the tower basket should be lined up with the bottom of the windowsill, producing as close to a 90° angle with the wall as possible **(Figure 12-13).** It is important that firefighters know how the doors operate on the particular tower basket that they are using (which way they open; **Figure 12-14).** When being utilized to have access to a balcony, it may be necessary to remove the railing on the balcony if it is structurally feasible to do so **(Figure 12-15).** This will eliminate the obstacle of trying to get up and over the railing.

FIGURE 12-11

The rescue litter can be secured to the top of the tower basket and then lowered to the ground with the aerial ladder.

FIGURE 12-13

The top railing of the tower basket should be lined up with the bottom of the windowsill, producing as close to a 90° angle with the wall as possible when positioning the basket for rescue.

FIGURE 12-12

Most manufacturers of tower ladders and snorkels provide special attachments for lowering a rescue litter with the basket.

FIGURE 12-14

It is important that firefighters know how the doors operate on the particular tower basket they are using.

FIGURE 12-15

When being utilized to have access to a balcony, it may be necessary to remove the railing on the balcony if it is structurally feasible to do so.

When working with fire escapes, the platform can be positioned level with the top railing of the fire escape landing. It is important to remember that more often than not, fire escapes are not structurally sound due to their exposure to elements, age, and lack of maintenance **(Figure 12-16)**.

Snorkels are platforms or baskets attached to articulating booms mounted on a truck. They are very useful for situations requiring the maneuverability of a master stream device at an upper level **(Figure 12-17)**. Snorkels are also useful in situations requiring movement around overhead obstacles. Disadvantages of the snorkel include the articulating action of the booms themselves. When the booms are raised, they

FIGURE 12-16

When working with fire escapes, the platform can be positioned level with the top railing of the fire escape landing. Careful consideration must be given to the condition and structural integrity of the fire escape.

FIGURE 12-17

Snorkels are useful for situations requiring the maneuverability of a master stream device at upper levels and in situations requiring movement around overhead obstacles.

will project where they are hinged, which can make it difficult to operate in tight areas. Because of their design, they will have a considerably diminished **scrub area** (area able to be covered by an aerial device) as opposed to the aerial ladder and tower ladder.

Some general rules in utilizing all aerial devices include: do not move an aerial device side to side when firefighters are on it unless it is directly needed to remove them from danger. Extending and retracting should never take place with personnel positioned on the ladder portion of the aerial device. A smooth, fluid maneuver is required when using the aerial device to remove a downed firefighter. It is best to do one motion at a time as opposed to attempting several simultaneously.

Extending and retracting should never take place with personnel positioned on the ladder portion of the aerial device.

Whenever possible, the aerial device should be located on a corner of the building. This will give the unit the ability to cover two sides of the building. In addition, all aerial devices need to be positioned as squarely to the building as possible. If the device is positioned at an angle other than 90°, only one side of the device will contact the building. Weight on the tip will cause twisting, which can lead to failure.

All aerial devices will operate in ways that will be specific to the manufacturer. The best way to determine their capabilities and operations is to practice with them.

Removing a Downed Firefighter Using an Aerial Device

One type of aerial device removal for a downed firefighter is to descend the aerial ladder with the downed firefighter lying horizontally across the rescuer's arms and the rails of the aerial ladder in a horizontal, cradle-type position. A high element of risk and danger are presented when making a removal in this manner, but it may be the only choice under the circumstances presented.

The rescuer on the ladder will receive the downed firefighter in the same manner as in a ground ladder removal, with the victim coming out face down or face up and the rescuer on the ladder putting one arm underneath the victim's armpit while the other is placed between the victim's legs at the inner thigh and groin area while grasping the rails of the aerial ladder, distributing the weight across the rescuer's arms **(Figure 12-18)**. Due to the design of aerial ladders, the rungs will not be available for the rescuer to grasp. Support of the downed firefighter's midsection will be challenging. When performing this maneuver there should be a second RIT team member backing up the rescuer descending with the victim. Another consideration when descending an aerial ladder with downed firefighters is the use of a safety tag line onto the victim as well as the rescuer. This can be anchored to the interior rescuers, letting the line out as the team member on the ladder descends with the

FIGURE 12-18
The rescuer on the ladder will receive the downed firefighter in the same manner as in a ground ladder removal, with the victim coming out face down or face up and the rescuer on the ladder putting one arm underneath the victim's armpit while the other is placed between the victim's legs at the inner thigh and groin area while grasping the rails of the aerial ladder, distributing the weight across the rescuer's arms.

victim. The use of these techniques should be avoided if the angle and height are too extreme. In such cases it is best to incorporate the use of the basket of a tower ladder or snorkel.

The use of a rescue litter presents the RIT with another option for removing the downed firefighter using an aerial ladder. Moving the rescue litter at a 90° angle down the rails of the aerial ladder will be difficult and dangerous because the load inside the litter will be top heavy. For that reason, it is recommended that the litter be brought down in a parallel position with the foot of the litter descending first. The rescue litter will more than likely be too wide to be lowered in between the rails of the aerial ladder and will not be wide enough to be transported on top of the aerial ladder rails unless some modification is made. Short pike poles or pry bars can be secured to a rescue litter basket by use of webbing and can then be used to slide the downed firefighter in the rescue litter down the aerial ladder by using the top of the ladder rails **(Figure 12-19).** A firefighter on a tag line at the tip of the device will control the descent as well take some of the weight as another firefighter slowly guides it down the ladder to the turntable area **(Figures 12-20** and **12-21).**

FIGURE 12-20

A firefighter on a tag line at the tip of the device will control the descent as well take some of the weight as another firefighter slowly guides it down the ladder to the turntable area.

FIGURE 12-19

Short pike poles secured to the rescue litter for removal down the aerial ladder.

FIGURE 12-21

The litter is lowered feet first as the pike poles allow it to "ride" on top of the rails of the aerial device.

Summary

The use of ladders and aerial devices can be a valuable tool for the RIT when operating on the fireground. Ladders proactively placed at windows on and above the fire floor can save a life if a firefighter needs to exit quickly. Their use for removal of a downed firefighter should only be considered when removal by an interior stairwell is not feasible. Removal of downed firefighters by use of a ladder or aerial device is a two-part process involving an interior and exterior team. Communication between these two teams will determine the success of the rescue.

One of the first steps that a RIT should perform as part of their size up is the availability of an aerial device and ladders that are available. An aerial device will have limitations and capabilities that are unique to the type of device that it is. The best way to overcome these limitations is to practice and train beforehand. Firefighters should be trained on and proficient in ground ladder techniques and should be familiar with aerial, tower, and snorkel devices and how they can be utilized in a fireground rescue.

■ KEY TERMS

Aerial ladder
Kicking out
Scrub area
Snorkel
Tower ladder

■ REVIEW QUESTIONS

1. ____________ is the safest method for removing a downed firefighter from an upper floor.
   a. Ground ladder
   b. Tower ladder
   c. Interior staircase
   d. Fire escape
2. When placing a ground ladder to a window, the tip should be placed
   a. above the window so it can be seen.
   b. at or below the sill to maximize the area for entry and egress.
   c. off to the side so that it is out of the way.
   d. above the window for easy access.
3. Ground ladders are rated to withstand a maximum load of
   a. 300 lbs.
   b. 500 lbs.
   c. 600 lbs.
   d. 750 lbs.
4. Multiple ladders are used
   a. to create greater stability.
   b. to distribute the weight load.
   c. when extreme angles are necessary.
   d. all of the above
5. Before raising a ground ladder the RIT member should
   a. tell Command.
   b. inform the interior RIT.
   c. check for overhead obstructions.
   d. none of the above
6. The three basic aerial devices are:
   1.
   2.
   3.
7. List two disadvantages of the aerial ladder.
   1.
   2.

8. The tip of the aerial ladder should be positioned _____ inches away from the building for rescue purposes.
   a. 2
   b. 4
   c. 8
   d. 10

9. Venting an upper-floor window is accomplished by moving the aerial ladder side to side through the window.
   True False

10. List three advantages of using a tower ladder to perform a rescue.
    1.
    2.
    3.

ADDITIONAL RESOURCES

Fire Department, New York City, *Ladder Company Operations, Use of Aerial Ladders.* Firefighting Procedures Volume 3, Book 2, 1997.

Hoff, Robert, and Kolomay, R., *Firefighter Rescue and Survival.* Tulsa, OK: PennWell Publishing, 2003.

IFSTA, *Fire Service Rescue,* 6th ed. Stillwater: Oklahoma State University, 1996.

Jakubowski, G., and Morton, M., *Rapid Intervention Teams.* Stillwater, OK: Fire Protection Publications, 2001.

McCormack, J., *Firefighter Rescue and Rapid Intervention Teams.* Indianapolis, IN: Fire Department Training Network, 2003.

Norman, J., *Fire Officers Handbook of Tactics,* 2nd ed. Saddlebrook, NJ: PennWell Publishing, 1998.

C H A P T E R

13 Roof Operations

Learning Objectives

Upon completion of this chapter, you should be able to:

- describe the maneuver that involves using a rope and roof ladder to remove a downed firefighter from a roof.
- discuss the purpose and use of the Munter hitch during roof removals.
- explain how the Halligan anchor lifting system provides a quick anchor while also incorporating a simple 2-to-1 mechanical advantage.
- illustrate how a trench cut can be used when retrieving a downed firefighter who has fallen through the roof into an attic space.
- list the techniques necessary to perform the rescue litter ladder slide method when removing and lowering a downed firefighter from a roof's edge.
- describe the parameters required to perform the ladder fulcrum method when lowering a downed firefighter from a roof.
- demonstrate the ladder fulcrum method when removing downed firefighters from roofs.

CASE STUDY

"On March 6, 2000, at approximately 1342 hours, the involved Fire Department was dispatched to a structure fire involving a 1 1/2-story, wood frame, single-family dwelling, measuring approximately 1,800 square feet, located 6.5 miles from the fire station. The victim, who was talking (in person) to the Fire Chief at the time, responded to the fire station in his privately owned vehicle (POV) to drive Engine 1 to the scene. The Fire Chief responded to the scene and was first on-scene. Four additional firefighters and three first responders responded in their POVs to the scene.

"The fire, which spread from burning trash, involved the north side of the dwelling, including the garage. Engine 1 arrived on-scene and the victim activated the pump. He delegated the driver/operator duty to another firefighter, and then stretched a 100-foot section of 1 1/2-inch hoseline to extinguish the external portion of the fire. After this was accomplished, the remaining firefighters arrived on-scene. The victim and one firefighter retrieved a 20-foot extension ladder and raised it to the garage roof. The victim and the firefighter, both wearing full bunker gear and self-contained breathing apparatus (SCBA), on air, and taking the 1 1/2-inch hoseline and an axe with them, climbed to the roof of the garage and began to perform roof ventilation with the axe. The firefighter climbed down the ladder to check for fire spread inside the garage area. He identified an area still burning and exited the garage, intending to retrieve the hoseline from the roof to continue extinguishment.

"After the victim completed roof ventilation, he took his SCBA facepiece off and suddenly collapsed. On-scene crew members initially thought he was looking down the ventilation hole. The firefighter standing on the ground below yelled to the victim to pass the hoseline down the ladder. After getting no response, the firefighter backed away from the edge of the garage and, again, asked the victim to pass the hoseline down the ladder. After backing up a few more feet, the firefighter saw the victim had collapsed. The firefighter yelled that a man was down and a radio call informed Dispatch of the situation.

"Crew members climbed onto the garage roof to aid the victim. Initial assessment by crew members found the assistant chief to be unresponsive, not breathing, and pulseless. CPR (chest compressions and assisted ventilations via mouth-to-mouth) began immediately. First responders on-scene retrieved a bag-valve-mask (BVM). However, attempts to ventilate via BVM were unsuccessful, and ventilations via mouth-to-mouth were resumed. The victim was then extricated from the roof and placed onto the ground.

"The ambulance was dispatched at 1358 hours. Medic units arrived on-scene at 1419 hours, finding the victim unresponsive, not breathing, and pulseless with CPR in progress. A cardiac monitor was attached to the victim, revealing ventricular fibrillation (V.Fib., a heart rhythm unable to sustain life), which was immediately defibrillated (shocked). The victim's heart rhythm reverted to asystole (no heartbeat). Advanced life-support measures, including intubation and intravenous therapy, were begun. The cardiac monitor again revealed V.Fib. and one additional shock was administered, without change in patient status. The victim's heart rhythm again reverted to asystole and CPR continued. The victim was then loaded onto a stretcher and placed into the ambulance, which began transport to the hospital at 1429 hours. Enroute, the cardiac monitor again revealed V.Fib. and two additional shocks were administered, without change in patient status. The victim arrived at the hospital's emergency department (ED) at 1446 hours. Inside the ED, CPR and ALS measures continued until 1508 hours, when the victim was pronounced dead by the attending physician."

—Case Study taken from NIOSH Firefighter Fatality Report # 2001-40,* "Firefighter Suffers Cardiac Arrest at Structure Fire—Illinois," *full report available online at http://www.cdc.gov/niosh/face200140.html.

FIGURE 13-1

Operating on the roof can be a very high-risk area on the fireground.

Introduction

Operating on the roof can be a very high-risk area on the fireground. Any given number of detrimental events can take place above the fire **(Figure 13-1).** Not only can firefighters fall into attic spaces because of construction failure, but RITs are also faced with firefighters becoming disabled on roofs because of medical problems as well. It becomes a very real possibility for a RIT to respond to a Mayday with a downed firefighter on top of a roof. If a Mayday occurs at the roof level it may be necessary to utilize an aerial device to establish a high-point anchor or safe working platform for the RIT **(Figure 13-2).** As we know, it is not always possible to get aerial devices positioned in the most advantageous place. This is especially true when dealing with setbacks, utilities, and large trees on residential lots.

FIGURE 13-2

If a Mayday occurs at the roof level it may be necessary to utilize an aerial device to establish a high-point anchor or safe working platform for the RIT.

Simple Removal Using Rope and Roof Ladder

Removing a firefighter from a roof can be simplified with the use of a rope bag, two large carabiners, some webbing, and a roof ladder. This technique does not provide any mechanical advantage for lifting the downed firefighter but assists in controlling his descent.

To perform a roof removal using rope and a roof ladder (Skill 13-1):

1. The first step necessary is to have a **roof ladder** (a ladder with folding hooks at the tip) on the roof with the hooks extended and firmly anchored across the **ridge board** (structural member at the highest point of the roof where rafters are attached). A roof ladder may already be in place if one was being used by the downed firefighter. The ladder *must* be positioned on the same side of the roof as the downed firefighter.

Safety

The ladder must be positioned on the same side of the roof as the downed firefighter when performing this maneuver.

SKILL 13-1
To Perform a Roof Removal Using Rope and a Roof Ladder

A An anchor point with a carabiner is established on a rung as close to the ridge as possible.

B A figure eight on a bight is then secured to the downed firefighter.

C Once weight is on the rope, a RIT member will control the downed firefighter's descent by using the Munter hitch.

D If ladder use is not possible, a second rope can be tied into a handcuff knot and placed on the downed firefighter's ankles and a RIT member on the ground can control the downed firefighter's body as it is lowered over the roof edge.

(continued)

2. An anchor point with a carabiner is established on a rung as close to the ridge as possible **(Skill 13-1A)**.
3. A figure eight on a bight is then secured to the downed firefighter. It is our recommendation to use a rated harness such as the MAST or one made from webbing **(Skill 13-1B)**.
4. A Munter hitch is placed into the carabiner located at the anchor point established in step 1.
5. The downed firefighter is lifted slightly to get the weight off of his ladder or safety belt, which is then disconnected. It is important that the downed firefighter be secured to the rope being brought up with

SKILL 13-1 (CONTINUED)

To Perform a Roof Removal Using Rope and a Roof Ladder

E If ladder use is not possible, a second rope can be tied into a handcuff knot and placed on the downed firefighter's ankles and a RIT member on the ground can control the downed firefighter's body as it is lowered over the roof edge.

the RIT prior to disconnecting his ladder or safety belt if operating on a steep roof.

6. Once weight is on the rope, a RIT member will control the downed firefighter's descent by using the Munter hitch. Other RIT members can guide the downed firefighter to the roof's edge **(Skill 13-1C)**.
7. Tandem ladders or even a single ladder with a rescuer can guide the downed firefighter to the ground with support and control being provided by the RIT team on the ground once over the roof edge. Again, ladders may not be able to be positioned to do this, as they must allow for the downed firefighter to be easily transferred from the roofline.
8. If ladder use is not possible, a second rope can be tied into a handcuff knot and placed on the downed firefighter's ankles and a RIT member on the ground can control the downed firefighter's body as it is lowered over the roof edge **(Skills 13-1D** and **13-1E)**.
9. The downed firefighter is assisted over the edge of the roof and lowered to the ground level, where he can be treated by EMS.

A change-of-direction pulley placed on the rope can further simplify this technique and allow the Munter hitch to be operated from the ground level if desired.

Halligan Anchor Lifting System

This **Halligan anchor lifting system** takes principles that have been introduced and expands them even further. A firefighter working on a roof may fall through a weakened section into the attic space. Because of fire conditions, the downed firefighter may not be able to be brought down from the attic and through the interior of the building. He may have to be raised back up to the roof to be rescued. This maneuver helps to establish a quick anchor and simple 2-to-1 mechanical advantage. Because the roof is weakened, RIT members must make certain to work off of a roof ladder that is properly placed. The RIT must also be able to reach the downed firefighter either from the roof level or by sending a rescuer into the attic.

To perform a Halligan anchor lifting system (Skill 13-2):

1. A Halligan bar is driven into the **roof decking** (the frame that roof materials cover), making certain that the pick end is at the ridge and the fork end is running lengthwise to the ground. After it is driven into the roof, a Halligan wrap (see Chapter 6) is utilized to establish an anchor for the rope **(Skill 13-2A)**.
2. A carabiner is then secured to the firefighter using a rated harness. In an extreme situation, the downed firefighter's SCBA shoulder straps can be secured together with a carabiner for lifting the firefighter out of the attic space.
3. The running end of the rope is placed in the carabiner attached to the downed

SKILL 13-2

To Perform a Halligan Anchor Lifting System

A A Halligan bar is driven into the roof decking, making certain that the pick end is at the ridge and the fork end is running lengthwise to the ground. After it is driven into the roof, a Halligan wrap is utilized to establish an anchor for the rope.

B The running end of the rope is placed in the carabiner attached to the downed firefighter. This will provide the 2-to-1 mechanical advantage. The running end now becomes the "haul" line.

C RIT members on the roof at the ridge board will take slack out of the running end of the rope and this will raise the downed firefighter out of the attic space to the roof level.

D Once on the downed firefighter is to the roof level, the rescuers can slowly allow slack into the line while lowering the downed firefighter to the edge of the roof.

firefighter. This will provide the 2-to-1 mechanical advantage. (The mechanical advantage may be slightly less than a true 2-to-1 because of friction created by the carabiner.) The running end now becomes the "haul" line **(Skill 13-2B)**.

4. RIT members on the roof at the ridge board will take slack out of the running end of

the rope and this will raise the downed firefighter out of the attic space to the roof level **(Skill 13-2C)**.

5. Once the downed firefighter is at the roof level, the rescuers can slowly allow slack into the line while lowering the downed firefighter to the edge of the roof **(Skill 13-2D)**.
6. Once at the edge, the downed firefighter can be lowered the rest of the way to the ground or consideration can be given to any of the ladder techniques already mentioned.

Trench Cut Method

The **trench cut** method is another simple alternative that can be considered to retrieve a firefighter through the roof. In this technique, the RIT extends the hole that the firefighter has fallen through by trench cutting from the hole down to the roof's edge. Some areas of the country refer to trench cutting as **strip ventilation.** It is routinely performed to head off fire that is spreading horizontally through a roof space. This can be done very quickly if a firefighter is well trained and skilled in the use of a power saw. Remember, it will be necessary for the firefighters trenching the roof to work off of roof ladders or a stable platform because the area is unstable.

To perform a roof removal of a firefighter using the trench cut method (Skill 13-3):

1. The firefighter that has fallen through to the attic is located and roof ladders are placed to distribute the weight load of the rescuers.

SKILL 13-3

To Perform a Roof Removal of a Firefighter Using the Trench Cut Method

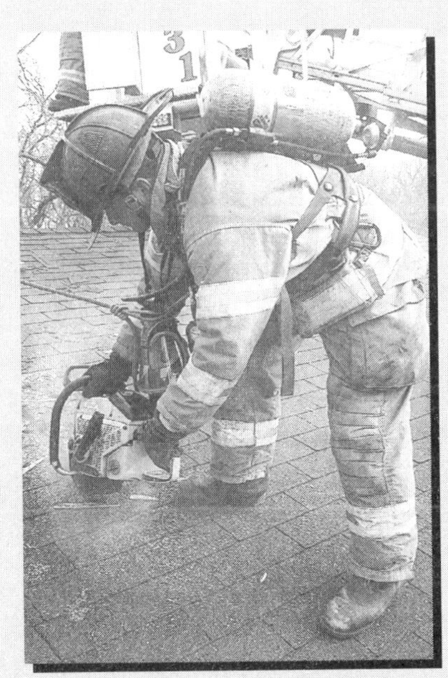

A Once the rafters are located, a cut to the inside of one rafter should be made parallel and run down to the roof edge.

B The roofing material should drop down into the attic as the cut is made.

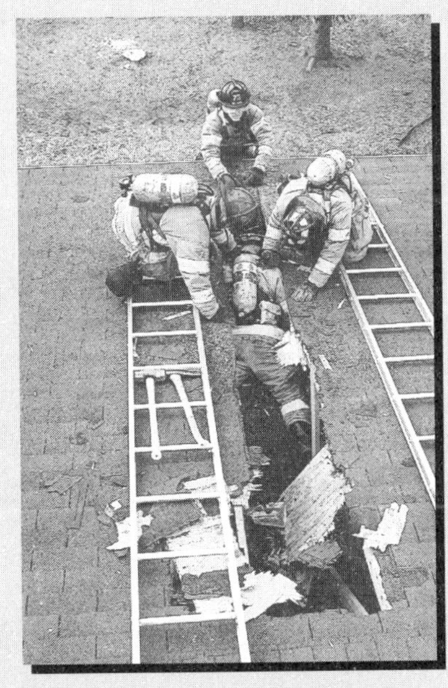

C Once the trench or strip is cut, the rescuers will be able to pull the downed firefighter to the roof's edge, where additional rescuers will be waiting on ladders.

2. With a power saw, a cut is made across the direction that the roof rafters are running to determine their location. Use caution and make certain that the saw blade is located away from the downed firefighter.

Use caution and make certain that the saw blade is located away from the downed firefighter.

3. Once the rafters are located, a cut to the inside of one rafter should be made parallel and run down to the roof edge **(Skill 13-3A)**.
4. The firefighter with the saw should return to place another cut parallel to the inside of the second rafter.
5. The roofing material should drop down into the attic as the cut is made **(Skill 13-3B)**.
6. Once the trench or strip is cut, the rescuers will be able to pull the downed firefighter to the roof's edge, where additional rescuers will be waiting on ladders **(Skill 13-3C)**.

Tandem ladders are again recommended to safely bring the downed firefighter to ground level. The ladders must be positioned just below the gutter line to allow the downed firefighter to be easily transferred to the ladders. This goes against the practice of extending ladders high above a roof line, but this is one of the very rare exceptions. This maneuver may also be hampered if the **fascia board** (the trim board that runs along the base of the roof) is not wide enough to safely support the ladders without extending past the roof line.

Rescue Litter Ladder Slide Method

Another maneuver that is available to the RIT is the **rescue litter ladder slide method** or the use of a **Stokes basket** (rescue litter) and a ladder to remove a downed firefighter from a rooftop. This technique can be used for the downed firefighter who may have serious injuries that require immediate spinal immobilization. The disadvantage to this technique is that it will take time to set up properly. For that reason, it may not be the number one choice of action when working in a hostile environment. The RIT will also have to consider positioning the downed firefighter and the SCBA in the rescue litter if being performed in hostile conditions.

To perform the rescue litter ladder slide (Skill 13-4):

1. The RIT will first need to prepare the head of the rescue litter. This can be accomplished by using webbing, rope, and carabiners. The simplest method is to attach a rope with a figure eight on a bight to the center of the top of the basket **(Skill 13-4A)**.
2. The rope is attached to the carabiner at the head and is run to an anchor that will utilize a Munter hitch to control the rate of descent. If preferred, a RIT member can utilize a figure eight or rappel rack to control the rate of descent in place of the Munter hitch. If no anchor is available, a body wrap can be utilized to control the descent of the litter **(Skill 13-4B)**.
3. Once the head of the litter has been secured, the downed firefighter will be placed in it. The downed firefighter's SCBA will have to be repositioned in order to allow the victim to be secured into the basket before lowering. This can be accomplished by removing the SCBA and placing it on top of the victim or off to the side in the litter. The victim is then secured in the basket by connecting the straps.
4. The RIT will ensure that the ladder tip is positioned just below the gutter line at an angle suitable to facilitate the litter sliding down the beams of the ladder. The ladder angle should be exaggerated to maintain control when guiding the litter down **(Skill 13-4C)**.
5. Two rescuers will guide the litter off the roof onto the beams of the ladder while tension is released from the guide line **(Skill 13-4D)**.
6. Rescuers at the ground level will lift the end of the litter up as it comes off the ladder. This will keep the victim and the litter parallel to the ground.

SKILL 13-4
To Perform the Rescue Litter Ladder Slide

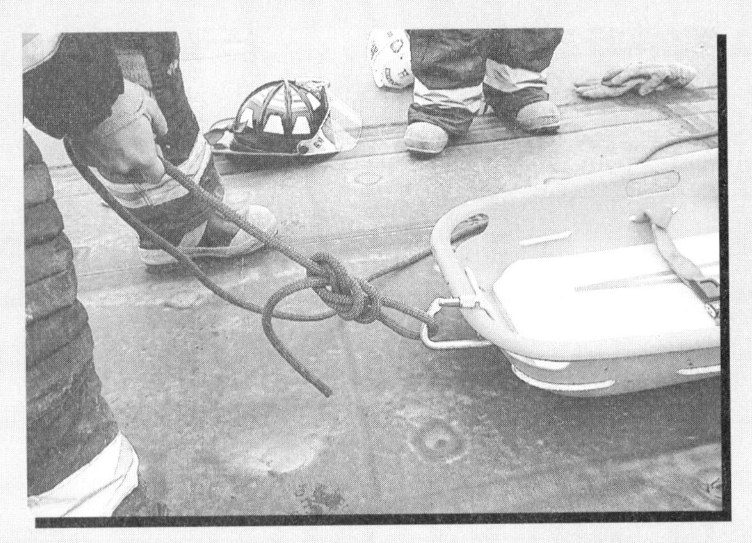

A The RIT will first need to prepare the head of the rescue litter by attaching a rope with a figure eight on a bight to the center of the top of the basket.

B If no anchor is available, a body wrap can be utilized to control the descent of the litter.

C The ladder angle should be exaggerated to maintain control when guiding the litter down.

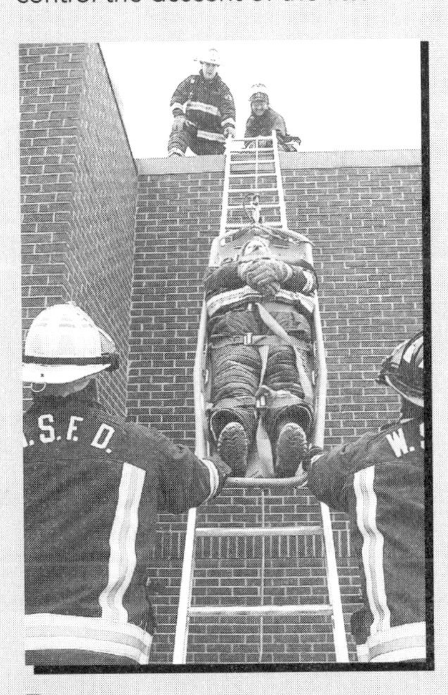

D Two rescuers will guide the litter off the roof onto the beams of the ladder while tension is released from the guide line.

Ladder Fulcrum Method

The **ladder fulcrum method** is only appropriate when there is time and fire conditions are under control. Working off a flat surface such as a flat roof is another requirement. The main advantage of this maneuver is that it allows the victim to be lowered in a horizontal position.

To perform the ladder fulcrum removal method (Skill 13-5):

1. The RIT will raise a ladder to the roof and ascend with the following:
   - rescue litter
   - rope
   - webbing
   - six carabiners
2. If still necessary, the downed firefighter's SCBA will have to be repositioned in order to allow the victim to be secured into the litter before lowering. This can be accomplished by removing the SCBA and placing it on top of the victim or off to the side in the litter. The victim is then secured in the litter by connecting the straps.
3. The base of the ladder will be set flush against the wall and have three to four rungs exposed over the roof line **(Skill 13-5A).**
4. The foot end of the litter will be attached to the second rung exposed over the roofline. This will be accomplished with webbing and carabiners. Rescuers should

SKILL 13-5
To Perform the Ladder Fulcrum Removal Method

A The base of the ladder will be set flush against the wall and have three to four rungs exposed over the roof line.

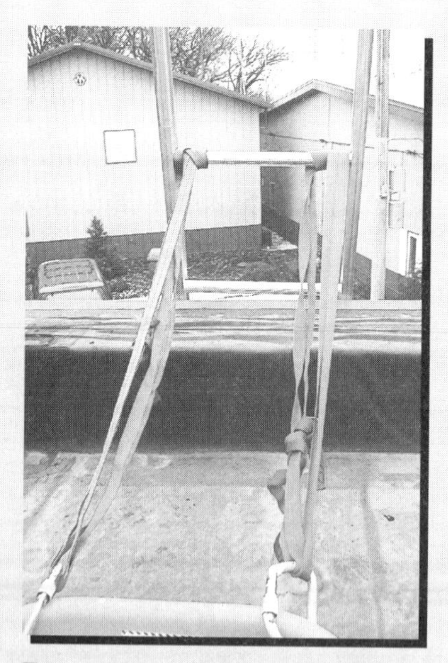

B The foot end of the litter will be attached to the second rung exposed over the roof line. This will be accomplished with webbing and carabiners. Rescuers should make certain that there will be enough slack in the webbing to allow the litter to pivot without being caught against the ladder.

(continued)

SKILL 13-5 (CONTINUED)

To Perform the Ladder Fulcrum Removal Method

C The head of the litter will have two ropes attached with a figure eight and carabiner. This provides the rescuers on the roof a method of controlling the rate and angle of descent for the litter. The rescuers on the roof will control these lines by utilizing a body wrap.

D The litter is then moved over the edge of the roof while the rescuers at ground level begin to walk the ladder down utilizing a hand-over-hand method.

make certain that there will be enough slack in the webbing to allow the litter to pivot without being caught against the ladder **(Skill 13-5B).**

5. The head of the litter will have two ropes attached with a figure eight and carabiner. This provides the rescuers on the roof a method of controlling the rate and angle of descent for the litter. The rescuers on the roof will control these lines by utilizing a body wrap **(Skill 13-5C).**
6. The litter is then moved over the edge of the roof while the rescuers at ground level begin to walk the ladder down, utilizing a hand-over-hand method. The base of the ladder should be tight to the wall at all times. It is important that the rescuers controlling the guide lines maintain tension to keep the litter level at all times until it is down on the ground. The RIT officer will have to be positioned at the roof's edge to relay critical information to the rescuers on the guide line once the litter is out of their field of vision below the roof line **(Skill 13-5D).**

Safety

The base of the ladder should be tight to the wall at all times and tension should be maintained on the guidelines by rescuers on the roof.

Summary

The roof is one of the most dangerous places to operate on the fireground. Any number of events can take place to cause the need for a RIT to perform a removal of a downed firefighter from the roof level. The safety of the RIT as well as time permitted to make a rescue from a roof will dictate what technique will be utilized for removal. The only way to find out what works best for a particular RIT is to train and practice with numerous techniques before they are needed on the fireground.

Firefighters should be trained on and proficient in the various roof rescue methods, such as the Halligan anchor lifting system, the rescue litter ladder slide technique, and the ladder fulcrum method.

As mentioned, the roof is a dangerous fireground location. Assuring proficiency in all areas of roof rescue operations is therefore a necessity.

■ KEY TERMS

Fascia board
Halligan anchor lifting system
Ladder fulcrum method
Rescue litter ladder slide method
Ridge board
Roof decking
Roof ladder
Stokes basket
Strip ventilation
Trench cut

■ REVIEW QUESTIONS

1. RITs respond to Maydays involving roof operations for reasons such as
   a. firefighters falling into attic spaces weakened by weight and fire conditions.
   b. heart attacks and other medical conditions.
   c. construction failure.
   d. all of the above

2. Simple removals that use a rope and roof ladder to remove a distressed firefighter from a roof do not provide any mechanical advantage for lifting.

   True False

3. An option that can be used when lowering a downed firefighter using a simple removal with a rope and roof ladder to help control their descent is the use of a
   a. Swiss seat applied to the pelvic area.
   b. safety sling applied to the extremities.
   c. handcuff knot applied to the lower extremities.
   d. barrel knot applied under the arms.

4. The Halligan anchor lifting system can provide a way for the RIT to raise a downed firefighter back up through an attic space to the roof while also supplying a
   a. simple retrieval system.
   b. 4-to-1 mechanical advantage.
   c. 2-to-1 simple mechanical advantage.
   d. Halligan raise.

5. Another alternative that can be considered to retrieve a downed firefighter who has fallen through the roof is for the RIT to extend the hole that the victim has fallen through to the roof's edge.

   True False

6. When trench cutting a pitched roof down to the roof's edge in order to remove a downed firefighter, it is a good idea to have rescuers remove the victim by the use of
   a. an extension ladder and pike pole.
   b. an extension latter extended over the roof line by five rungs.
   c. tandem ladders placed to assist in bringing down the distressed firefighter safely.
   d. both (a) and (b)

7. Ladders must be positioned __________ when performing the trench cut roof removal technique.

8. A RIT can use a rappel rack to control the rate of descent in place of a Munter hitch when lowering the rescue litter down the ladder.

   True False

9. The ladder tip should be extended at least four rungs above the roofline when performing the rescue litter ladder slide roof removal.

   True False

10. What are three requirements needed to perform the ladder fulcrum removal method?

    1.
    2.
    3.

■ ADDITIONAL RESOURCES

McLees, M., “Rapid Intervention Rope Bag,” *Firehouse,* October 2001.

Mittendorf, J., *Ventilation Methods and Techniques.* El Toro, CA: Fire Technology Services, May 1991.

Sitz, T., “Rapid Removal of an Unresponsive Firefighter from a Peaked Roof,” *Fire Engineering,* March 2003.

CHAPTER

14 Enlarging Openings for Removals

Learning Objectives

Upon completion of this chapter, you should be able to:

- explain when a breaching or enlarging an opening technique can be used and what considerations need to be taken into account.
- identify the tools that may be needed to breach a wall or enlarge an opening.
- discuss the characteristics of the following:
  - wood-framed or sided walls
  - masonry walls
  - metal sided walls
  - stucco walls
  - lightweight steel construction
  - glass block
- explain the procedure used to enlarge an existing opening in wood-framed or lightweight steel construction.
- demonstrate the procedure for breaching glass block.
- discuss different aspects of forcible entry techniques required for doors.
- execute the technique needed to make a "doggy door" cut when forcing a door.
- explain the procedure used to enlarge an existing door opening.
- explain the dangers associated with overhead doors.
- explain how to cut open a roll-up overhead door, sectional overhead door, and tilt-up overhead door.
- give details on how burglar bars are attached to buildings and how they can be removed.
- demonstrate the cradle cut technique for managing a circular saw.

Case Study

"One firefighter was killed and eleven others injured, one critically, fighting a residential fire in San Francisco, California, on March 9, 1995. Three firefighters were trapped when an overhead garage door closed behind them without warning. The fire spread rapidly from its origin in a lower-level bedroom. Sixty mile per hour winds accelerated the spread of the fire from the rear of the structure towards the front. The fire was initially attacked by entering the structure through the open overhead garage door.

"The cause of the door closing is not known. Several of the injuries were due to smoke and hot debris pushed onto the rescuers by the extreme wind conditions as they fought to save the trapped firefighters.

"In an effort to open the garage door a firefighter pulled on the garage door handle with no result. Five members, including the Battalion Chief, then attempted to raise the door but there were not enough hand holds available on the bottom of the door to allow sufficient force to be applied.

"Two firefighters began to work on the door with axes in an attempt to gain entry. An attempt also was made to gain entry with a Chicago Door Opener, but because of the resilience of the plywood, the door could not be opened. Even with a hose line cooling the rescuers, they were driven away from the door several times as the products of combustion were blown onto them. The force of the wind was blowing the smoke and flames horizontally toward the rescuers and across the street.

"A chain saw and a multi-purpose saw were being used on the door. Both tools would only run for short periods of time before stalling due to the smoky conditions.

"The work with the axes paid off when a hole was made in the lower right corner of the door near a vent. The senior firefighter put a hand through the hole in an attempt to escape. The rescuers ripped parts of the plywood off by hand until he was able to crawl out. He informed the rescuers that two others were still trapped inside.

"The rescuers then managed to get the gas-powered tools to run long enough to cut a hole in the left side of the door. Through this hole the missing personnel could be seen. The door finally failed and the five rescuers were able to raise it."

Some recommendations and key points emphasized in this technical report are that:

"1. Overhead doors require special attention.

This incident illustrates the danger of overhead doors to fire personnel. Overhead doors, including fire doors, are generally very heavy. Once closed they may be impossible to reopen under fire conditions. If entry is made through an overhead door, steps must be taken to ensure that the door will not close.

2. Getting out of a fire safely is more important than getting in.

Companies entering a building should always seek secondary exits. These exits must be maintained at all costs. A secondary exit at the fire was obstructed and not visible because it was blocked by a vehicle. This is often overlooked in residential fires because the risk is not recognized. In many situations a hoseline will help protect the exit path and keep a swinging door from closing. Windows are often acceptable as emergency exits for firefighters. In some situations the secondary exit points must be guarded, even at the expense of slowing the fire attack, to ensure firefighter safety.

3. Extreme weather conditions require reassessment of routine response levels.

Extreme wind conditions can speed the spread of fire and quickly turn a routine fire into a critical situation.

4. A thirty-minute SCBA will not deliver thirty minutes of breathing air.

There are many factors that affect SCBA duration. Both physical and emotional stress cause an increase in the consumption of oxygen and therefore air. A person's physical size and conditioning are also major factors in air supply duration. Each firefighter should know how they react to stress with an SCBA on. These reactions affect the duration of the air supply. A certain level o f stress can be relieved if training is sufficient to make firefighters comfortable and confident with their equipment.

5. Delayed reporting can greatly increase the potential for loss of life and property damage.

CASE STUDY (CONTINUED)

6. *Protective equipment is only effective if it is used. Eleven firefighters were injured in this incident. As many as five of the injuries could have been prevented if the firefighters had worn all of the protective equipment supplied to them. Several of these injuries occurred during the rescue efforts; however, failing to use SCBA and other protective equipment in these situations is likely to put the rescuer in need of rescue.*
7. *Importance of a 360-degree size-up.*

   The dangerous size of this fire was not apparent from the front of the building. A complete size-up should include a view from every possible side of the incident scene. If decisions have to be made without all the facts, every effort should be made to gain additional information and ensure the safety of crews until a complete size-up can be accomplished.
8. *Use natural forces as advantages whenever possible.*

   In extreme wind conditions, changes in tactics should be considered."

—*Case Study and information taken from,* "Entrapment in Garage Kills One Firefighter," *United States Fire Administration Technical Report Series, Report #084.*

Introduction

Enlarging openings can create an entry or egress point to retrieve a downed firefighter. When a Mayday is initiated from the interior of the structure, the RIT will attempt to respond to the downed firefighter's location. At any time, the RIT can decide that the means of removing the downed firefighter may be more efficient if an opening were provided on the exterior of the structure rather than dragging the victim to the original entry point.

Complicated floor plans and changing fire conditions can hinder the ability for a RIT to efficiently remove a downed firefighter within many structures. It is wise for the RIT to assess the interior situation and consider a breach by an exterior team to aid in removal.

FIGURE  **14-1**

At any time, the RIT can decide that the means of removing the downed firefighter may be more efficient if an opening were provided on the exterior of the structure rather than dragging the victim to the original entry point.

Enlarging Openings and Breaching Walls

Enlarging openings or breaching walls **(Figure 14-1)** is a decision that should be contemplated early in the rescue efforts. It is important that the timing of these operations coincide with the interior rescuers. It does no good for the exterior RIT to make or enlarge an opening if the interior RIT brings the downed firefighter to another point. It is a good idea for the interior rescuers to communicate their intentions early on in order for the outside team to begin breaching or enlarging an opening while the interior team is packaging and moving the downed firefighter to the opening to avoid any delay.

When enlarging openings, it is best for the exterior rescuers to utilize an existing opening such as a door or window. By doing so it is probable that they will make fewer cuts and deal with less surface area, which saves time. If there is no preexisting opening available at the area, one will have to be made into the side of the structure. If this is the case, it is important for the exterior team leader to maintain communication with the interior RIT in regard to their progress and their arrival into the room or area were the wall is being breached. The interior and exterior rescuers should be extremely careful when deciding to breach or enlarge an opening in the same room that they are in with the victim. There are a few reasons for this. One reason is that opening up or enlarging an opening on the exterior of the building in the same room with the rescuers may draw fire conditions to them due to the influx of fresh air. Another reason is that RIT members and the victim can get injured from saws and penetrating tools that are coming through the wall with great force. Even in light smoke conditions, it may be difficult for the interior rescuers to see the penetrating saws and tools coming through the wall.

Safety

If the interior team members request a breach or an enlarged opening in the same room that they are in, they should communicate with the exterior rescuers to ensure that the victim and the interior rescuers are not on the wall that is being opened.

If a preexisting opening is to be enlarged, it is possible for an exterior rescuer to enter through that opening and act as a safety measure to help prevent the interior rescuers from being injured by chain saws and other penetrating tools as they are dragging the victim into the area of the breach. Having a rescuer on the interior also will help provide direction to the exterior rescuers on the progress and placement of their breach and the effectiveness of their tools.

The RIT will always need to consider the time that may be required to complete a proper, safe, and adequate opening and will need to make certain that adequate resources are available to make it possible. It is also important that the RIT on the exterior have full control of the utilities, ensuring that they have been shut off to avoid any cuts through live electrical service lines running through the walls.

Enlarging Openings: Forcible Entry

Breaching walls and enlarging openings should be considered part of the RIT's proactive fireground behavior. If crews are operating in a sector with no doors or windows for egress, consider making one **(Figure 14-2).** Enlarging openings equals forcible entry!

Fire behavior and communication with interior crews must be considered prior to the removal of any windows or doors!

FIGURE 14-2

If crews are operating in a sector with no doors or windows for egress, consider making one.

A

B

C

FIGURE 14-3

Stairwell grates (A), burglar bars (B), and plywood-covered windows (C) can prevent a firefighter in trouble from being able to escape.

The RIT must be well trained in the techniques of forcible entry to help establish egress points. Remove any objects such as burglar bars, wire mesh, or plywood that may be covering windows and would impede an egress **(Figures 14-3A, 14-3B, and 14-3C).** The following listed items are examples of some of the possible forcible entry responsibilities that can be taken into account as long as the RIT is not jeopardizing fire behavior or the interior suppression efforts.

- forcing and clearing windows
- removing security bars
- forcing doors
- cutting or removing fences that may impede RIT access
- forcing padlocks on all exterior doors (including cellars)
- removing security gates and window gates
- removing grating that may be above window wells or below-grade staircases

These are just a few of the responsibilities of the RIT that may be present at a fire scene while companies are working inside. These obstructive materials and security issues should be addressed around the entire building in order to provide a safe environment as well as to prepare for a rapid egress if necessary.

The RIT will be faced with several types of wall construction materials that include wood frame, masonry, metal, drywall, glass block, and an assortment of other materials. Effectively breaching and enlarging openings will require a RIT to be familiar with the proper tools and techniques needed to accomplish the task. The stability of the wall being breached or opening being enlarged must also be taken into consideration. Will the wall's integrity and stability still be present after being breached, or will it cause instability or possible collapse? Whenever RITs involve themselves with breaching and enlarging openings, they should realize that the majority of these walls will be load bearing and that any holes or breaches that are made in them must be kept to a minimal size. These walls, once opened, should be carefully looked at in order to avoid damaging them excessively, which could possibly lead to collapse.

Tools for Breaching Walls or Enlarging Openings

In order for exterior RIT members to accomplish breaches and enlarge openings, the proper tools should be assembled for the job in relation to the type of wall or preexisting opening that has been selected. Some of the basic equipment that an exterior breach requires is listed as follows, but other equipment may be necessary:

- circular saw, with proper blade for the material addressed
- chain saws
- pike pole
- plaster hook
- sledge hammer
- maul
- flat-head and pick-head axes
- halligan bar
- eye protection
- ladders

Wood-Frame Walls

Wood-frame or sided walls are the most common type that the RIT will encounter, especially in areas of residential construction. The wood-frame wall is also the easiest to breach or enlarge when it comes to exterior breaches. Exterior wood-frame walls on most structures include **sheathing** that is installed on the outside of wood framing, which is what gives rigidity or "shear" strength to the wall. If we started with the outside layer of an exterior-wood frame wall we could list the materials from the outside in that the RIT may encounter as follows. This allows us to see what the RIT members and their tools will be penetrating, cutting, or enlarging.

FIGURE  14-4

A chain saw, which has a much greater depth to its cut, is preferred for enlarging openings in wood-frame structures.

Exterior Wood-Frame Wall

1. Siding
2. Construction paper or weatherproofing
3. Sheathing
4. Floor joist
5. Insulation
6. Vertical studs
7. Interior finish (paneling, drywall)
8. Interior flooring.

When breaching or enlarging openings in wood-frame walls, whether it is a solid breach or a cut-down from the windowsill, the RIT can consider using a circular saw with a carbide tip wood blade or a chain saw. A chain saw, which has a much greater depth to its cut, is preferred **(Figure 14-4).** If the rescuers are using a circular saw they will have to ensure that a depth cut of 3 1/2 in. is provided in order to cut through the wood material on the exterior wall.

Masonry Walls

Masonry walls may pose a slightly more challenging breach for the RIT than wood-frame walls but can easily be penetrated with the proper technique and tools. Masonry walls include those that are made of brick, concrete, stone, and veneers. Masonry walls are commonly used as exterior walls of residential structures and commercial buildings. When RITs are breaching or enlarging openings in masonry walls, they should be conscious that the opening being made can affect the structural integrity of the wall, possibly causing a collapse. What appears to be solid masonry in residential and commercial structures is many times a wood-frame building with a brick exterior, better known as a **brick veneer.** A brick veneer wall will not contain a true **header course** (which provides a structural bond for two vertical sections of masonry) in its construction and is only one brick wide. The bricks will be held in place to a wood-frame wall by the use of metal ties. Brick that is structural will carry its own weight load. Typically, every sixth course will be turned to its end to form a header course and the wall will be at least two bricks wide **(Figure 14-5).** Materials and components of a masonry wall that may be encountered by the RIT from the outside in when breaching a masonry wall are:

Masonry Wall

1. Brick veneer, mortar, and metal ties
2. Air space
3. Flashing at the foundation
4. Sheathing
5. Vertical studs
6. Floor joist
7. Interior finish

The easiest way to breach a masonry wall is by starting with an existing opening and enlarging it from there. The use of sledge hammers, pneumatic hammers, air chisels, and hydraulic tools works quite well for solid masonry and stone and should be available to the RIT. These tools can also be used for brick veneer until the sheathing is reached, which is made of wood, impregnated fiberboard or foam-insulating-type material, in which case the rescuers can then use power saws as they would when performing a breach or enlargement of an opening involving wood-frame walls. When breaching masonry walls, integrity and stability are very important in preventing brick or stone wall collapses. In order to help prevent this, the RIT should provide an opening that is a pyramid or triangle shape in order to avoid curtain falls and other types of collapses with brick and stone **(Figure 14-6).**

The exterior RIT should consider the thickness and metal involvement such as **rebar** (metal bars used to reinforce concrete) that may be present in a masonry wall. The opening of masonry walls can be quite difficult and time consuming.

FIGURE 14-5

Brick that is structural will carry its own weight load. Typically, every sixth course will be turned to its end to form a header course and the wall will be at least two bricks wide.

FIGURE 14-6

When breaching masonry walls, integrity and stability are very important in preventing brick or stone wall collapses. In order to help prevent this, the RIT should provide an opening that is a pyramid or triangle shape in order to avoid curtain falls and other types of collapses with brick and stone.

The RIT should first try to create a small opening that can be slowly enlarged as progress is being made to the size of the breach. If the exterior rescuers are faced with concrete block walls that are filled with additional concrete and metal rebar, they should consider another point of breach because trying to breach this type of wall will require a considerable amount of time.

Metal-Sided Walls

The metal-sided wall is another obstacle to the RIT when requested to breach or enlarge openings. Metal-sided walls can be found mainly in commercial buildings but also in mobile homes **(Figure 14-7).** On residential structures it can be seen as aluminum siding. Many garage and storage buildings will have metal-sided walls. A main concern with breaching and enlarging openings involving metal-sided walls is the flexibility of the material. It can be easily cut with a circular-type saw with a metal cutting blade. The other difficulties and dangers of metal-sided walls is their sharp edges after they are cut, in which case the rescuers should make sure that these edges are protected in some way to avoid injury during the removal of the downed firefighter and his rescuers. Special tools such as a torch or **slice tool** (torch that uses metal rods and pure oxygen to generate heat) or hydraulic spreaders can be considered when breaching the metal-sided wall **(Figure 14-8).**

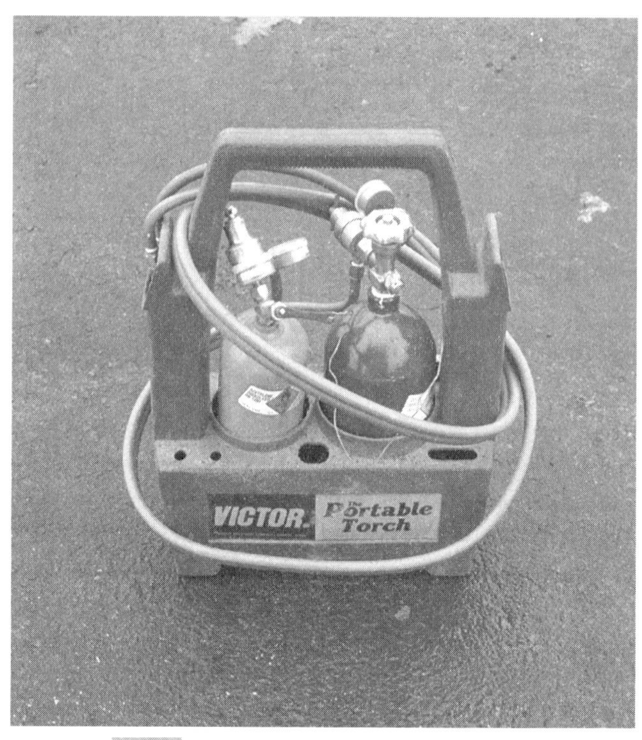

FIGURE  14-8

Torch or "slice tool."

Stucco and Frame Walls

Stucco is a form of concrete consisting of Portland-based material, sand, and water. It is common in older homes **(Figure 14-9).** Stucco is placed over wood sheathing in three layers. A layer of weatherproofing or housewrap is placed over the wood sheathing first to protect it from any moisture. Metal lathing is nailed on next and

FIGURE 14-7

Metal-sided walls can be found mainly in commercial buildings.

FIGURE  14-9

Stucco is a form of concrete consisting of Portland-based material, sand, and water. It is common in older homes.

provides a space for the first coat to be pushed through to form a solid layer and anchor the other layers. This is followed by a second coat and then a third in which the finish pattern (smooth, troweled, stippled) is applied **(Figure 14-10).** Stucco differs from the synthetic material being used on new construction. The material that looks similar to stucco on newer construction, **EIFS—External Insulating Finish System,** is softer and easier to penetrate as it is backed by a Styrofoam-like material **(Figure 14-11).** If an exterior wall is to be breached through stucco and frame construction it is best to use a circular saw that has a carbide tip or compound wood blade in order to complete the cuts necessary from the windowsill downward as described. This type of blade can provide the necessary cutting action to cut through the stucco as well as the wooden studs behind it. This blade is preferred over a masonry blade, which cannot cut as fast and also has the tendency to bind with this type of material. Certain manufacturers of chain saws have chains that are being used allow the chain saw to cut through the stucco and the wood at the same time. Stucco may also be applied over masonry, where the final coat is applied directly over block or concrete—this is a much more challenging wall to breach but can be done with the proper equipment.

FIGURE 14-10

Stucco is placed over wood sheathing in three layers. A layer of weatherproofing or housewrap is placed over the wood sheathing first to protect it from any moisture. Metal lathing is nailed on next and provides a space for the first coat to be pushed through to form a solid layer and anchor the other layers. This is followed by a second coat and then a third, in which the finish pattern (smooth, troweled, stippled) is applied.

FIGURE 14-11

The material that looks similar to stucco on newer construction, EIFS—External Insulating Finish System, is soft and easy to penetrate as it is backed by a Styrofoam-like material.

Lightweight Steel Construction

Steel is a common material being used in new construction and renovations **(Figure 14-12).** It is a good possibility that it could be encountered on any building. In most new commercial construction the building is held up with structural steel and lightweight gauge steel is used for partitions. 18 gauge is commonly used on load bearing walls. In preengineered steel buildings, the primary load-carrying members are spaced from 24 inches (as opposed to 16 inches with wood) up to 8 feet, which enables larger open spaces in buildings. With steel construction, wall studs must be installed directly over the floor joists because the track that replaces the **sole plate** (the horizontal structural member along the bottom of a wall) in wood construction is not load bearing **(Figure 14-13).**

FIGURE 14-12

Steel is a common material being used in new construction and renovations.

FIGURE 14-13

With steel construction, wall studs must be installed directly over the floor joists because the track that replaces the sole plate in wood construction is not load bearing.

FIGURE 14-14

Lightweight steel studs, rafters, and joists are in the shape of a "C" as opposed to being solid, which gives them their strength.

Lightweight steel construction is competitively priced, resulting in less cost than wood, and carpenters do not have to worry about the quality of consistency of the material. Rather than using nails, components of the building are put together using screws. Steel building components can weigh as much as 60 percent less than wood. Its strength, which comes from its shape, results in less structural members being required. Studs, rafters, and joists are in the shape of a "C" as opposed to being solid **(Figure 14-14).** Wood still is used in steel construction to provide solid nailing surfaces for doors, windows, and trim. Generally, these openings will be blocked with wood 2 × 4s. A circular or reciprocating saw will be best to cut through this material.

Enlarging the Opening

After all the equipment has been assembled and the location of the breach or enlargement has been chosen, the exterior rescuers can begin. As always, the exterior rescuers should make certain that their SCBAs are worn properly and should avoid having loose or unbuckled straps. This will help avoid any shifting of the SCBA while working, which may cause a rescuer to fall off-balance into a saw or tool or to become caught into the operations of these tools.

Rescue Tip

When working on the opening, it is a good idea for the RIT members to be on-air with their SCBAs, even though they are located outside the building. The volume of smoke can become very thick, diminishing visibility and breathing ability to a minimum as the opening is made. The facepiece will also provide eye protection from flying debris.

The enlargement of existing openings at the ground-floor level is where the majority of breaching or enlarging operations will take place, although they can be performed on an upper level if necessary.

To enlarge an existing window opening in wood-frame or lightweight metal construction (Skill 14-1):

1. The first step in enlarging the window opening is to clear the window of glass. Personnel should stand to the upwind side in order for any issuing smoke, heat, or fire to move away from the rescuer. Make sure that the window is cleared of all remaining pieces of glass and broken pieces of its sash. The RIT member should also make sure that all obstacles are removed such as drapery, blinds, and decorations.
2. The exterior RIT will begin to make the cuts to the existing window opening. The first cut to the windowsill should be down from the bottom edge of the windowsill directly down to the floor area. When the saw begins to bind up on the floor, remove it from the cut **(Skill 14-1A).**

Safety

The rescuers doing the cutting should remember to operate the saw in a safe and secure way, making sure that the initial cuts are always made off to the side and away from themselves in order to reduce the chances of injury.

3. The second cut is the bottom cut. This cut is started just off the floor level. The bottom cut is done second to avoid any binding of the chain saw blade from the weight of the cut section **(Skill 14-1B).**
4. The final cut to be made will be the second vertical cut down from the windowsill. This will also meet with the cut made at the floor level **(Skill 14-1C).** Once this cut is completed the entire section of material below the windowsill can be removed down to the floor level. The RIT members should make sure to get control of this piece of material in order to prevent it from falling inward and possibly injuring someone. Once the material is removed it should be placed out of the way **(Skill 14-1D).**
5. After the enlarged opening has been completed and the interior rescuers with the victim are ready to exit, the RIT on the exterior will need to provide a rescue litter or long board to remove the downed firefighter away from the structure **(Skill 14-1E).** This will be easier than trying to attempt various types of carries or drags that could cause tripping and falling due to debris and tools. It is important for the RITs on the exterior to make sure that everything is prepared in regard to providing a clear path for the victim and the rescuers. All debris and tools involved in the rescue should be moved out of the way.

Another alternative to this technique involves two additional cuts to be made that will extend the opening.

As the exterior rescuers are making the horizontal cut at the bottom near the floor below the windowsill, they will continue the cut past the windowsill an additional few feet. Extension cuts are provided for the purpose of enlarging the opening beyond the width of the window. This can provide up to an additional 12 to 15 square feet of working space for the rescue effort **(Figure 14-15).**

Glass Block

Glass block is a material that is being used extensively by building designers to give buildings modern aesthetics and more natural light. It has also been used in the past to also allow light into

SKILL 14-1

To Enlarge an Existing Window Opening in Wood-Frame or Lightweight Metal Construction

A The first cut to the windowsill should be down from the bottom edge of the windowsill directly down to the floor area. When the saw begins to bind up on the floor, remove it from the cut.

B The second cut is the bottom cut. This cut is started just off the floor level. The bottom cut is done second to avoid any binding of the chain saw blade from the weight of the cut section.

C The final cut to be made will be the second vertical cut down from the windowsill.

D The RIT members should make sure to get control of this piece of material in order to prevent it from falling inward and possibly injuring someone.

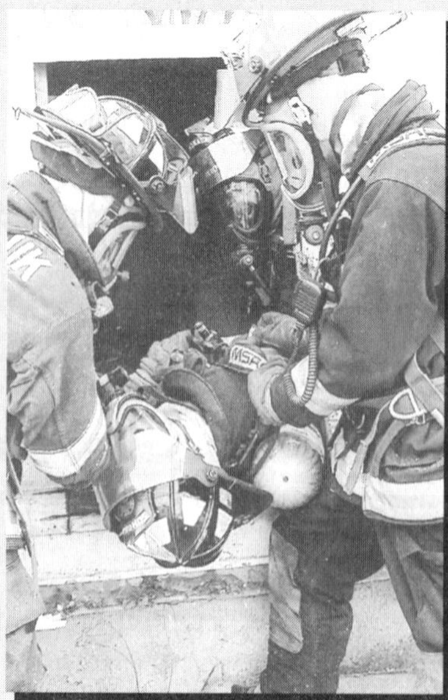

E After the enlarged opening has been completed and the interior rescuers with the victim are ready to exit, the RIT on the exterior will need to provide a rescue litter or long board to remove the downed firefighter away from the structure.

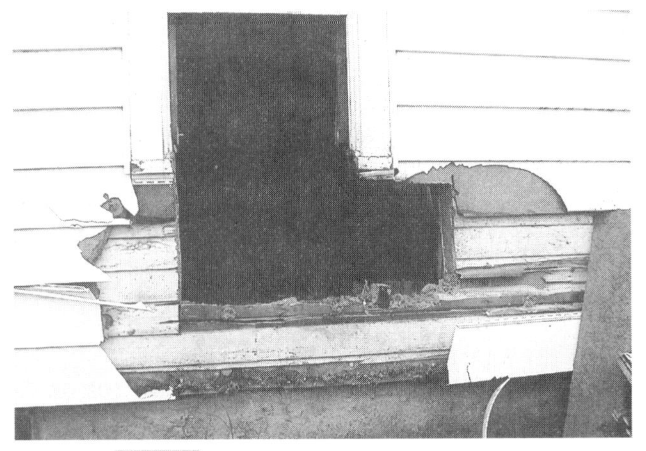

FIGURE 14-15

Extension cuts can be made for the purpose of enlarging the opening beyond the width of the window.

certain areas of a building while providing a high level of security **(Figure 14-16).** To breach glass block will take additional time compared to regular glass and must be complete to avoid any danger of injury.

To remove or breach glass block (Skill 14-2):

1. If necessary, glass block can be removed by using a sledge hammer to first remove the bottom row of blocks. This is best done by striking the center of the blocks and then breaking them out of the mortar joint **(Skill 14-2A).**
2. A row of blocks on either side should be removed next, working from the bottom up. Once these two rows are removed it may be possible for the rest of the blocks to fall in a curtain fashion, so caution should be exercised.
3. Remaining blocks should be removed in a diagonal fashion to create an opening as large as possible **(Skill 14-2B).** Rescuers need to make sure that all sharp edges of glass are removed along with the steel reinforcing mesh that is sometimes used to strengthen the mortar.

Doors

The RIT may also be involved in enlarging openings by using an existing door on the exterior of the structure. In order to begin the process of enlarging a door opening, it will be necessary for the RIT on the exterior of the structure to remove the door. Again this is a forcible entry technique that the RIT should be experienced in. The RIT

SKILL 14-2

To Remove or Breach Glass Block

A If necessary, glass block can be removed by using a sledge hammer to first remove the bottom row of blocks. This is best done by striking the center of the blocks and then breaking them out of the mortar joint.

B Remaining blocks should be removed in a diagonal fashion to create an opening as large as possible.

FIGURE 14-16

Glass block is used to allow light into certain areas of a building while providing a high level of security. What problems can windows such as these present to a firefighter in trouble?

will need to have extensive knowledge and techniques in forcible entry to remove doors very quickly **(Figure 14-17).** Attacking the hinges of a door will prove to be quick and efficient for a RIT. In order to gain quick access to begin the cuts necessary for enlarging and opening of the door, the rescuers may have to cut external hinges with a circular saw when the hinges are exposed on the outside of the structure. Most inward-opening doors can be forced by driving a Halligan bar as a fulcrum between the door and the jamb **(Figure 14-18).** Outward swinging doors can be forced by driving the adz end of a Halligan bar between the door and the jamb and then prying outward. A good tool for RIT members to have as part of their equipment for exterior operations is a **rabbit tool** (handheld hydraulic forcible entry tool), which can be used to force doors **(Figure 14-19).** The RIT may be faced with bars or other devices that are often used on the inside of outward-swinging doors to

FIGURE 14-17

The RIT should be highly skilled in the area of forcible entry techniques.

FIGURE 14-18

Most inward-opening doors can be forced by driving a Halligan bar as a fulcrum between the door and the jamb.

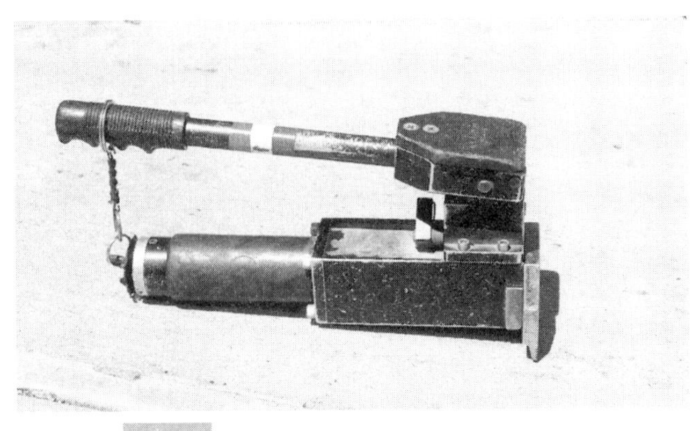

FIGURE 14-19

Rabbit tool.

provide increased security. This is especially true when speaking of commercial buildings. These devices sometimes can be recognized by bolt heads on the exterior of the door. These can be removed by using a circular saw to cut through the bolt heads, separating them from the shanks. The device will then fall off the door, allowing it to open.

Another technique that is applicable to forcing outward-swinging doors is to perform what is known as the **doggy door cut.** This cut is placed into the door to allow the bottom portion to swing open.

To perform the doggy door cut for forcing an outward-swinging door (Skill 14-3):

1. Identify that the door is outward-swinging by the presence of exposed hinges. Attempt to open the door to make certain that it is in fact locked.
2. With a circular saw, make a cut from one side of the door jamb to the other. It is imperative that this cut is located below the locking mechanism.
3. Once the cut is complete, insert a Halligan bar between the door and jamb while prying outward. The door should swing open freely **(Skill 14-3A).**
4. Once inside, the RIT can unlock and open the top portion of the door.

These are just some of the forcible entry techniques that may be required of a RIT in order to provide emergency egress for personnel that are working on the interior of a structure. This is one

SKILL 14-3

To Perform the Doggy Door Cut for Forcing an Outward-Swinging Door

A Doggy door forcible entry technique.

of the main responsibilities of the RIT while performing its proactive behaviors on the fireground.

After the RIT on the exterior has removed the door, members will be able to begin the cuts that will be necessary to enlarge the doorway if needed.

To enlarge a door opening (Skill 14-4):

1. The first cut that should be made by the rescuers is the cut along the floor area at the bottom of the open doorway coming off of its jamb. This low cut should extend at least 2 feet off either side of the jamb **(Skill 14-4A).**
2. The next cut that will be made is the high horizontal cut, which should be started at least three-fourths of the way up of the

SKILL 14-4
To Enlarge a Door Opening

A The first cut that should be made by the rescuers is the cut along the floor area at the bottom of the open doorway coming off of its jamb. This low cut should extend at least 2 feet off either side of the jamb.

B The next cut that will be made is the high horizontal cut, which should be started at least three-fourths of the way up of the existing door height or door opening. This horizontal cut will be the same length as the lower one.

C The third cut will be the vertical cut that will be started from top to bottom to meet the two horizontal cuts.

D After the cuts have been performed it is important for the RIT members to secure this large piece of material in order to prevent it from falling and possibly injuring anyone.

(continued)

SKILL 14-4 (CONTINUED)

To Enlarge a Door Opening

E After the removal of the material, the enlarged door opening should be cleared of all debris and tools to guarantee a clear path for the interior rescuers and downed firefighter.

existing door height or door opening. This horizontal cut will be the same length as the lower one **(Skill 14-4B)**.

3. The third cut will be the vertical cut that will be started from top to bottom to meet the two horizontal cuts **(Skill 14-4C)**.
4. After the cuts have been performed it is important for the RIT members to secure this large piece of material in order to prevent it from falling and possibly injuring anyone **(Skill 14-4D)**.
5. After the removal of the material, the enlarged door opening should be cleared of all debris and tools to guarantee a clear path for the interior rescuers and downed firefighter. Again, a rescue litter or long board should be ready to receive the downed firefighter **(Skill 14-4E)**.

Breaching/Opening Exterior Overhead Doors

Other excellent locations for the purpose of creating enlarged openings and breaches for the RIT on the exterior are the presence of overhead doors. These will be found in both commercial and residential applications. Overhead doors coming down under fire conditions or not being able to be opened have caused numerous firefighter injuries and fatalities over the years.

Most overhead doors in residential applications utilize springs under tension to help make the weight of the door manageable for a person to lift. Overhead doors in commercial applications will also utilize springs to assist in lifting as well as electric motors or chain hoists. Without these tensioned springs, the door would require its total weight to be lifted by the operator. When exposed to extreme heat, the spring may become damaged and no longer provide the mechanical advantage provided when it was in place. Electric door openers may also have a locking mechanism in place that will not allow an overhead door to be moved when it loses power unless the lock is disengaged. Electrical components of the door opener can also become damaged and fail from heat or the application of water. Overhead doors can definitely be a danger to firefighters and must be secured once opened. This can be done by damaging the tracks that the door rolls on with a tool or by securing vise grips or a similar tool in the door track to prevent the door from rolling back down **(Figure 14-20)**. A pike pole or similar tool bracing the door up is not acceptable to secure an overhead door; the weight of the door itself can cause the tool to break or be pushed out from underneath. Once an overhead door is secured open, still be aware that the door itself is located at the ceiling level and can thus be a hazard if it drops straight down with the tracks.

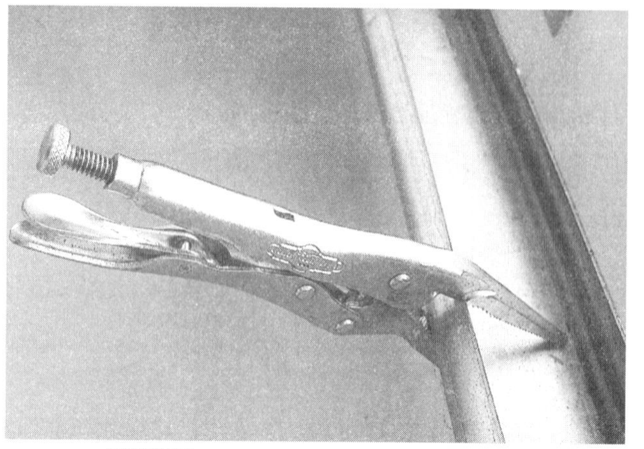

FIGURE 14-20

Overhead doors can be secured by damaging the tracks that the door rolls on with a tool or securing vise grips or a similar tool in the door track to prevent the door from rolling back down.

FIGURE 14-21

Roll-up overhead door.

Overhead doors may be easier to penetrate than an exterior wall or may provide a closer location as a means of egress for interior crews. As with any maneuver that introduces air into a fire, considerations must be given as to how conditions will change once an overhead door is opened prior to it being done.

Forcing overhead doors can be accomplished in two different ways: attacking the locking mechanism and then lifting the door open or cutting your own opening in the door. The type of door and time allotted are factors affecting which method should be used.

Roll-Up Doors

One of the most common types of commercial overhead doors is the **roll-up overhead door.** The doors are constructed out of 1–3-inch steel strips that interlock and are flexible at their joints to allow the door to roll around a drum that is located at the top of the door **(Figure 14-21).**

To cut open a roll-up door (Skill 14-5):

1. The RIT can use a circular saw to make a cut vertically downward at both ends of the door from shoulder height **(Skill 14-5A).**
2. Once these two cut have been made, another is made down the middle of the door **(Skill 14-5B).**
3. The pick of a Halligan bar is buried into the slats and used to pull the slats out of their channels **(Skills 14-5C and 14-5D).**

An inverted "V" or **teepee cut** is another variation of this technique that can be utilized to cut open a roll-up door. The teepee cut allows the door to also remain in the down position, eliminating the overhead weight hazard of having it in the open position. Two cuts are placed in an inverted V pattern and the cut section is pulled out **(Figure 14-22).**

The RIT should have a charged hoseline if there is a possibility of approaching or altering fire conditions in order to protect the rescue efforts.

Sectional Overhead Doors

If the RIT is faced with a **sectional overhead door,** it can be approached with the same considerations as the rolling steel door. Sectional doors are usually made up of steel, aluminum, fiberglass, or wood sections that are hinged to one another **(Figure 14-23).** A sectional door will usually consist of four to six sections that are 12 to 18 inches in width. When the door opens, it rolls in a track that is mounted at the sides. When open the sectional door will present an overhead hazard that also must be monitored. A sectional door can roll out of the tracks when stressed.

A **box cut** can be utilized to cut open a sectional door.

SKILL 14-5

To Cut Open a Roll-Up Door

A The RIT can use a circular saw to make a cut vertically downward at both ends of the door from shoulder height.

B Another cut is made down the middle of the door.

C The pick of a Halligan bar is buried into the slats and used to pull the slats out of their channels toward the center of the door.

D View of door with slats removed.

To perform a box cut of an overhead door (Skill 14-6):

1. Once the decision is made to breach the door, a vertical cut from top to bottom is made. A pry bar or Halligan may need to be utilized to lift the bottom of the door slightly off the concrete slab of the garage to finish the cut all the way down. The blade guard on the saw may also require adjustment **(Skill 14-6A).**

FIGURE 14-22

Inverted "V" or tepee cut.

FIGURE 14-23

Sectional overhead door.

FIGURE 14-24

Tilt-up overhead door.

2. Next, at the point where the vertical cut stops, start a horizontal cut. Do not make the cut at the joint space between each section because there are steel hinges that are on the other side of the door that if cut would allow the sections to fall downward and possibly bind the saw blade **(Skill 14-6B).**
3. Another vertical cut is completed on the opposite side **(Skill 14-6C).**
4. The RIT can help pry the sectional door outward and away from the opening.

Tilt-Up Doors

Tilt-up doors are commonly found in older residential homes. They are one-piece slabs that are constructed of metal or wood and are hinged at the top by the sides and middle to allow them to pivot open or closed. These doors swing out at the bottom to open **(Figure 14-24).** A box cut is also the recommended method of cutting an opening into a tilt-up overhead door. A tilt-up overhead door has no way of being secured into the open position that is both safe and reliable—these doors can present a significant hazard.

Exposure from heat to the springs and hinges on overhead doors will weaken them and make them incapable of supporting themselves. If this is the case, the RIT should avoid any breach or opening as it would become an inappropriate means of egress for the interior crews.

SKILL 14-6

To Perform a Box Cut of an Overhead Door

A First, a vertical cut from top to bottom is made. A pry bar or Halligan may need to be utilized to lift the bottom of the door slightly off the concrete slab of the garage to finish the cut all the way down. The blade guard on the saw may also require adjustment.

B Next, start a horizontal cut at the point where the vertical cut stops.

C Another vertical cut is completed on the opposite side, allowing the RIT to remove the material from the opening.

Burglar Bars

Unfortunately, in our society the fear of crime has prompted people to take measures to protect their safety and property interests. Many people install security bars over their windows to deter criminals from breaking into their property. These also keep firefighters from being able to egress through that window in an emergency. Burglar bar removal should be addressed as part of the RITs proactive fireground behavior. Burglar bars are often attached to a building in one of two ways: they may be set into the masonry of the building (which is common on older buildings),

FIGURE 14-25

Burglar bars set into masonry.

FIGURE 14-26

Burglar bars to the building with lag bolts.

or they may be attached to a building with lag bolts or carriage bolts **(Figures 14-25** and **14-26).** If carriage bolts are used, keep in mind that they are secured on the inside of the wall with a nut.

Burglar bars can be removed several different ways:

- If the bars are set into the masonry, striking them with a sledge hammer several times should loosen them enough to be pulled away from the wall. Hydraulic tools such as a rabbit tool, spreaders, or cutters can also be used.
- If they are set into the building with lag bolts, attacking the lag bolts will be the preferred method. The adz end of the Halligan can be used to pry them away from the building or can be used in conjunction with a striking tool to shear off the heads of the bolts.
- A torch or slice tool can be very efficient if used to cut the bars themselves or at their attachment point. A circular saw with a carbide metal blade can be very effective for removing burglar bars. Most firefighters will shy away from using the circular saw, however, because the saw will need to be lifted to make the cuts. This can be done easily and safely if a firefighter utilizes the cradle technique for operating the saw.

To perform the cradle technique for operating the circular saw (Skill 14-7):

1. The saw, which is already running and at idle, is placed across the rescuer's forearms with the blade parallel to the ground **(Skill 14-7A).**
2. The throttle should be controlled with the outside hand. The cuts will be made with

Safety

Proper eye protection is put in place by the rescuer.

SKILL 14-7

To Perform the Cradle Technique for Operating the Circular Saw

A Proper eye protection is put in place by the rescuer. The saw, which is already running and at idle, is placed across the rescuer's forearms with the blade parallel to the ground.

B The rescuer will control the saw with his body weight as opposed to extending the arms. This prevents the rescuer's arms from having to hold the weight of the saw.

the saw at full throttle. The rescuer's forearms will provide a shelf for the saw.

3. The rescuer will control the saw with his body weight as opposed to extending the arms. This prevents the rescuer's arms from having to hold the weight of the saw **(Skill 14-7B).**

Summary

RITs performing breaches and enlarging openings should have a high degree of proficiency with forcible entry techniques—such as the doggy door cut, tepee cut, and box cut—and tools required, such as a power saw and slice tool. The importance of having an understanding of building construction and the properties of the materials involved—such as stucco, brick veneer, glass block, and wood frame—cannot be stressed enough prior to beginning any breaching or enlarging operations. Rescuers should realize that all options need to be considered when looking for the quickest and most efficient route of removal for a downed firefighter.

■ KEY TERMS

Box cut
Brick veneer
Doggy door cut
EIFS (External Insulating Finishing System)
Glass block
Header course
Rabbit tool
Rebar
Roll-up overhead door
Sectional overhead door
Sheathing
Slice tool
Sole plate
Stucco
Teepee cut
Tilt-up door

REVIEW QUESTIONS

1. When is an exterior breach or enlarged opening required?
   a. when a complicated floor plan exists
   b. when conditions are rapidly changing
   c. when time is a factor
   d. all of the above
2. What are two advantages of using an already existing opening for enlargement or breaching?
   1.
   2.
3. What component gives a wood-framed wall its stability?
   a. vertical studs
   b. sole plate
   c. sheathing
   d. gusset plates
4. Brick veneer walls have a true header course of brick and will support their own weight.
   True False
5. Masonry walls should be breached in a __________ shape or pattern.
6. Two concerns of metal sided walls are:
   1.
   2.
7. In preengineered steel buildings, the primary load-carrying members are spaced from _____ to _____. This allows for _______________.
8. Lightweight steel construction gets its strength from its
   a. size.
   b. shape.
   c. position.
   d. type of connector used.
9. When enlarging an existing opening, the second cut is made horizontally at the floor level to _____________________.
10. Bolt heads on the exterior of an entry door may indicate
    a. the presence of an alarm.
    b. the presence of an additional locking device.
    c. a door that has been priorly damaged.
    d. none of the above
11. When performing the doggy door cut, the door should be cut
    a. near the top.
    b. below the locking mechanism.
    c. below the hinge.
    d. along the jamb vertically.
12. What should be done to secure a roll-up or sectional overhead door in the open position?
13. Why should vertical cuts be made from the bottom up when performing a box cut on a sectional door?
14. Tilt-up overhead doors are commonly found in
    a. newer residential construction.
    b. older residential construction.
    c. commercial buildings.
    d. strip malls.
15. Describe the two different ways that burglar bars are attached to a building.

ADDITIONAL RESOURCES

Brannigan, F., *Building Construction for the Fire Service,* 3rd ed., Quincy, MA: NFPA, 1992.

Crawford, J., *Enlarging Openings for Removal.* 2002. Available online at http.//rapidintervention.com.

Firefighter's Handbook, 2nd ed. Clifton Park, NY: Thomson Delmar Learning, 2004.

Gustin, B., "Forcing Overhead Doors." *Fire Engineering,* November 2004, pp. 67–74.

McCormack, J., *Firefighter Rescue and Rapid Intervention Teams.* Indianapolis, IN: Fire Department Training Network, 2003

Mittendorf, J., *Truck Company Operations.* Saddlebrook, NJ: PennWell Publishing, 1998.

United States Fire Administration, *Security [Burglar] Bars Special Report.* United States Fire Administration Technical Report Series, Report # 138, February 2002.

CHAPTER

15 Rapid Intervention and the Thermal Imaging Environment

Learning Objectives

Upon completion of this chapter, you should be able to:

- discuss the basic principles on which thermal imaging is based.
- explain how mass and density affect the image visualized on the thermal imager.
- clarify the three different types of infrared energy emitters.
- discuss the drawbacks and limitations of thermal imaging.
- identify how thermal imaging can be utilized in searching for trapped firefighters.
- explain the effects that image inversion has on a search with the thermal imager.
- recognize the importance of using thermal imaging to detect convected heat currents.
- define the parameters that make up a systematic search method using the thermal imaging camera.
- explain what is meant by a TIC lead search.
- explain what is meant by a landmark search.

CASE STUDY

"On June 15, 2003, a 39-year-old male career lieutenant (Victim 1) and a 39-year-old male career fire fighter (Victim 2) died while trying to exit a commercial structure following a partial collapse of the roof, which was supported by lightweight metal trusses (bar joists). The victims were part of the initial entry crew searching for the fire and possible entrapment of the store manager. Both victims were in the back of the store operating a handline on the fire that was rolling overhead above a suspended ceiling. A truck company was pulling ceiling tiles searching for fire extension when a possible backdraft explosion occurred in the void space above the ceiling tiles. Victim 1 called for everyone to back out due to the intense heat. At this point, the roof system at the rear of the structure began to fail, sending debris down on top of the firefighters. Victim 1 and Victim 2 became separated from the other firefighters and were unable to escape. Crews were able to remove Victim 2 within minutes and transported him to a local hospital where he succumbed to his injuries the following day. Soon after Victim 2 was removed, the rear of the building collapsed preventing further rescue efforts until the fire was brought under control. Victim 1 was recovered approximately 1 1/2 hours later."

***—Case Study taken from NIOSH Firefighter Fatality Report # 2003-18,* "Partial Roof Collapse in Commercial Structure Fire Claims the Lives of Two Career Firefighters—Tenessee," *full report available online at* http:// www.cdc.gov/niosh/face0318.html.**

One of the key points brought forward by NIOSH investigators to minimize the risk of similar occurrences from this incident was that:

1. "Fire departments should consider using a thermal imaging camera as a part of the size-up operation to aid in locating fires in concealed areas."
2. "Thermal imaging cameras are being used more frequently by the fire service. One function of the camera is to locate the fire or heat source. Infrared thermal cameras assist firefighters in quickly getting crucial information about the location of the source (seat) of the fire from the exterior of the structure, so they can plan an effective and rapid response with the entire emergency team. Knowing the location of the most dangerous and hottest part of the fire may help firefighters determine a safe approach and avoid structural damage in a building that might have otherwise been undetectable. Ceilings and floors that have become dangerously weakened by fire damage and are threatening to collapse may be spotted with a thermal imaging camera. The use of a thermal imaging camera may provide additional information the Incident Commander can use during the initial size-up. Thermal imaging cameras (TICs) should be used in a timely manner, and firefighters should be properly trained in their use and be aware of their limitations."

Introduction

The use of thermal imaging technology has enhanced search and rescue activities for downed firefighters as well as civilians. The technology of these cameras has provided a new edge and limitless possibilities on the fireground. Through the proper use of thermal imaging technology, rescues can take place more quickly and more efficiently while establishing an increased level of accountability and safety on the fireground.

History of Thermal Imaging Technology

Thermal imaging technology has been available and used by the military since the 1950s. As the technology became declassified, it was passed down to the fire service. The first fire departments began to use **thermal imaging cameras (TICs)** in the late 1980s. Early TICs were large, cumbersome, and paled in comparison to the TICs available to the fire service today **(Figure 15-1).**

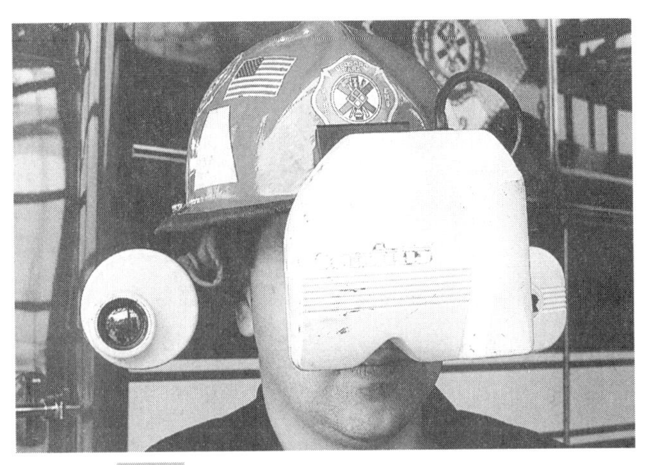

FIGURE 15-1

Early TICs were large, cumbersome, and paled in comparison to the TICs available to the fire service today. One of the first readily available TICs used by the fire service was mounted to a special helmet.

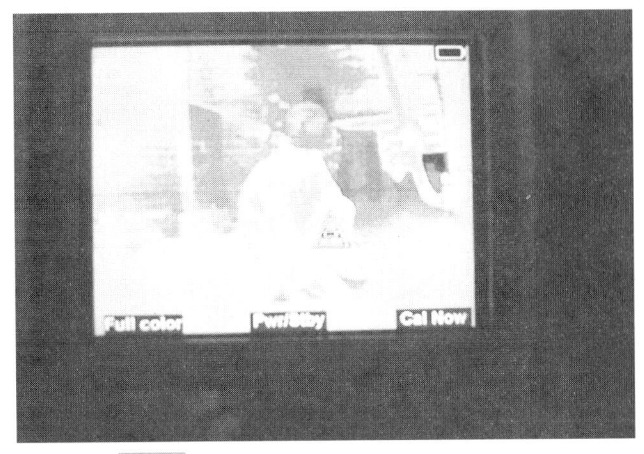

FIGURE 15-2

Visual representation of the TIC viewing screen.

The use of TICs is now an important function for the RIT, not only in the areas of search and rescue, but also in identifying potential and additional hazards on the fireground. Once the understanding of this technology is acquired, the RIT will begin to understand the camera's capabilities and limitations. Training conducted on a regular basis using TICs is the only way to understand this technology properly in order to reap its benefits to the fullest extent.

How Do Thermal Imaging Cameras Work?

TICs are of no use unless the user understands the images being represented on the screen. Thermal imaging technology is based on **infrared energy.** All objects located in an area above absolute zero temperature (**0 degrees Kelvin**) will emit infrared energy to some degree. Heat is classified as a form of infrared energy. Our unaided eye is not able to see this energy. Visible light is also considered infrared energy but it is on a different wavelength than heat. The amount of heat energy given off by objects is different. The picture on the viewing screen of a TIC is a visual representation of temperature differences within an area. Even though visual impairment has taken place for the firefighter, the heat of combustion continues to exist. It is the heat that the technological function of thermal imaging capitalizes on. This is why a firefighter can get a visual representation of an area on the viewing screen when products of combustion have brought visibility levels near zero **(Figure 15-2).**

Even though visible light (which is blocked by smoke) is not required for use, the TIC cannot see through objects. A user may be able to obtain a skeletal view of a stud wall behind drywall, but this will only occur if heat or differentiation in temperatures exists behind the drywall in order to illuminate the objects or wood framing behind it. Mass and density of an object will have a direct effect on the image visualized on the screen of the TIC. The technology of thermal imaging cannot see through the density of solid objects provided in any given space within a structure.

There are three types of infrared-energy emitters, and they can be used to help explain how mass and density affect the image visualized on the screen.

Passive emitters are inanimate objects whose temperature will vary depending on the environment and time limit that it is exposed. Basic physics tells us that heat moves from hot to cold. Heat will move to a cold object until the object is the same temperature as the surrounding environment. A passive emitter will absorb heat in the same manner.

Active emitters are objects that generate their own thermal energy, such as human beings

FIGURE 15-3

A direct source emitter will give off the most thermal energy and will be easily detectable with the TIC. Fire is classified as a direct source emitter.

and animals. These objects can be hidden or **masked** very easily when searching with the TIC. The density of a firefighter's turnout gear or debris (passive emitters) covering a firefighter may prevent his body (active emitter) from being picked up by the TIC. Gear that is wet can also mask an image on the camera. It is important that the user understands this and is able to recognize shapes or objects that may be a part of a downed firefighter protruding from underneath a debris pile.

A **direct source emitter** will give off the most thermal energy and will be easily detectable with the TIC. The fire itself can be classified as a direct source emitter **(Figure 15-3).**

Drawbacks and Limitations

The biggest present drawback with TICs is that they may be too expensive for a large number of departments that are regulated by budgetary constraints. Although the price of the technology has come down since its introduction into fire service use, it is still in the neighborhood of ten thousand dollars or more for a TIC. Many departments that could not have otherwise afforded a TIC have been able to do so through community fundraising and grants. If a department does not presently have a TIC, it is highly recommended that these options be aggressively pursued. Not having this technology readily available can mean the difference between life and death on the fireground.

It should also be mentioned that TICs like any other piece of firefighting equipment, have limitations for use. The most important limitation is that the use of thermal imaging cannot replace a secure and basic foundation in firefighter search techniques.

Note

The use of thermal imaging cannot replace a secure and basic foundation in firefighter search techniques.

The use of TICs can result in firefighters developing an overconfidence and a dependability on the use of the camera. The camera allows firefighters to see in otherwise near-zero visibility environments, which can cause them to stand or walk into a structure without recognizing prominent reference points or keeping aware of obstacles such as holes in floors. This tunnel vision of concentrating on the camera and not the surroundings can also lead to disorientation if the camera fails and the firefighter must now exit. The components of the camera are intricate. The camera is being used in a hostile environment. These intricate components are subjected to impact, heat, and water exposure once inside the fire building. Condensation and fogging will alter images seen. In addition, the TIC is powered by a battery. These are all factors that can lead to its failure. The TIC should be used to supplement the firefighter's search techniques, not replace them **(Figure 15-4).**

TICs will not see through glass or water. Shiny objects such as mirrors, glass, aluminum, steel, and water will reflect infrared energy. Glass and water are transparent to the unaided eye but are opaque to infrared wavelengths. Reflective surfaces within a structure such as doors or windows may be unable to be determined as exits or openings due to high heat and contrasting information that is provided to the camera.

Each TIC will vary in the size of the viewing screen available according to manufacturer specifications. It should be understood that the field of view and depth perception is very limited

FIGURE 15-4

The TIC should be utilized to supplement the firefighter's search techniques—not replace them.

with a TIC. Again, using the TIC to aid search techniques and not replace them will help to avoid this problem.

As mentioned, density will have a direct affect on the capability of the TIC. Certain materials and construction features provide different levels of protection, mass, and density. As the mass increases—such as when building materials are layered over one another or old materials are covered with new materials through remodeling—the capabilities of the TIC decrease.

Using the Thermal Imaging Camera

The recognition of thermal layering within a room is a very important aspect when interpreting **contrast** or heat differential on the TIC screen. Firefighters operating the TIC must be able to recognize thermal contrast.

TICs allow the RIT to identify heat patterns that are available through the existence of thermal layers in a given room and aid in monitoring the true stage of fire conditions that may not be visible due to smoke. The direction and velocity of **convected heat currents** can be visualized on the TIC to help firefighters determine the location and extent of a fire **(Figure 15-5).** The monitoring of changes in thermal layering and convected heat must be continual to avoid the RIT getting into trouble. If a thermal layer is recognized by the TIC it will appear as a high-intensity contrast with a wavelike motion at any given height within a room or area. The level of the thermal layer is of utmost importance to the search team because it indicates the given time available to search for a downed firefighter or to not search at all.

Contrast is simply the images that are provided for us through the camera's lens onto the visual screen that we are looking at. The images appear in different contrast because of the differences in temperatures from one object to another within an area. It is difficult for the TIC to provide a clear image if there is no heat contrast within the area. If this were the case, the images would barley be able to be seen or they would simply appear as either all black or intensely white. Dark or black contrast indicates that the object is producing or holding very little heat in comparison to the surroundings, whereas white contrasts are objects and figures that are giving off more heat than the surroundings. The less heat emitted by an object, the darker the image

FIGURE 15-5

Convected heat currents unseen to the eye can be detected with the proper use of a TIC.

will appear on the screen. The amount of heat within a space and the effect of heat on objects located in that space will determine the ease of distinguishing one object from another on the viewing screen.

Firefighters should also be aware that a downed firefighter may not be easily discernable on the viewing screen inside of a fire environment. Because of the high ambient temperature, the downed firefighter may appear the same color as the surroundings or even darker. This is known as **image inversion.** This is very unlike what is seen on the camera screen when the TIC is trained on in an environment using nontoxic theatrical smoke or normal-temperature environments. Thus, to locate a downed firefighter will require the TIC operator to be able to interpret shapes of objects as opposed to relying on only color shade variations on the screen.

Training with the thermal imaging camera should be instituted in the classroom so that all team members understand the imaging applications of the technology. Training in the classroom should provide understanding of the unit's operation, which includes turning it on and off, battery charging and exchanging procedures, contrasting procedures, and maintenance. The classroom should also provide training on using the camera in zero visibility as well as determining the operational power and time length of the batteries.

Search and Thermal Imaging

Search time can be reduced drastically with the proper use of a TIC. A systematic approach is necessary when incorporating the TIC as part of a search.

When the RIT decides to use thermal imaging for the purposes of search, the camera operator should establish a systematic approach when viewing through the camera at the entryway. The RIT member that is viewing with the camera should place himself in the entryway to the area and begin viewing the space prior to entry. This gives the camera operator and the team members an immediate impression of the overall space, providing them specific information to help plan for their search. It is important for the camera operator to communicate the information that he is seeing to the rest of the team. In order to view an area properly, start at the entryway to the area and apply a specific pattern of scanning to the room. This will help to recognize its hazards as well as the victims you are looking for. When you are scanning the room you can receive the information needed to establish a plan of direction and pattern of search.

When scanning an area to search, perform a quick scan of the floor area first—the downed firefighter may be located near the door or you may determine that the floor is not structurally sound to advance any further. When scanning with the TIC at an entryway to an area, point the camera at the top right or left at the ceiling area and scan slowly across it to the opposite corner from which you started. The reason for scanning the ceiling area after a quick scan of the floor is to determine any possible threatening conditions that may be over the intended rescue team's search area before entry into the space. By also scanning the ceiling you will be able to tell if heated gases, fire, or potential collapses are present. After the ceiling area has been viewed bring the camera back to the original corner of the ceiling area you started with and bring the camera downward from that corner to about midlevel on the wall and then began scanning at that level the same way you did when viewing the ceiling.

Repeat the process at the floor level as well, pointing the camera from one side of the area or space to the other. A good pattern to use is as follows. The camera operator should start at the entryway and point the camera to his extreme right or left corner at the floor area, moving down the wall to the next corner at the floor area, and then begin moving the camera along the floor area of the opposite wall across from the camera operator. The camera operator will then move the camera to the midsection of that wall, viewing the floor area, while moving the camera back toward himself at the entryway. The camera operator will then point the camera to the opposite extreme left or right corner and proceed down the wall to the next corner at the floor area and proceed down the opposite wall to its midsection and back to himself as he had done before. This ensures that the entire space at floor level will have been viewed and that the floor layout has revealed its obstructions, partitions, and hazards, and possibly the downed firefighter, before the search begins **(Figure 15-6)**.

By using a planned approach when viewing with the thermal imaging camera, the RIT will be able to cover all the visible areas between the walls, ceiling, and floor. The team can then begin the actual hands-on searching, which will require looking into areas of obstruction, behind and under furniture, and around stored items within a given area. This can also be directed by the operator of the camera. The camera operator should ensure that the camera is scanning the entire area. The reason for this is that the camera screen does not provide any peripheral view for the operator. The camera's tendency is to force the viewer into a tunneled approach, which makes the operator and possibly the team members unaware of other surrounding conditions if the camera is not continually scanned. Also, if the area is not properly scanned you may miss or overlook hazardous conditions as well as the downed firefighter you are looking for. Whenever the RIT has decided to use thermal imaging for the purposes of search, there are five main objectives to be kept in mind by all team members along with the operator of the TIC:

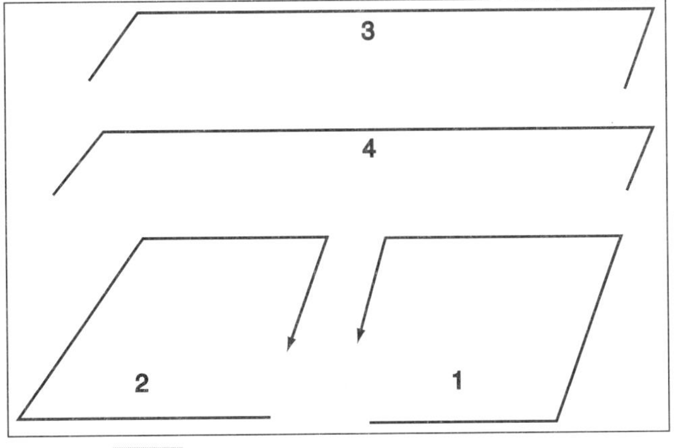

FIGURE 15-6

Systematic order for scanning a room with the TIC.

1. Always scan for life hazards that can affect the RIT as well as the victims the team is looking for.
2. Always be on the lookout for changing fire conditions or approaching conditions that would jeopardize the RIT or the victim.
3. Always try to remember to systematically scan an area, looking for unsafe conditions.
4. Always be aware of your surroundings and avoid tunnel vision, or paying attention only to the screen of the camera.
5. Always scan the floor and ceiling of an area you are about to enter for its integrity and fire conditions.

Methods

Thermal imaging can allow RITs to increase the speed of their operations to the point that the rescue team may find themselves moving too quickly, in turn possibly getting themselves into trouble. Thermal imaging will always be a great benefit but the downside is that if the operators cannot interpret the images properly or the operators and the team members abandon basic search principles, they can find themselves in situations of becoming lost or trapped along with the victim they are looking for. One way to help avoid this problem is to have a member of the RIT other than the camera operator to keep the team in check by concentrating on basic search procedures and slowing the team down when it is moving too fast. This member will have responsibility for the accountability of the team members as well as keeping track of the operator of the TIC. This member can slow the team down to allow time for understanding and recognizing fire conditions as well as possible threatening hazards that might be missed by a fast-moving team.

TIC Lead Search

In the **TIC lead search**, the thermal imaging operator leads the search team through the area to be searched while relating information to the team members in regard to direction of travel and

FIGURE 

In a TIC lead search, the camera operator will provide direction to the team members to reach the victim.

obstacles encountered. When a victim is located, the camera operator will provide direction to the team members to reach the victim **(Figure 15-7).** The camera operator should be monitoring fire and heat conditions while the team members prepare the victim for removal. If a search line rope is being utilized, the camera operator is to stay on the line, directing the other team members who are tethered on the main line to the victim. If the camera should fail, the members tethered off the main line can return with the victim to the camera operator positioned on the main line. The camera operator would then continue to lead the rescuers with the downed firefighter to the point of egress.

Landmark Search

Another method of search when using the TIC is the **landmark search** used when moving through a structure. With this type of search, the camera operator will identify a specific landmark when scanning the area and bring the team to that point as the rescuers search along the way. This can be done with or without a search rope but it is suggested that one be used anytime that the RIT is activated, especially if the search involves a large area or complex floor plan. Once the landmark is reached, the camera operator begins to scan the area in a systematic approach to locate another landmark and proceeds to move the team again to that point as they search along the way. When utilizing the landmark search technique it is important for the rescue team to be aware of the amount of time and penetration made into the building in regard to their air supply. When using landmark-type searches with thermal imaging, the team can move very quickly and directly to items of interest or victims while not knowing or realizing the amount of time and distance they have covered. This is why it is so important for one member, as stated earlier, to maintain team safety and basic search procedures.

Searches should be disciplined within a specific plan and these plans should be trained on and practiced so that members of the team are successful and accountable to each other and the rescue effort. Because the use of thermal imaging is very helpful in searching operations, it is important for the teams using this technology understand and comply with the responsibilities of accountability and continually monitoring conditions around themselves. To do otherwise may actually get the rescuers into trouble. TICs and their usage should be outlined in a department's standard operating guidelines **(Table 15-1).**

Thermal Imaging Standard Operating Guidelines

Purpose

To provide a guide for the deployment and maintenance of thermal imaging cameras (TICs) for the XXX Fire Department while establishing a procedure to provide for their most effective use in a manner that promotes the maximum level of safety.

Scope

These guidelines will be used by all XXX Fire Department personnel engaged in operations using TICs. These guidelines cover the deployment, operation, and maintenance of the TIC. Further information on the maintenance and features of the TIC on each fire company is available in the manual provided that is specific to that particular camera. Personnel should study the information provided with the camera and be familiar with the advantages and limitations of the unit before using it.

Safety

- The TIC is a tool to be used to augment but not replace the use of sound strategy, tactics, and standard structural firefighting and rescue procedures.
- Water, plastics, glass, and shiny surfaces are all an effective barrier for the TIC and may cause a reflective image.
- **The TIC is a mechanical device and can fail;** personnel must plan for this possibility by always employing standard and safe fire-fighting practices such as maintaining contact with a hoseline, wall, etc.
- **The thermal imager is NOT intrinsically safe and could produce a spark;** personnel must consider the possibility of a potential explosive atmosphere before placing the unit in service.

Uses

TICs can be used to aid personnel in the following situations (the list is not exhaustive):

- Initial Size-Up
- Primary and Secondary Search
- Location of Fire and Extension
- Monitoring of Conditions Indicative of Flashover
- Overhaul
- Rapid Intervention Operations
- Hazardous Material Situations
- Preplanning
- Fire Inspections
- Urban Search and Rescue
- Assisting Law Enforcement

Limitations

A. Depth perception—TICs only provide a two-dimensional view of the environment.
B. Shiny surfaces will reflect infrared energy, causing an untrue image on the viewing screen.
C. TICs will not see through glass or water.
D. Condensation and fogging will distort images on the viewing screen.
E. Active emitters can be easily masked if clothing is wet.
F. Active emitters can be easily masked over by a passive emitter.
G. Mass and density will have a direct effect on the image.
H. Thermal inversion can occur when changing location within an environment.
I. TICs are high-technology, battery-operated devices and can fail.
J. TICs are not intrinsically safe.

Deployment

It is the company officer's responsibility (or person riding in that capacity) to carry the TIC on their person into a structure regardless of size-up information given on all fire-related calls.

A. In Fire Conditions: the TIC must be used in conjunction with established safety practices.
    - Crews using the TIC must use established search procedures such as left- and right-hand searches, entering with a hoseline, and/or the use of safety lines if required.
    - When entering with a crew, the person operating the imager shall not advance ahead of or lose contact with other personnel involved in the search operations.

(continued)

TABLE 15-1

Example of Thermal Imaging Standard Operating Guidelines. *Source: Adapted from several existing policies.*

- Areas should be scanned with the TIC in a systematic fashion to allow a full view of the area.
- **At no time should a TIC operator be leading only by use of the camera—conventional search techniques must not be abandoned.**
- The TIC operator should constantly monitor for changes in thermal contrast and movement of convected heat currents.
- The officer in charge should consider additional personnel for searches when personnel are engaged in active fire suppression along with a search.

B. Secondary and Final Searches:
- The TIC shall be employed for all secondary and final searches of the structure. **No exceptions.**

C. Non-Fire-Suppression Searches:
- Department personnel should consider the use of the TIC in any situation where a person could be obscured or not able to be of assistance to themselves.
- Use in confined spaces or other hazardous situations must conform to all safety procedures.

D. Rapid Intervention Operations:
- All rapid intervention teams operating at an incident shall have at least one TIC available for their use.

E. Detection of Abnormal Conditions:
- The TIC should be employed in any situation where FD personnel need to detect abnormal conditions such as overheated motors, shorted electrical equipment, out of range conditions in heating and cooking appliances, machinery, etc.
- Department personnel are advised to employ the TIC in any situation in which direct visual inspection is limited or detection of temperature is useful.

F. Overhaul:
- The TIC should be employed throughout the overhaul process and for follow-up inspections of the property.
- After structure fires, post-fire inspections shall be done with the TIC before the property is turned back over to the owner.

Maintenance

A. Storage—TICs shall be stored either in their case or approved apparatus docking station (charger). The TIC should be protected from moisture, and temperature extremes and be secured within the fire apparatus.

B. Daily Inspection—Each morning the TIC shall be removed from its storage case, turned to the ON mode and operated. Make certain to check that the battery level is fully charged.

C. Post-Use Maintenance—After each use, the TIC shall be cleaned with the use of mild soap and water. The TIC should be dried with a soft cloth. At no time should a TIC be put back into its storage case wet. The camera shall be inspected for damage and the batteries shall be replaced/recharged.

D. Any malfunction or damage shall be reported by filling out a repair order form with a concise description of the problem encountered. The TIC and corresponding work order shall be directed to the on-duty shift commander. Once at this level, all repairs on TICs will be coordinated through the Support Services Division.

Definitions

Active Emitter—object that generates its own continuous source of thermal energy, examples are humans and animals.

Masked—blocking of the thermal emission from the sensor of the TIC. Objects can be easily masked when wet or covered by dense objects.

Passive Emitter—objects whose thermal emissions will be dependent upon the environment and length of time that it is exposed.

TIC (Thermal Imaging Camera)—device that shows a representation of temperature differences among objects.

Thermal Inversion—color change on the thermal imager viewing screen due to changes in ambient temperature. Can occur when the camera location is changed within a fire area.

TABLE 

Example of Thermal Imaging Standard Operating Guidelines. *(continued)*

Summary

TICs can vastly improve a rescue effort, but they do not ensure success. Some important aspects that should be remembered by the RIT utilizing thermal imaging is that the tool or camera has its limitations, as does every tool within the fire service. Remember these important points:

- The camera cannot see through objects.
- The camera receives and interprets infrared energy from relative differences in surface temperatures.
- The camera cannot look through walls or floors for victims unless there is extreme heat absorbed into these surfaces and even then they will provide only little information.
- Heat moves from hot to cold objects.
- Glass and water are transparent to eyesight but opaque to infrared energy.
- Mass and density of a given object determines the amount of absorption and infrared energy given off by that object, which provides us with thermal contrast.
- Always turn the camera on before you use it in a particular environment, allowing time for warm up and checking systems.
- Rescuers should place spare batteries in their pockets.
- There are two different ways you can incorporate thermal imaging into search—to either lead the search or direct it.
- When you are directing a search, your team is in front of you at all times. The TIC operator provides monitoring and accountability.
- Always check the ceiling of a given structure for the existing thermal conditions.
- Never become tunnel-visioned into a narrow field of view; constantly scan.
- Thermal imaging can cause RITs to proceed too far, too fast, and to deep within any given structure if proper techniques are not exercised.
- TICs should be protected from dropping, high heat, and water submersion.
- RITs should always be aware of the door they came in and a second way out through conventional search skills, not thermal imaging.
- Conditions always change; be aware of your surroundings, not just the TIC screen.
- All searches by the RIT will be in a crawling position, not walking, whether thermal imaging is used or not.

These are just some of the main observations and important points to think about when utilizing the TIC. What these points stress are that the operators of these cameras as well as the team members should have an in-depth understanding of how thermal imaging works. Understanding these points will allow the rescuers to properly interpret the images when looking at the thermal imaging screen. Thermal imaging technology is one of the most valuable and advanced technological tools within the fire service to date. As valuable as thermal imaging is, it still cannot provide or guarantee its abilities unless firefighters are well-trained in the fundamentals of fire suppression, RIT operations, and search in order for a high degree of success to be attained with the camera. Firefighters should always be aware and understand the limitations and dangers of the fireground to avoid injuries and misuse of the camera's capabilities.

KEY TERMS

0 degrees Kelvin
Active emitter
Convected heat currents
Contrast
Direct source emitter
Image inversion
Infrared energy
Landmark search
Masked
Passive emitter
Thermal imaging cameras (TICs)
TIC lead search

REVIEW QUESTIONS

1. Thermal imaging technology has been available and used by the military since the
   a. 1950s.
   b. 1970s.
   c. 1990s.
   d. 1940s.
2. Thermal imaging technology is based on
   a. magnetic energy.
   b. reflective energy.
   c. infrared energy.
   d. electrical energy.
3. Heat is classified as a form of infrared energy.
   True False
4. ______________________ of an object will have a direct effect on the image visualized on the screen of the thermal imaging camera.
   a. Size and shape
   b. Mass and density
   c. Density and color
   d. Weight and size
5. Passive emitters are inanimate objects whose temperature will vary depending on the
   a. region and location of exposure to heat.
   b. weight and movements during exposure to heat.
   c. environment and time limit of exposure to heat.
   d. none of the above
6. The biggest present drawback with thermal imaging cameras is that they may be too expensive for a large number of fire departments with budgetary constraints.
   True False
7. Thermal imaging cameras can be used for water rescue.
   True False
8. Downed firefighters may not easily be discernible when visualized through the thermal imaging camera because of high ambient temperatures, which may show the firefighter in the same contrast as well as appearing in even darker contrast than the surrounding fire environment. This is referred to as
   a. blackout.
   b. image inversion.
   c. contrast variance.
   d. white out.
9. When scanning an area to search, you should perform a quick scan of the ______________ before advancing into an area.
   a. floor and ceiling
   b. walls
   c. windows and doors
   d. none of the above
10. When scanning a given area, the leader will identify key spots or objects while directing the team to that point as the rescuers search along the way. This is known as a
    a. direct search.
    b. survey search.
    c. landmark search.
    d. perimeter search.

ADDITIONAL RESOURCES

Bastian, J., "The Five Don'ts of Thermal Imaging," *Firehouse,* June 2003.

Eisner, H., "The ABCs of Thermal Imaging," *Firehouse,* November 2000.

Firefighter's Handbook, 2nd ed., Clifton Park, NY: Thomson Delmar Learning, 2004.

Norman, J., *Fire Officers Handbook of Tactics,* 2nd ed. Saddlebrook, NJ: PennWell Publishing, 1998.

Woodworth, S., "Thermal Imaging for the Fire Service, Part 1: The Basics of Thermal Imaging," *Fire Engineering,* July 1996.

Woodworth, S., "Thermal Imaging for the Fire Service, Part 3: Thermal Characteristics," *Fire Engineering,* November 1996.

Woodworth, S., "Thermal Imaging for the Fire Service, Part 4: Thermal Imaging Devices," *Fire Engineering,* February 1997.

Woodworth, S., "Thermal Imaging for the Fire Service, Part 6: The Search," *Fire Engineering,* August 1997.

CHAPTER

16 Search Techniques and Search Rope Systems

Learning Objectives

Upon completion of this chapter, you should be able to:

- define what is meant by a large-area search and the challenges that it presents.
- talk about the types of operations that are required for large-area searches.
- explain the procedure of radio-assisted feedback when used for locating a downed firefighter.
- discuss the basic parameters that must be in place prior to deploying the RIT to perform a large-area search.
- give details on how to perform a simple oriented search.
- describe aspects of what a main line rope-assisted team search entails and clarify what the roles and responsibilities of the RIT members are when performing a main line rope-assisted search.
- discuss some of the basic techniques and procedures used in main line searches using rope.
- identify some of the equipment needed for large-area, rope-assisted searches.
- list the advantages and disadvantages of performing an oriented search using a hoseline.
- explain what can be done to determine and overcome air supply limitations when performing large-area searches.

CASE STUDY

"On December 18, 1999, a 47-year-old male battalion chief (the victim) was fatally injured during a paper warehouse fire. Firefighters were dispatched to the fire and upon arrival they immediately ordered all employees to evacuate the approximately 300,000-square-foot warehouse. The fire was located in the paper-bale section and was causing the structure to fill with a haze of white smoke. The incident commander (IC) assumed overall command and ordered an interior fire attack. He also ordered the battalion chief (the victim) from Car 106 to take command of interior operations. The firefighters battled the fire for approximately 52 minutes before the IC and the victim decided conditions were deteriorating and they should go to a defensive attack. The IC ordered all firefighters to evacuate the structure; however, several firefighters' radios malfunctioned and they did not receive the evacuation order. Some of the firefighters with the malfunctioning radios eventually ran out of air, became disoriented, and needed assistance to exit. The victim also became disoriented and did not exit. After learning that all the firefighters except for the victim had exited, the IC ordered the two initial rapid intervention teams to enter and search for the victim. Both teams entered but eventually ran low on air and were forced to exit without the victim. Additional RITs were formed and found the victim approximately 1 1/2 hours after the initial dispatch. He was transported to a nearby hospital where he was pronounced dead."

As a result of this incident, NIOSH investigators concluded that to minimize similar occurrences, "fire departments should

1. *ensure that the department's standard operating procedures (SOPs) are followed and refresher training is provided.*
2. *ensure that all firefighters performing firefighting operations are accounted for.*
3. *ensure that proper ventilation equipment is available and ventilation takes place when firefighters are operating inside smoke-filled structures.*
4. *ensure that when entering or exiting a smoke-filled structure, firefighters follow a hoseline, rope, or some other type of guide.*
5. *ensure that firefighters are equipped with a radio that does not bleedover, cause interference, or lose communication under field conditions.*
6. *ensure that the assigned rapid intervention team(s) (RIT) complete search and rescue operations and are properly trained and equipped.*
7. *ensure consistent use of personal alert safety system (PASS) devices at all incidents and consider providing firefighters with a PASS integrated into their self-contained breathing apparatus (SCBA).*
8. *develop and implement a SCBA preventative maintenance program to ensure that all SCBAs are adequately maintained."*

—Case Study taken from NIOSH Firefighter Fatality Report # 1999-47,* "Warehouse Fire Claims the Life of a Battalion Chief—Missouri," *full report available on-line at http://www.cdc.gov/niosh/face9947.html.

Introduction

Searching for a downed firefighter in a large area is one of the most challenging and dangerous tasks that a RIT can undertake. Chaos is an immediate reaction and must be controlled by a good rescue plan and superb search techniques. This chapter will discuss the operations and techniques required for large-area searches.

Large-Area Searches

We must first define a large-area search. A large-area search can take place at a commercial-type occupancy or a residential home. In many cases, features such as square footage and open areas will be similar. New residential and business construction is becoming progressively larger **(Figures 16-1** and **16-2)**. This causes concerns in

FIGURE 16-1

Large, open areas with high ceilings and sprinkler systems will create "cold smoke" situations that can cause firefighters to become easily disoriented.

areas of accountability, air supply, and search techniques.

A solid foundation in the basic skills of search techniques will be a prerequisite for firefighters operating as a RIT. Procedures will have to be well planned and flexible to meet the demands of each scenario. Emotional discipline and control will also be a priority. Searching for a downed firefighter is different in many ways than searching for a civilian, but the basic skills of search must still be executed. One of these differences is that the RIT members should have accurate information to lead them to a general area of where to search for the lost firefighter, provided an accurate accountability system is in place.

FIGURE 

New residential construction is being built with large, open spaces and high ceilings, which can cause problems similar to those encountered in commercial spaces.

Upon Arrival

The RIT members should begin collecting information on the fireground to accurately handle a situation as soon as they arrive. In the early chapters of the book we discussed many responsibilities for the RIT. Those responsibilities should be understood before a large-area search begins.

A rescue of a downed firefighter in a large area will require many RITs with different responsibilities. One of the most important fundamentals when responding to rescues of this nature is for the RIT to have a solid plan. Well-written operating procedures will establish a base of actions to build from, which in the end will provide the ability to establish plan A and a back-up plan B.

It is important to acquire information that could give the RIT a specific location of a downed firefighter before the Mayday has occurred. The RIT will also be responsible for understanding the mode of fire attack and predicting possible fire extension. Knowing these factors will enhance the development of a plan to retrieve a victim. All of this information should be gathered by the RITLO from the IC's tactical worksheet.

There may need to be as many as four to six teams made up of five to six firefighters to complete a large-area search. Designated responsibilities for each team and individual member are important.

Supplies

When searching in larger structures, a rope-assisted search will be necessary to locate and retrieve a downed firefighter. Search rope will provide a safety factor for the rescuers. Rope will allow the RIT to move and branch off into areas within a large area to cover it quickly while remaining in contact with each member of a team. Main lines and tether lines to branch off further will help keep accountability of rescuers while they are searching.

Thermal imaging cameras (TICs) are an essential tool for the large-area search. A TIC-led search can provide the ability to see a downed firefighter from a substantial distance, which could reduce the amount of time required for the rescuers to locate the victim as well as help conserve the RIT's air supply.

The activation of PASS devices can also improve the outcome of a large-area search. When distressed firefighters activate their PASS devices, they increase their chances of being found in a timely fashion. A RIT may be able to move quickly through a building to the sound of a PASS device. It should be noted that this technique can sometimes be confusing due to the sound of the PASS echoing off walls. However, if the sound is becoming louder it is obvious that the RIT is getting closer and if the sound begins to diminish, the RIT is moving further away. The PASS alarm is at least helpful in moving toward the general location of the downed firefighter.

There are other techniques that can be used if a PASS cannot be heard. **Radio-assisted feedback** can be attempted when two radios are placed close together and keyed up to produce a high-pitched sound that may enable the RIT members to hear feedback from the downed firefighter's portable radio **(Figure 16-3).**

FIGURE  16-3

Two radios can be placed close together and keyed up to produce a high-pitched sound that may enable the RIT members to hear feedback from the downed firefighter's portable radio.

The Basics

Before any search can begin it is important for certain responsibilities to be addressed:

- All forcible entry problems and blocked exits should be opened into the structure.
- Additional hoselines should be deployed to protect all search teams, especially in the areas of advancing fire.
- Proper ventilation should be addressed to open the structure and increase visibility, enhancing the search.
- RIT members should possess full knowledge and understanding of their SCBA equipment and know their limitations. Each member should be fully aware of all emergency procedures involving the SCBA.
- An air management plan with proper accountability must be in place.

The proper tools, as noted earlier, should also be available in order to make the search possible. Some basic equipment that should be available to the RIT includes:

- hand lights
- irons (Halligan and flat-head axe or sledge hammer)
- pike poles
- large-area search kit (150-ft rope bag with carabiners)
- thermal imaging cameras
- portable radios

Large-area search is a race against time involving the rescuer's air supply as well as the rescue of a downed firefighter. A large-area search for a downed firefighter becomes a high-priority primary search. RITs should have the ultimate attitude that a secondary search is out of the question. We have to remember that the RIT is in the mind-set of searching for the victim firefighter and not necessarily the extension of fire. Even though the RIT is focused on reaching the victim, the responsibilities of providing ventilation, closing doors, noticing fire extension, and communicating should still be undertaken to provide safety and more time. A second company with a charged hoseline should provide protective cover for the RIT's efforts to search quickly.

RITs should maintain radio contact while providing progress reports on a regular basis. Communicating search results, areas searched, and the need for additional help are high priorities for successful outcomes. Radio communications should also be provided for relaying fire conditions, structural stability, locations of search crews, and whether a new plan needs to be considered.

There are several types and methods of searching large areas. The type of search used depends on the conditions present. Searches will be faster or slower with greater or less accuracy depending on those conditions. For instance, the light smoke search takes place when smoke conditions allow rescuers to go into areas that still have visibility. Conditions are not extremely bad at this point. In this situation, usually the rescuers can simply crouch down below the smoke level while using hand lights to visualize a large area or room. The RIT must take advantage and be thorough in the search while good visibility is still available. As conditions deteriorate, rescuers will gain entry into areas and find themselves with visibility diminishing. In this case they will have to secure themselves to a wall and incorporate basic search procedures such as right- and left-handed searches while using a tool to extend out to the front and to the side to cover an area.

Hand lights can become orientation points when placed on the floor in entryways to larger areas. The use of personal ropes or tethers may have to be considered to cover more area and to provide accountability of each RIT member. The search activities should stop occasionally to listen for the downed firefighter's SCBA and PASS device. Rescuers should continue to use their hand lights to scan below the smoke level at the floor area in hopes of spotting the reflective material of a downed firefighter's gear. Adjoining rooms should be searched progressively. Rescuers should remain in the same area, providing each other with a point of reference.

Standard Searches

Each search technique has advantages and disadvantages. It should be understood that several techniques and variations can be involved in one rescue. A standard search will utilize walls. Rescuers proceed left or right into a room or one rescuer may go left and the other will go right and meet somewhere in the middle of a room. Once the room search is completed the rescuers will then move on to the next room. Techniques such as crawling along a wall while another rescuer holds on to the first rescuer's boot will *not* suffice **(Figure 16-4)**. Being directly behind one another leaves a large amount of area uncovered. Firefighters performing a wall search should extend off one another toward the center of the

FIGURE 16-4

Crawling along a wall while another rescuer holds on to the first rescuer's boot is *not* a productive search technique.

FIGURE 16-5

Firefighters performing a wall search should extend off one another toward the center of the room to cover more area.

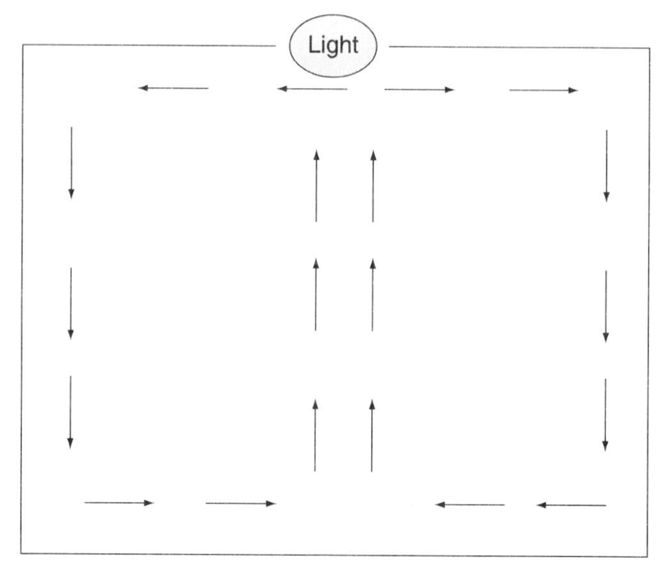

FIGURE 16-6

Simple oriented search technique.

room to cover more area **(Figure 16-5)**. The advantages of this particular concept are that it maintains team continuity. The drawback is that all members are searching, including the team leader, with no single person concentrating on the surroundings. By following from wall-to-wall, the original point of entry also can become an opening to a maze, with the RIT members not concentrating on points of entry or egress and becoming disoriented.

The Simple Oriented Search

The **simple oriented search** starts by leaving a firefighter with a powerful light at the entry to provide a point of reference that can be seen underneath smoke conditions. The second firefighter should move swiftly but consciously by proceeding down the wall while still looking back occasionally, making certain that the hand light is still visible. The rescuer will then proceed to the next wall and cross over to the center of the room. At this point, the rescuer should not continue across the entire room but should cross over to the center and proceed toward the hand light at the entry point. The rescuer should then proceed on the opposite wall of the initial search on entry. The search is then executed in the same manner by going down the wall and crossing over to the center of the room while proceeding back toward the light at the entry point. This provides good coverage of a medium-size or small open-area room in place of a complete right- or left-handed search along a wall **(Figure 16-6)**.

The methods just mentioned involve the basic principles of search techniques and have a slightly different approach because the rescue members are dealing with the search of a downed firefighter and not a civilian. It should be mentioned that the use of thermal imaging in all procedures provides a more thorough and safer approach for the RIT and the downed firefighter.

Main Line Rope-Assisted Team Search

When the RIT is involved in a team search, the rope is used as a guide. If the proper procedures are followed, this can be a very safe search method. The **main line rope-assisted team search** concept is simple: a rope is anchored to a stationary point outside the hazardous environment and RIT members lead out the search line as the team crawls into the structure following the line. The rope can incorporate a guide system to indicate which way is in and which way is out. This is sometimes accomplished by tying a series of knots into the line periodically, depending on protocol and needs of a department. This can also provide specific information in regard to telling how deep into the building the RIT may be. This is accomplished by counting the knots. For example, if using 20 feet between knots, two knots = 40 feet, three knots = 60 feet, and so on. However, it is important to realize that the number of knots put into a main line should be limited to avoid making the system confusing and to allow it to deploy from the bag without any difficulty.

There are many different procedures involving the use of main search lines. Some departments utilize short ropes or tethers to attach searchers to one another. Some use ropes or webbing from 15 to 20 feet in length attached to the main line while searching in patterns. The RIT members should be trained in the specific use and deployment of these lines and make certain they are familiar with the procedure their department chooses. No attempts should be made to try to hook on to the main rope line by using devices such as small clips or by applying any type of hitches or knots to line. Instead the tethers should be hand-held on the line by another team member or team leader while the search is being conducted. It is also important to remember to only search 180 degrees to either side of the main line in order to avoid any 360-degree movement over or around the main line, which could cause entanglement to personnel and other objects in the immediate area.

Five to seven RIT members will be required to conduct a large-area search using a rope system. Several positions with specific responsibilities will have to be integrated:

- The *team leader* will be responsible for the overall operations of the search team. This will include control and deployment of the main line. This individual will also be responsible for directing the search on the interior and maintaining communications with the control/entry supervisor. Tools essential for the team leader will include the main line search rope, two-way radio, hand light, and TIC.
- The **control/entry supervisor** is responsible for tracking the team's progress, and most important, monitoring and logging the time in and out of the structure as it pertains to the team's air supply. The control supervisor should have a worksheet and stopwatch to monitor and record the team's air supply. The control supervisor will also be responsible for accountability and monitoring conditions from outside the RIT's entry point. The team leader will communicate progress reports back to the control/entry supervisor regarding the RIT's progress and air supply status. The control/entry supervisor's main responsibility is to act as a safety officer for the interior rescuers. This is an ideal position for the RIT operations chief.
- The remaining members of the RIT will need to bring in forcible entry/exit tools, hand lights, two-way radios, personal rope or tethers, and an additional air supply.

An additional RIT should be in a standby position at the point of entry with the control/entry supervisor ready to assist the first RIT as soon as it is requested. Backup teams will be responsible for relieving the initial entry teams during extended operations. Multiple entry points with multiple teams may become a feasible plan of action depending on the location of the downed firefighter and type of structure.

The search rope system should always be anchored to an outside area, whether the conditions are tenuous or not **(Figure 16-7).** At any time visibility can go from crystal clear to heavy smoke conditions, which will require the rope to provide the RIT with an immediate way out. It

FIGURE 16-7

The search rope system should always be anchored to an outside area, whether the conditions are tenuous or not.

should also be stressed that the control/entry supervisor at the entry point will ensure the security of the main guide line to its attachment.

The RIT members should check that their radios are on the correct frequencies prior to initially making entry. Multiple RITs and entry points can cause confusion. Each should operate on a separate frequency if possible. All tools and equipment should also be checked once more prior to entry. As the RIT enters into the large area, the control/entry supervisor at the door should record the time the company went inside as well as the lowest SCBA cylinder pressure. The RIT will only be as good as the cylinder with the lowest pressure. The RIT and the control supervisor should have a set amount of time allowed for the team to be inside. Once this time is reached, the control/entry supervisor should alert the interior rescuers to make their way back out. A good baseline parameter to work off of is the application of the **10-10-10 rule.** This rule suggests ten minutes to get in and work, ten minutes to get out, and ten minutes remaining for a margin of safety. The ten-minute margin of safety can be manipulated depending upon conditions, depth into a structure, and the progress of rescue operations. It should also be pointed out that the ten-minute margin of safety will immediately be decreased if an individual member of the RIT has an air supply that is not near full.

The RIT should work from a search rope bag that can be shouldered while controlling the rope as it is paid out from the bag **(Figure 16-8).** The rope bag should never be attached to the SCBA of a rescuer. This will prevent the rope bag from being entangled or caught on something, trapping the rescuer. Another important reason for not attaching the bag to a rescuer is that if the need arises to leave the rope and the bag, the RIT will be able to anchor it quickly and follow the line back out. A good idea is to provide a quick-release-type strap such as a buckle-and-clasp, seat-belt-type connection to the rope bag so it can be disengaged quickly if necessary. This also limits the possibility of entanglement problems occurring with the team member holding the rope bag, in which case a quick-release method provides a safe method for freeing oneself. Another important aspect in providing this feature is that all rope-assisted searches being deployed as a main search line from a rope bag should be left in place and anchored preferably when the team exits the structure, if possible. Deployed search ropes should never be collected while traveling out of the structure.

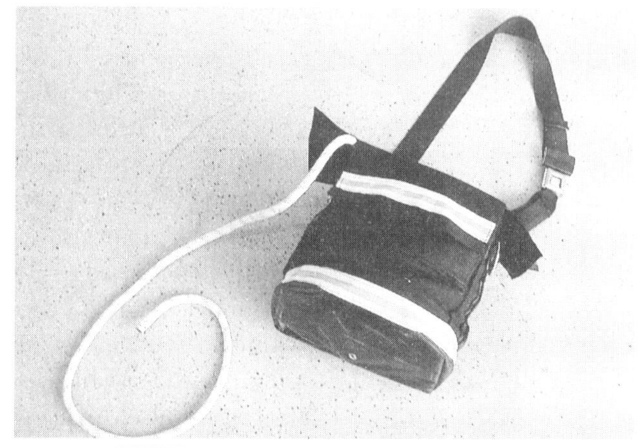

FIGURE 16-8

The RIT should work from a search rope bag that can be shouldered while controlling the rope as it is paid out from the bag.

FIGURE 16-9

The RIT members behind the team leader should remain several feet behind on the line to allow the team leader to stop occasionally and draw the line tight to ensure accountability and any need for a change of direction.

The RIT members behind the team leader should remain several feet behind on the line to allow the team leader to stop occasionally and draw the line tight to ensure accountability and any need for a change of direction **(Figure 16-9).** RITs should occasionally stop and listen for activated PASS devices, sounds of breathing apparatus, and striking noises that are given off by the victim.

When changing direction, the main line should be tied off and secured to prevent the rope from moving as the team travels through the building. This is important during egress so that the RIT is not going over areas that it did not originally cover. Hazards such as balconies or holes in the floor could cause problems if the line is not secured. If the rope is not anchored properly, the line can be pulled back or through the hazard area, putting rescuers in harm's way as well as changing the route to safety. Tying the line off and leaving the rope bag when the RIT reaches the point of no return will allow the next team to resume the search from where the previous team left off. A hand light can be left next to the bag to allow the second team to find its way quickly. When being deployed, search rope should be kept tight and several inches off the floor. This will make it easier to locate and follow back out in a swift manner if necessary.

As stated earlier, a main line could have a series of knots tied into it in order to identify how deep the RIT has traveled within a structure. One common application of these knots is to apply them in the rope at every 20 to 25 feet. Any tethers being utilized by individual members working off the main line may not exceed the length of the knot system that it is supplied into the rope. For example, if a butterfly knot is tied into the rope at every 15 feet, then all tethers, whether they are made of rope or webbing, shall not exceed 15 feet. The reason for this is simple. When a rescuer is searching off the main line using a tether, the rescuer will be searching in a semicircular pattern off to one side of the main line while another rescuer is searching on the other side of the main line, with the team leader stationed at the knot in the main line **(Figure 16-10).** This causes the rescuers on tethers to always end up at the next knot in the main line if they have attached themselves to the previous knot. The team leader may have to act as an anchor point on the main line in order for other members to search off it with their tethers **(Figure 16-11).** If the knot system is being used, the rescuers—after having extended themselves from the main line using tethers—shall return to the

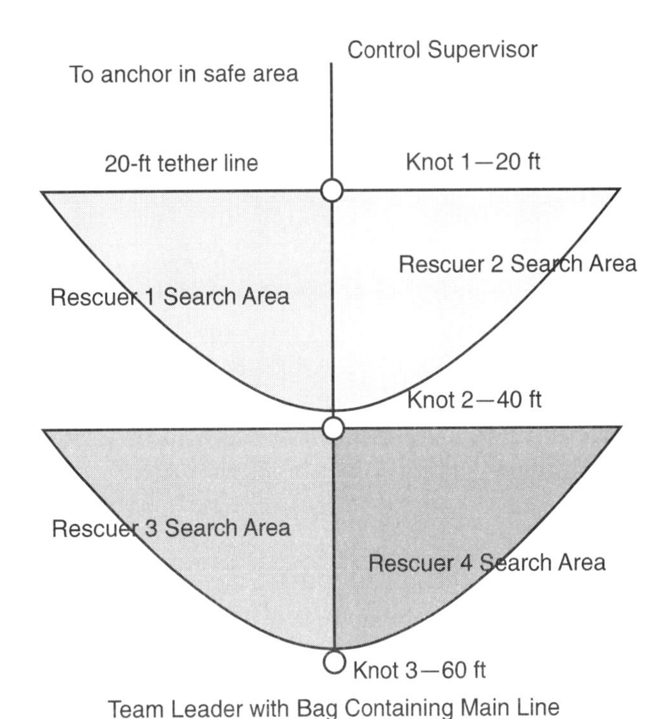

FIGURE 16-10

Semicircular main line search.

FIGURE 16-11

The team leader may have to act as an anchor point on the main line in order for other members to search off it with their tethers.

main line after completing a specific area and advance to the next knot. The advantage of using this type of system is that it can help determine how deep the RIT is into a structure and allows members to reference that information to the remaining air in their SCBAs to determine a point of no return.

It is possible for several RITs to be working at one time at different entry points into a structure. This can be extremely difficult for communication because it may require additional radio frequencies for the separate team searches being conducted. Communications and order by multiple control/entry supervisors will be the key in successful multiple searches happening at one time.

When the victim is found, the RIT should call for immediate assistance from an additional RIT while an air supply and injury assessment is made on the downed firefighter. If the search results in the use of multiple teams traveling up and down the main guideline, they should have a prearranged system to avoid confusion and congestion. It may be necessary to establish which side of the line the RIT members should be on when going in and out of the structure. A system such as choosing the right side of the main line to enter and the left side as an exit can help avoid overcrowding.

A system useful for high-rise apartment buildings and large-area offices as well as open areas is the use of a 150-foot length of rope with a series of knots tied into it at 50 feet and 100 feet. One knot is tied at 50 feet and two knots are tied at 100 feet. The RIT of four or five members will be placed on the main line rope, which is anchored outside of the hazardous environment. The officer or team leader will shoulder the rope bag and the TIC to lead and direct the search **(Figure 16-12)**. Behind the team leader will be two firefighters (position 2 and position 3) with forcible entry/exit tools and any other specialized tools that may be needed. Tool selection will be largely dependent on the situation and type of structure being dealt with. Behind these firefighters will be a firefighter with an additional SCBA or RIT emergency air supply unit (position 4), which could be utilized by the members of the RIT or the downed firefighter.

FIGURE 16-12

The officer or team leader will shoulder the rope bag and the TIC to lead and direct the search.

This setup allows for easy maneuverability down hallways and into office areas or apartments by allowing the team leader to deploy the main line rope along walls and scan into areas with the TIC.

When approaching an opening into an apartment or room, the team leader will allow the position 2 firefighter to observe the scan of the room **(Figure 16-13)**. The position 2 firefighter will enter and search the room or compartment. The rope does not enter the room with the position 2 firefighter. The position 4 firefighter is left at the doorway of that room to monitor and assist the position 2 firefighter. The team leader and position 3 firefighter proceed to the next room or apartment while deploying additional rope down the corridor. Upon reaching the

FIGURE 16-13

When approaching an opening into an apartment or room, the team leader should allow the position 2 firefighter to observe the scan of the room.

next doorway, the position 3 firefighter will observe the team leader's scan of the room and proceed to perform a search of that room while the leader monitors from the doorway.

Upon completion of searching the first room, the position 2 and position 4 firefighters proceed down the rope to the team leader. At this point, the position 2 firefighter will be left at the door and the leader and position 4 firefighter will continue deploying the rope. When at the next room, the position 4 firefighter will search the room after the leader scans with the TIC.

This process continues, with the position members taking turns on the search of rooms—excluding the team leader, who always stays to the front of the search with the camera. This will allow for maximization of the team's air supply and work efforts. The fifth rescuer (position 5) should be used as an entry supervisor at the point of original entry **(Figure 16-14)**.

The knots used in this particular rope system are basic and allow the rescuers to determine their depth into the structure. As with all rope-assisted searches, the team leader will be responsible for securing the rope anytime there is a change of direction in the search. If there is no object to secure the rope, it will become necessary for a member of the rescue team to be stationed at that point to secure the line. When the appropriate search and distance has been covered, the rope bag should be secured in its place by the team leader. The team leader will make certain that all members exit quickly down the line while a new team moves up to search further.

There are many other choices for the RIT in regard to the techniques that can be used to control the main line rope as well as the search. One simple method is to have the team leader move forward, holding onto the rope and using the other arm to sweep along the floor from one side to the other with a tool. Another technique that can be used is to position a rescuer on each side of the main line while moving directly behind the team leader, sweeping the floor on each side of the line while advancing. This particular technique will involve two or three members staying on the main line and moving very quickly through a space while maintaining communication between each member. These types of search will not be as thorough for larger areas but offer another alternative to search an area quickly. These techniques can be improved by the use of tethers, which will enable the RIT to extend outward to cover additional square footage. The more involved a search becomes, the more time it will require. It is possible that many different variations and techniques will be taking place at any given time during one rescue. Be inventive, but be accountable to yourself and your team at all times.

We have covered the basic methods of the main line search to protect against becoming separated or disoriented within a large area. Using a search line does not guarantee safety for the RIT. One of the disadvantages of a main line

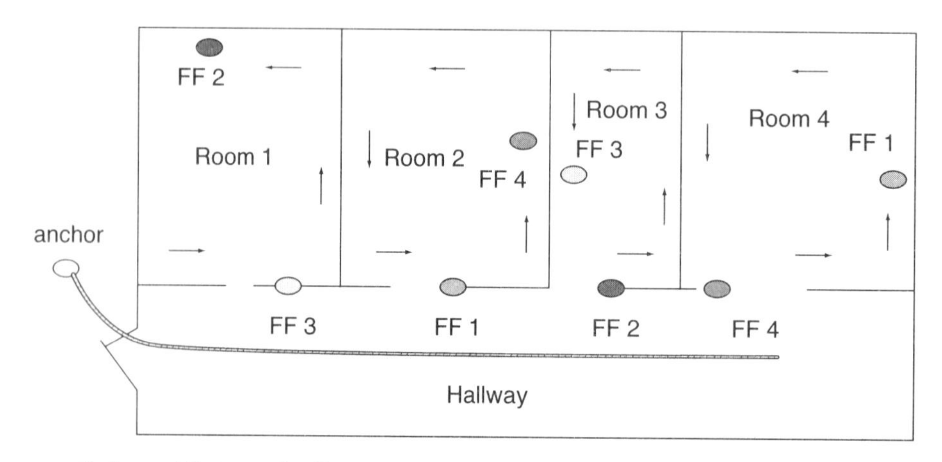

Order and Patterns for Rope-Assisted Search to Maximize Air Supply Working Time

FIGURE 16-14

Procedure for searching off a main line rope.

search is that the RIT is relying on the rope to find its way back out to the original entry point. Strong consideration should be given to the length and depth of any RIT's involvement by limiting the length of main search lines to 150 feet. The other possible disadvantage is that damage can occur to the main line such as burning or it being cut. The possibility of entanglement also exists if the rope is not deployed and managed properly **(Figure 16-15)**. Conventional methods and tools are still of importance and can help keep RIT members from becoming victims themselves.

Rope Systems

When main line rope is chosen, it should be able to be felt easily with a gloved hand and be easily differentiated from personal ropes or tethers. There are numerous rope bags available to store the rope in, but the main concern for the RIT is how the rope is deployed out of the bag. It is important that rope is able to deploy easily from the bag but is still constricted enough at its opening to not allow the rope to come out uncontrollably. The length of the rope recommended is not to exceed a 150-foot maximum. This is due to the time constraints of air supply. The RIT should not be allowed to search so far that it is not able to get back out. A large carabiner should be ap-

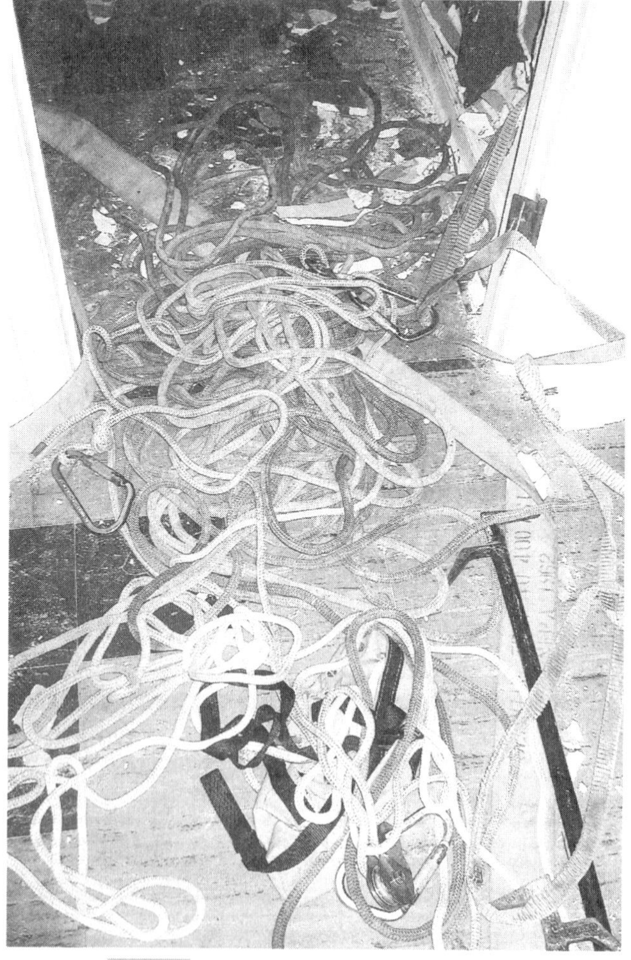

FIGURE 16-15

Rope must be managed properly once deployed to prevent a disaster such as this from occurring.

plied to the main guideline, placed at the starting point of the rope that is exiting the bag. This allows the rope to be applied to an exterior object in a safe area outside of where the RIT is entering. The opposite end of the main guideline should be secured to the interior bottom of the bag so that the rope cannot get separated from that bag.

Some departments employ the use of large carabiners when attaching tethers to main line rope-assisted searches. The use of large carabiners with tether lines for the purposes of attaching to the main line is recommended due to the ease of operation with a gloved hand as well as their ability to travel over a knot system if it is applied into the main line. The carabiner connection allows for quick connecting and disconnecting. The large carabiner is also able to pass over knots easily without becoming snagged when the rescuer is traveling up and down the main line **(Figure 16-16).** When these types of systems are utilized, a minimal amount of travel up or down the main search line is preferred. It is recommended that a team member or team leader hold the carabiner that is attached to the main line in order to prevent traveling while still allowing a 180 degree search to be conducted on either side of the main search rope.

The preferred type of large carabiner is the **captive carabiner (Figure 16-17).** This type has a rod across the inside of the carabiner below the gate, creating an enclosed area. The personal rope or tether is passed through this enclosed area with a knot affixed to it. By doing this, the rope is unable to be separated from the carabiner, avoiding the possibility of the rescuer becoming separated from his personal search rope or the main guideline.

A large carabiner can be applied on the main line or into a knot system on the guide line very quickly. When a firefighter travels up the main guide line into the building, he can arrive at a butterfly knot and apply the large carabiner into the knot and begin to search off of the main line in a fan or semicircle pattern. When this is the method being used, an attendant should be placed at the point of connection to ensure its security as well as prevent entanglement or wrap-around problems of the tethered line. The attendant or team member should also make certain that the disengagement of the large carabiner from the butterfly knot is established. After completing the search on that particular side of the main guide line, he can disconnect the large carabiner from the knot and apply it back onto the main search line and continue to travel

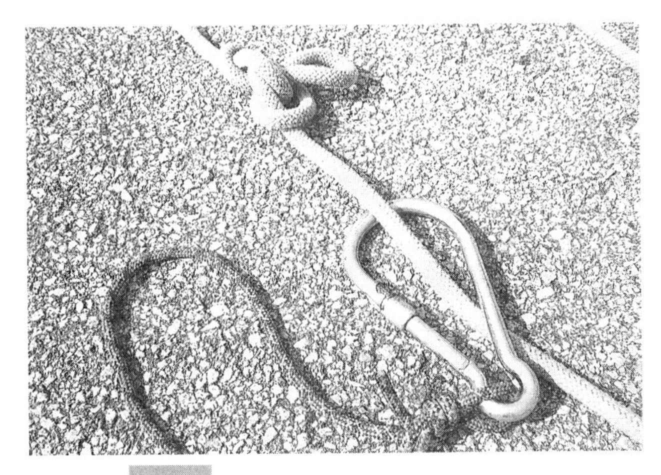

FIGURE 16-16

The large carabiner should be able to pass over knots easily without becoming snagged when the rescuer is traveling up and down the main line.

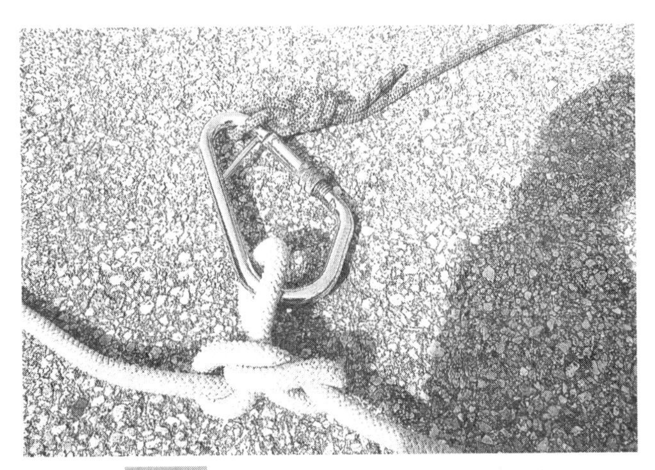

FIGURE  **16-17**

A captive carabiner.

up the line to the next knot and repeat the procedure.

If the rescuer chooses to search in between the knots, the team leader will hold the large carabiner to the main line while the individual rescuer searches the perimeter and then returns back to the team leader on the main guide line. They travel up the line, passing over the butterfly knot if they choose, and begin to search at a different point on the line, repeating the same procedure. These are just a few examples in regard to tethers and main line search ropes.

In order to store this system properly and be able to deploy it when needed it is important to keep the system organized. A rapid intervention large-area search kit should incorporate one large bag with the main search rope with its knots premade into the line at every 20 feet (or other distance determined). It should also have the necessary personal ropes or tethers contained with it. This includes the large carabiners already attached to the ends. The following is an example of a complete large-area search kit for RITs to conduct successful searches of an organized nature.

Large-Area Search Rope System

- 150 feet of 7/16- to 1/2-in. main line search rope
- Two 20-ft lengths of 8 mm personal ropes or tethers (webbing) with large carabiners attached to their ends and secured to pouch bags
- Two pouch bags
- Four large captive carabiners
- Door wedges
- One large rope bag with shoulder straps (seat-belt style to allow easy disconnect in case of entanglement) with an exterior equipment pocket for the personal rope
- Marking devices such as bright colored chalk to mark off searched areas
- One clipboard for documentation purposes for the control/entry supervisor
- Stop watches for tracking SCBA air time
- Hand lights
- Additional carabiners for change of directions

Oriented Searching in Structures: Hoseline or Rope

Searching with a hoseline can be just as effective as search rope and should be considered by the RIT when dealing with large-area searches. One advantage that a hoseline provides is the rigidity and stability of its presence. Search ropes can get pulled from side to side as rescuers are searching which may move the rope or cause it to change direction without the rescuer's knowledge if not controlled and kept tight. A hoseline will move very little when it is charged and lying on the ground. However, it must be remembered that a hoseline is difficult to move. It will also require manpower that may not be available.

The RIT should consider using an existing hoseline that is already present, especially if the lost firefighter was part of the crew operating it. Not only does it cut down on the time needed to find the area of the lost firefighter, it is also in a position to provide water protection for the RIT and victim. The team leader will maintain a position on the hose while the other rescuers will be searching off it. Rescuers should use a 15- or 20-foot piece of rope or webbing to apply a girth hitch to the hoseline while they search off the hose until they meet an obstruction or reach the end of their tether. At this point, the rescuer moves back to the hoseline. The rescuers can hand their tether to the team leader, who moves the anchor further up the line. This process is continued every few feet.

Upon entering the structure, the RIT members should move rapidly up the hoseline to the nozzle if they do not have a specific location of the downed firefighter. This allows them to fan the area around the nozzle by extending their reach through the use of their tethers. They will then move back down the hoseline and perform their searches off the line until they are back to their entry point. When using the hoseline method, it is best to have a rescue team consisting of five members. This allows for one member, the team leader, to stay on the line while the other members search by twos on each side off the line. If more RITs are available, they should be sent up the hoseline to approximately the halfway point on the line and should begin to work backwards. This will enable time to be saved in regard to the overall search. The same principles can be applied to a main rope search line to accomplish these procedures. When using a hoseline for the purposes of large-area search, there obviously will be no knots available, just the presence of the hose and its couplings at every 50 or 100 feet (depending on the hose lengths used).

Air Supply Limitations

Fire departments should develop SOPs to account for tracking the RIT members and their air supply. A good rule of thumb for the purpose of large-area search is to assign a specific time frame within which they will be allowed to work. Starting with a thirty-minute bottle will allow a probable ten minutes of operating time while allowing ten minutes of egress time, which would also allow a ten-minute safety factor. Time frames will have to be adjusted depending on the average air consumption rates of the firefighters.

All rescuers that enter the structure should report to the control/entry supervisor and have their SCBA cylinder pressure recorded. The team will be limited by the RIT's lowest cylinder pressure. The following should be kept in mind when considering the RIT's air supply:

- All RIT members involved in large-area searches are to be aware of their air levels and their abilities.
- All members of the RIT preparing to search in large areas for a downed firefighter will base their activities on the team member with the lowest cylinder pressure.
- All members of the RIT should exit a structure before their low-air alarms are activated. This will guarantee the best possible chance for egress.
- Proper monitoring by the control/entry supervisor of the team's air supply is crucial for safety during search and rescue efforts.
- All members of the RIT that lose communication with the control/entry supervisor shall exit the structure with a minimum of 1000 psi.

Components of the Search Rope System

Rope systems help the interior RIT maintain direction, depth of search, and location and provide means for immediate egress. The following list and brief descriptions should be apparent in any rope search system.

Depth—The system used must provide a means of knowing how far into a structure the team is located. Knot systems can be put into the main line rope every 10, 15, or 20 feet. Knots can be applied into the rope at these intervals or additional knots can be tied to relate to the amount paid out of the rope bag into the building. An example is one knot equals 10 feet, two knots equals 20 feet, and so on.

Orientation—A point of reference for rescuers to establish direction on the main line must be established, such as utilizing a knot indicating the depth while adding an additional spacer knot or ring app into the line to help establish which way is in or out. It is advisable to keep this application as simple as possible. These types of techniques are usually used to aid the rescuer's ability to feel direction by touching the knots and identifying their placement in order to know which way leads in and which way leads out of the structure.

Search Patterns—Patterns should be standardized using the 12 and 6 o'clock positions when searching off of a main line. Patterns should provide the ability for two rescuers to conduct a full 360-degree search. Tethers should never exceed the length of the knot system being used. For example, if a 15-foot tether is being used by a rescuer, the main line rope should also have its knots placed into the rope at 15-foot intervals.

Progress Reports—Searching team members should identify items and points of reference in their respective areas and provide information to team leaders and Command on a continuous basis. These reports should include conditions, actions, and needs.

Control Member/Buddy System—Rescuers performing searches on main line ropes with tethers or tag lines should utilize search patterns that involve one rescuer searching off one side while their partner searches off the other. A team leader is to provide tension on the line to prevent it from drifting, which can be caused by the tethered rescuers pulling the main line to one side or another.

Entry and Exit Accountability—RITs operating and deploying search lines should have communications with sector officers and IC at all times. A ten- or 15-minute rule should be applied to the interior rescuer's air supply in order to allow for egress at any time and up to that time. All members will exit if for any reason they cannot determine the status of their air supply at any given time in order to avoid unknowingly running out of air.

Air Supply/PAR Reports—These reports go hand in hand when asking for air supply information from the interior rescuers. It is also a good idea to ask for PARs through control sectors and IC at the same time air supply information is requested or reported. The team is only as good as the lowest air supply of any member of the RIT entering the structure.

Location—Interior rescuers should always provide their location and depth within the structure in relationship to the rope and knot system employed. If a multiple knot system is being used it is important to not only relay the number of knots traveled, but to also know the number of feet involved.

Searched and Unsearched Areas—Control/entry supervisors and IC, through communications of interior RITs, should be able to establish a general picture of the search operation through drawing. Sketches should be provided to other RITs about to enter the structure. This helps establish a tactical plan to complete a better search.

Summary

Large-area searches present both challenges and dangers to the RIT. Basic search skills as well as techniques specific to large areas must be understood by all RIT members to ensure personal safety and increase the possibility of enacting a successful rescue. Basic search skills include assessing the fireground upon arrival and assigning specific responsibilities to each RIT member. Proper tools for large-area searches should be priorly assembled and available, such as two-way radios, hand lights, and TICs.

There are several types of searches—such as the standard search, the simple oriented search, and the main line rope-assisted team search—that should be understood by the RIT. Each involves specific procedures and equipment that all RIT members should be familiar with.

Large-area searches using rope systems are common, and RIT members must be proficient in the different methods of utilizing the main search line, especially those employed by their particular department.

Rope-assisted searches usually require five to seven members, each of whom plays a specific role: team leader, control/entry supervisor, or rescuer. As in all fireground situations, clear communication between all members is essential for both safety and a successful outcome.

KEY TERMS

10-10-10 rule
Captive carabiner
Control/entry supervisor
Main line rope-assisted team search
Radio-assisted feedback
Simple oriented search

REVIEW QUESTIONS

1. A solid foundation in the basic skills of search techniques and procedures involves being
   a. well planned and flexible.
   b. disciplined and controlled.
   c. accurate and accountable.
   d. all of the above
2. How many team members, preferably, are needed to make up a RIT involved in large-area searches?
   a. two to three
   b. five to six
   c. one to two
   d. none of the above
3. Before any search can begin, especially those being conducted in large areas, the RIT along with the IC should address
   a. opening the structure, eliminating forcible entry problems, and opening blocked exits.
   b. deployment of additional hoselines to protect search teams.
   c. proper ventilation, which can help increase visibility.
   d. all of the above
4. Large-area searches are a race against time and are limited by the RIT's air supply.
   True False
5. Standard searches usually involve the use of walls for orientation.
   True False
6. Simple oriented searches utilize
   a. a powerful light at the entry point for reference that can be seen under smoke conditions.
   b. The use of walls and crossing over to the center of the room and proceeding toward the hand light at the entry point.
   c. looking into the rooms and deciding if they are to be searched or passed by.
   d. both (a) and (b)
7. A main line rope-assisted search involves a rope that is anchored to a stationary point just inside the hazardous environment and the RIT moving into the building, laying out the line as they search.
   True False
8. When utilizing short ropes or tethers for the purposes of searching by extension from a main line search rope, an appropriate length for these tethers would be
   a. 15 to 20 feet.
   b. 20 to 30 feet.
   c. 30 to 40 feet.
   d. none of the above
9. A good baseline for a control entry supervisor and a team leader regarding air supply and when to return to the point of entry is to
   a. listen for the low-level alarm device on any given member of the team.
   b. apply the 10-10-10 rule: ten minutes to get in, ten minutes to get out, and ten minutes for safety.

c. observe the cylinder gauge and begin to exit when it is at 500 psi or the low-level alerting device activates.
d. call for additional air supplies to be brought to the members of the team at the halfway point of the search.

10. When a downed firefighter is found by the RIT, members should
a. call for immediate assistance and notify the RIT operations chief.
b. assess the downed firefighter's air supply and injuries.
c. begin resuscitative efforts immediately.
d. both (a) and (b)

11. When approaching an opening such as a doorway into apartment or room, the team leader should allow the team member searching to observe the viewfinder of the TIC and scan the room for orientation purposes along with specific points of interest that may need to be investigated.
True False

12. Knot systems utilized in main line search ropes help identify what major feature of orientation?
a. room size
b. the width of an area
c. the depth of penetration into a structure or area
d. time into the structure

ADDITIONAL RESOURCES

Firefighter's Handbook, 2nd Ed. Clifton Park, NY: Thomson Delmar Learning, 2004.

Hoff, Robert, and Kolomay, R., *Firefighter Rescue and Survival.* Tulsa, OK: PennWell Publishing, 2003.

Norman, J., *Fire Officers Handbook of Tactics,* 2nd Ed. Saddlebrook, NJ: PennWell Publishing, 1998.

CHAPTER

17 Rapid Intervention and the Collapse Environment

Learning Objectives

Upon completion of this chapter, you should be able to:

- identify key factors to minimize the hazards to firefighters regarding fireground collapse.
- describe the differences between balloon-frame, platform-frame, and lightweight truss construction.
- identify the characteristics of steel that make it vulnerable to collapse.
- discuss the various indicators of structural collapse.
- explain the three ways that a wall could collapse.
- recognize the parameters necessary to establish a collapse zone.
- describe the priorities of the RIT once a collapse occurs.
- explain the five stages of a fireground collapse incident and the RIT's role in each.
- describe four ways of mitigating hazards at a collapse incident and the tools and equipment that might be utilized.
- clarify the five ways that a building collapse will occur and the characteristics of each.
- identify the various symptoms that may be associated with post-traumatic stress disorder.

CASE STUDY

"On January 10, 1999, three firefighters became trapped when the second floor of a nightclub collapsed during an interior fire attack. One male firefighter (the victim) died and two other firefighters were injured as they battled the late-morning blaze. Arriving on the scene of a two-story taxpayer building (commercial occupancy on the first floor and living quarters on the second), firefighters reported heavy smoke emitting from the second-floor windows and eaves with fire showing in a secondary front doorway that led to the second floor. The fire quickly spread up the walls of the first floor to an area above the false ceiling over the first floor. The fire also spread up the walls to the attic area above the second floor.

"As firefighters prepared entry through the main front door, other firefighters began applying water to the fire that was emitting through a secondary front door that led to the second floor. Gaining entry through the main front door, two firefighters from Engines 2550 (a captain and a firefighter) and three firefighters from Engine 2552 (lieutenant [injured], engineer, and a firefighter [victim]) advanced two 1 1/2-inch charged lines and began applying water to the fire. Upon entering the smoke-filled structure, they noticed that the ceiling (drywall drop ceiling) was down in some areas, and they could see fire going up the walls and across the ceiling. As firefighters from Engine 2552 advanced their line, the engineer was struck in the head by falling debris that knocked off his helmet and he was forced to exit. Minutes later, a third firefighter from Engine 2550 joined his crew inside the structure. A firefighter (injured) from Engine 2552 also entered at the same time, relieving the victim on his line. The captain of Engine 2550 stated that as he continuously surveyed the interior conditions, it had appeared to him there had been a partial roof collapse (referring to the second floor as the roof). He then exited the structure and went to the Command post to give the incident commander (IC) a report on the interior conditions, explaining that he felt there had been partial roof collapse (second floor). Before the IC could make any changes, the captain returned to the interior of the structure to find his crew. Just as he located his crew, the second floor collapsed, trapping three firefighters from Engine 2552. The captain of Engine 2550 and his crew escaped without injury. The rescue team quickly freed one firefighter and had to use hydraulic jacks, air bags, and cribbing to free the lieutenant and the victim. All three firefighters were transported to a local hospital where the victim was pronounced dead."

NIOSH investigators concluded that, to minimize the risk of similar occurrences, fire departments should:

- *"use extreme caution and recognize potential hazards that could exist when fighting a fire in a balloon-framed structure.*
- *implement an emergency notification system to rapidly warn all persons who might be in danger if an imminent hazard is identified or if a change in strategy is made.*
- *ensure that firefighters wear protective clothing whenever they are exposed or potentially exposed to hazards.*
- *ensure that a separate incident safety officer, independent from the incident commander, is appointed.*
- *ensure that when firefighters are performing an interior attack with the possibility of a ceiling collapse, they should establish a collapse shelter.*
- *provide the incident commander with a Command aide.*
- *ensure that once a rapid intervention team (RIT) is established they remain the RIT throughout the operation.*
- *develop and implement a preventative maintenance program to ensure that all SCBA's are adequately maintained."*

"Additionally, building owners should:

- *ensure that all modifications/renovations to buildings are in compliance with current building codes (i.e., any renovation or remodeling does not decrease the structural integrity of supporting members)."*

CASE STUDY (CONTINUED)

—Case Study taken from NIOSH Firefighter Fatality Report # 1999-03, **"Floor Collapse Claims the Life of One FireFighter and Injures Two—California,"** *full report available on-line at http://www.cdc.gov/niosh/face9903.html.*

Introduction

Gravitational forces are always at work to bring a building down. When fire takes possession of a portion of a building, changes begin to take place with the materials and components that provide the structure with its ability to stand. Most times, these changes are able to be detected before collapse occurs; other times these changes cannot be readily detected. The collapse of the World Trade Center is an example of changes that were not readily detectable **(Figure 17-1)**. Over the years there have been countless numbers of tragedies of firefighters killed or seriously injured in the line of duty with regard to collapsed structures. Appropriate preplanning, a good knowledge of building construction, and proper application of risk management principles are key factors when trying to minimize the hazards presented to firefighters.

FIGURE  17-1

The collapse of the World Trade Center was unpredictable and well beneath the time limits of flame exposure that engineers stated it could withstand. The FDNY "Bravest" lost that day will never be forgotten.

The Collapse Environment and the Rapid Intervention Team

The RIT's relationship to the collapse environment is a relatively new area of understanding and purpose in rapid intervention. One of the reasons for this is the vague understanding of what the responsibilities are of a RIT that is on the scene of a collapse event. What should be the expertise of these immediate rescuers and what are they expected to do for the downed firefighter in the collapse scenario?

In order to come to grips with the responsibilities of RITs in the rescuing of downed firefighters involved in collapse, the RIT will need to understand its limitations and abilities when functioning in this type of environment. The most important realization that should be well understood by the RIT and IC is that the collapse environment involving the trapped firefighter and its relationship to the RIT's response is not equivalent to the expertise of a technical rescue team, nor should it ever be considered as such. This chapter's purpose is not to provide in-depth coverage of structural collapse at the technical responder level. The operating procedures that are established for the RIT in the collapse environment should be specific and provide for the immediate response of a technical rescue team at any collapse situation.

Safety

The RIT's response is not equivalent to the expertise of a technical rescue team when dealing with building collapse.

This chapter deals with the immediate response of RIT companies and their involvement in prompt search and rescue efforts when retrieving downed firefighters buried underneath debris. It is also understood that the responsibility, as stated earlier, for the IC and the RIT company will be to immediately call for a technical rescue team to all collapse environments. The initial actions taken by a RIT can provide a definite impact on the outcome of a collapse situation.

Collapse Indicators

The RIT has a vital responsibility in the area of reconnaissance of the fire building and fireground. RIT members are the additional eyes and ears of the safety officer and IC. Through good reconnaissance of the fire building, the RIT can better protect the firefighters and predict the possible collapse of a building before it occurs. The RIT has a responsibility for reporting any collapse indicators to the IC and safety officer.

The one thing that should be realized by everyone is that there is almost always an indicator that a collapse is about to occur. Unfortunately, these indicators are sometimes not seen or noticed until it is too late. The RIT company should be well aware of and be able to communicate the indications that a collapse may occur.

During the RIT's size-up of the fireground, members should be aware of the type of construction of the building. By understanding the building's construction features, they can determine how fire may react with the structural components of that building.

Wood-frame construction is very common and is used for all types of occupancies. Thus, the RIT should be well versed in its characteristics. The most common forms of wood-frame construction are **balloon frame, platform,** and lightweight wood truss.

Balloon-frame construction is likely to be encountered with wood frame buildings that were constructed prior to 1950 **(Figure 17-2).** The exterior walls are built with vertical 2 × 4 studs that are continuous for the height of the building **(Figure 17-3).** These walls would be built on the ground and then raised into place. Once upright, all four walls would be joined together. Because the studs that frame these walls are continuous for the total height, channels are formed from the bottom of the wall to the top, with no horizontal members for firestopping. Without the benefit of any firestopping, it is very probable that a cellar or basement fire in this type of construction will extend up into the living areas and attic space. Fire burning inside balloon-frame walls can be very detrimental to the stability of the structure. Vertical ventila-

FIGURE 17-2 Balloon-frame construction is likely to be encountered with wood-frame buildings that were constructed prior to 1950.

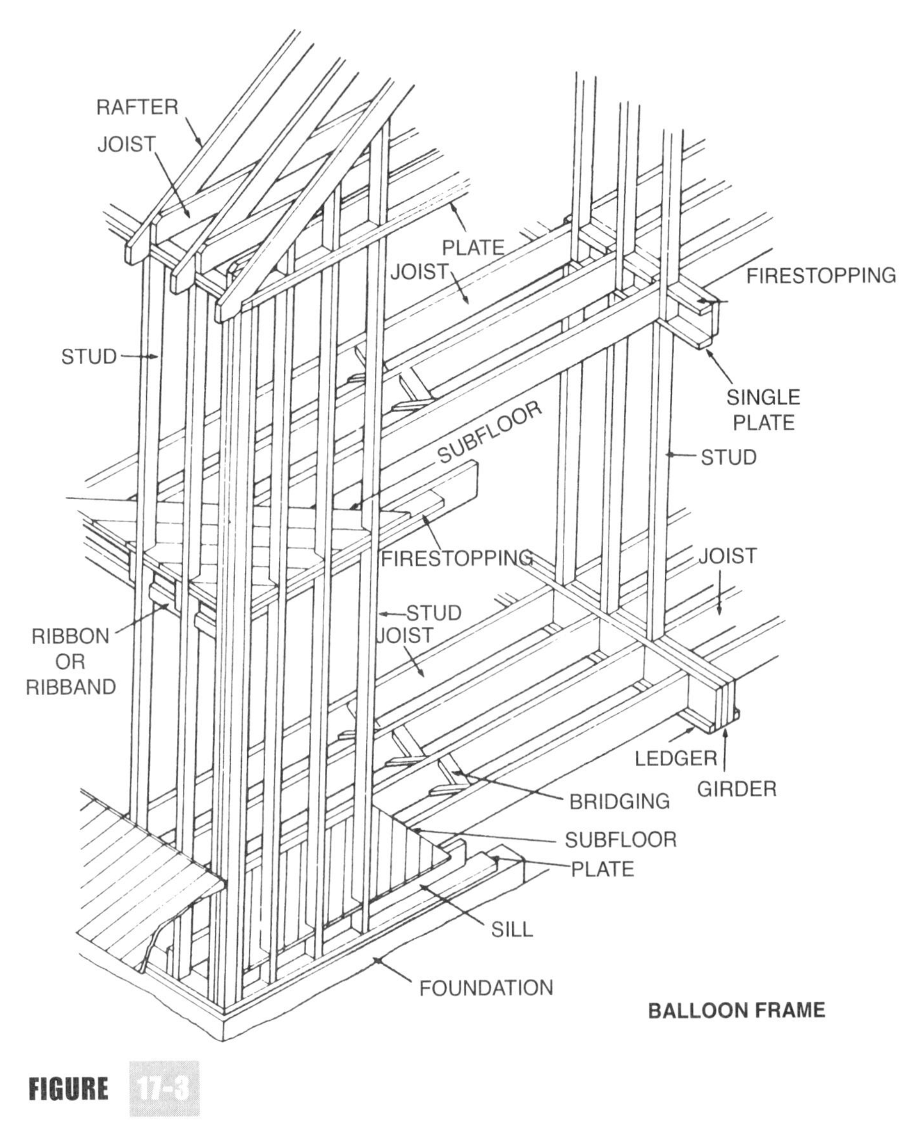

FIGURE 17-3

The exterior walls of balloon-frame construction are built with vertical 2 × 4 studs that are continuous for the height of the building.

tion, aggressive fire attack, and overhaul will be paramount to controlling fire in balloon-frame construction.

Platform-frame construction consists of the floor (floor joists and subflooring) being built as a platform for the floor above. The walls are constructed on the platform, which in turn the joists and subflooring for the floor above are placed on top of. The end result is that the wall studs are only the length of one story. The platform of the floor above provides a horizontal break in the channel from the ground to the attic, which helps to control fire extension **(Figure 17-4)**.

Lightweight wood truss construction is being used more and more with new construction. Lightweight truss construction is very attractive to builders because it offers cost savings, easier access to run utilities and ventilation, and can support a weight load equivalent to a solid structural member under normal conditions **(Figure 17-5)**. Buildings that contain lightweight truss construction are susceptible to collapse from fire exposure in a very short amount of time **(Figure 17-6)**.

Lightweight truss construction consists of top and bottom members that run parallel. These

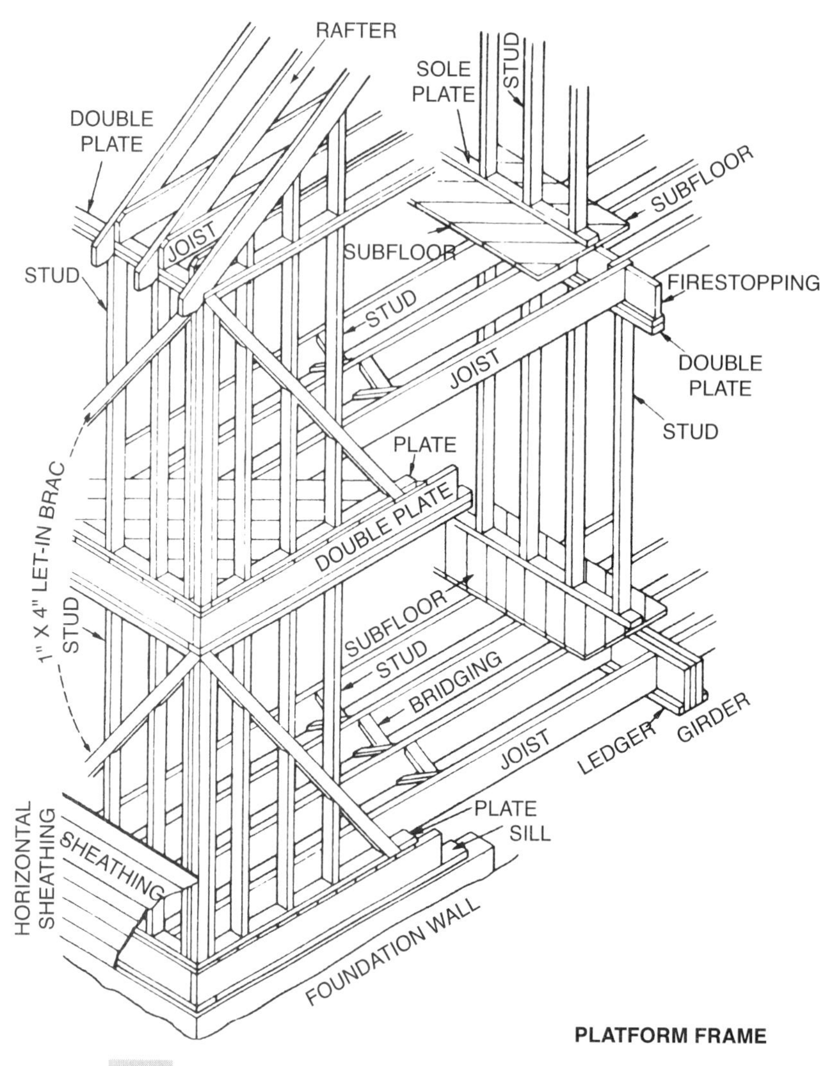

FIGURE 

Platform frame construction.

FIGURE 

Lightweight truss construction is very attractive to builders because it offers cost savings, easier access to run utilities and ventilation, and can support a weight load equivalent to a solid structural member under normal conditions.

FIGURE 17-6

Buildings that contain lightweight truss construction are susceptible to collapse from fire exposure in a very short amount of time. Most fast-food chain restaurants are lightweight construction.

are referred to as *chords* and are made of wood. These chords are cross-connected for support by wood that forms a web-like pattern. All wood usually consists of 2 × 4s or 2 × 3s. The wood members are connected together with a fastener made of stamped sheet metal containing spikes (gusset plates).

Unlike conventional construction, lightweight wood truss construction does not obtain its strength from the size of the materials used but rather from compression and tension of the materials used in its construction. The top chord is supported by load-bearing walls. It acts as a bridge between these walls. With this being under a load, the top chord is being placed under compression while the bottom unsupported chord provides tension.

Conventional construction techniques do not rely on a sum of the total members for structural stability, whereas lightweight truss construction does. Because of the bottom chord providing tension, a failure of any one connection point (gusset plate) will cause the load of that truss to be transferred to another, which may already be weakened, thus causing a collapse of multiple trusses.

Another hazard of lightweight trusses exists with the connection points of the wood members themselves. These are often referred to as *gusset plates* or *gang nails*. They are usually made of 18-gauge stamped steel and penetrate the wood only 1/2 to 3/8 of an inch **(Figure 17-7).**

A fire that is burning under these trusses and in between floors directly affects the building's structural integrity.

The **lightweight wooden I beam** is common throughout newer residential construction and commercial spaces **(Figure 17-8).** The center "beam" portion is made up of chips and wood particles held in place with glue and is noted for its quick failure when attacked by heat or direct flame. The RIT will have to make sure that its reconnaissance of the structure and the progress of the fire has not affected these construction features.

FIGURE 17-7

Gusset plates are usually made of 18-gauge stamped steel and penetrate the wood only 1/2 to 3/8 of an inch.

FIGURE 17-8

Lightweight wooden I beam.

FIGURE 17-9

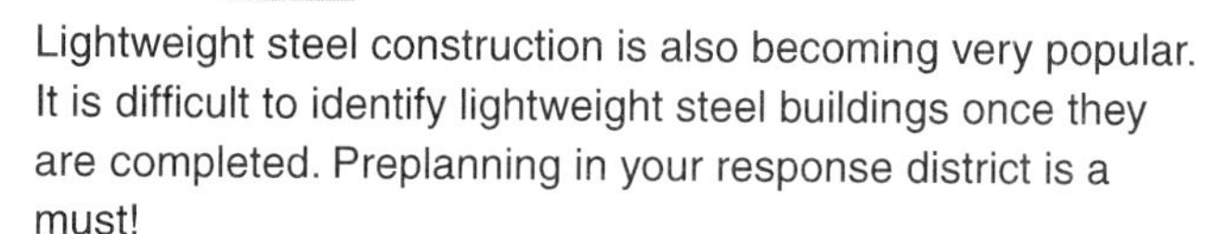

Lightweight steel construction is also becoming very popular. It is difficult to identify lightweight steel buildings once they are completed. Preplanning in your response district is a must!

FIGURE  **17-10**

Steel bar joist roof assemblies are often unprotected, exposing them to high-heat conditions.

Lightweight steel construction is also becoming very popular **(Figure 17-9).** Steel building components can weigh 60 percent less than wood. Its strength, which comes from its shape, results in less structural members being required. Buildings that are constructed of steel components that are not protected can expand or twist at temperatures as low as 1200°F. Steel bar joist roof assemblies are often unprotected **(Figure 17-10).** It is important to get large amounts of water on exposed steel to cool it down and prevent it from further distortion. Steel beams expanding and pushing walls outward, causing walls to bulge in appearance, can predict possible collapse.

Wood and steel are also being used together to form structural components **(Figure 17-11).** Anytime steel members are mixed with wood construction features, the steel will act as a conducting medium, moving the heat directly into the wood material. This will cause connection points to loosen and fall away, allowing the truss to collapse. The RIT should recognize this and communicate the possibility of truss failure to the safety officer and IC.

Commercial buildings often have exposed structural members over large, open areas. Many void spaces may also be present **(Figure 17-12).** The construction features of these buildings allow fire to quickly spread from area to area and floor to floor. High-heat conditions and pyrolysis to structural members will inevitably lead to structural failure.

FIGURE 17-11

Anytime steel members are mixed with wood construction features, the steel will act as a conducting medium, moving the heat directly into the wood material.

FIGURE 17-12

Many void spaces may be present in commercial structures. The construction features of these buildings allow fire to quickly spread from area to area and floor to floor. High-heat conditions and pyrolysis to structural members will inevitably lead to structural failure.

Fire conditions must be monitored very closely by the RIT. The presence of a continued fire-suppression attack involving heavy fire with conditions that do not improve is a possible indicator that collapse may occur. Fire located on two or more floors of a building indicates that extremely high heat conditions exist and that the impingement of these conditions on the structural components of the building will most definitely have an affect on its ability to stand.

The RIT should also monitor ventilation efforts. Smoke that is showing through walls defines the fire as being under pressure as it pushes through openings available either through existing construction features or openings being created by the fire itself. Heavy smoke under pressure could indicate the possibility of **backdraft** (or smoke explosion), which can lead to collapse of the structure or portions of it.

Any visible fire that is showing through walls or sudden exposed cracks in exterior walls is a reliable indication the building has been seriously compromised **(Figure 17-13).** This also allows oxygen to enter the building, intensifying the fire and allowing it to burn using additional materials of the structure. The

FIGURE 17-13

Sudden exposed cracks in exterior walls are a reliable indication that the building has been seriously compromised.

RIT should also notice exterior walls that were once unaffected by the fire initially and then began showing signs of new cracks or bulging, which may be the result of interior forces pushing against the exterior walls of the building. Cracks appearing on the exterior walls of buildings involved in fire indicate movement of some type that will lead to collapse. Any increase in the size of the crack is also an indicator of possible collapse.

The RIT should take notice of any bulges or leans in walls. Anytime bulging or leaning is noticed it is a good possibility that structural members on the interior of the structure are moving or distorted, causing force to be applied to the exterior walls of the building.

The RIT should establish an awareness of the added loads to a building. This includes the addition of water streams pouring into the structure from fire-suppression activities **(Figure 17-14).** Water and building loads drastically affect the ability of the structure to fall or stand. Keep in mind that water weighs approximately 8.64 pounds per gallon and several thousand gallons may be flowing from multiple master stream devices if being used. Heavy fire streams can break apart mortar joints, causing the wall to become weaker in regard to its load-bearing capacity. **Dead loads** such as HVAC units and communications equipment should be recognized and acknowledged **(Figure 17-15).** The largest factor that may presumably cause building collapse is the addition of firefighters, their equipment, and the excess of water from fire streams **(Figure 17-16).**

FIGURE 17-15

Dead loads such as HVAC units and communications equipment should be recognized and acknowledged by the RIT.

FIGURE 17-14

Water from master stream devices will drastically affect the ability of the building to fall or stand.

FIGURE 17-16

The largest factor that may presumably cause building collapse is the addition of excessive numbers of firefighters in certain areas.

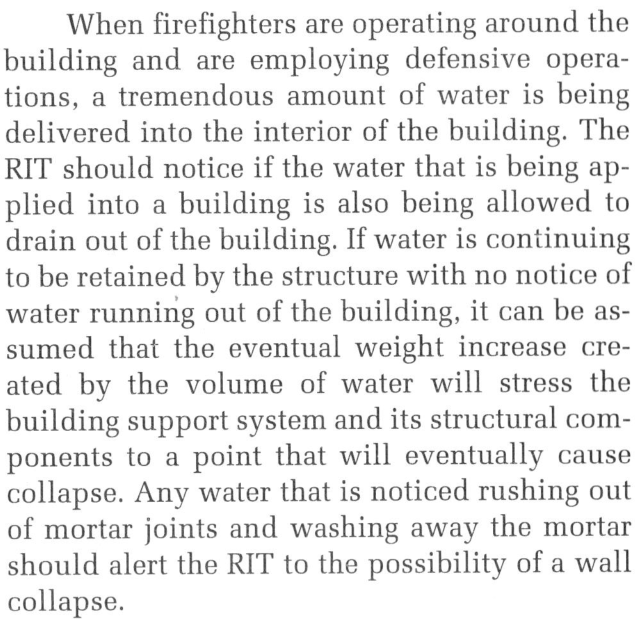

When firefighters are operating around the building and are employing defensive operations, a tremendous amount of water is being delivered into the interior of the building. The RIT should notice if the water that is being applied into a building is also being allowed to drain out of the building. If water is continuing to be retained by the structure with no notice of water running out of the building, it can be assumed that the eventual weight increase created by the volume of water will stress the building support system and its structural components to a point that will eventually cause collapse. Any water that is noticed rushing out of mortar joints and washing away the mortar should alert the RIT to the possibility of a wall collapse.

There are many more indicators that could possibly provide the RIT with the information of a possible collapse before it happens on the fireground. Previous history of fire or structural damage to a building, interior collapses, abnormal noise, excessive loads on the interior or exterior, pitched or sagging floors, floors burned through, and collapsed stairs or floors can all be measurable signs of the possibility of collapse **(Figures 17-17, 17-18, 17-19)**.

FIGURE 17-17

Previous history of structural damage can be an indication for an increased collapse potential.

A

B

FIGURE 17-18

The deadly bowstring truss roof of this building may not be detectable from the ground level due to its being hidden by a parapet wall. Trusses are vulnerable to collapse under fire conditions due to heat buildup and open void spaces.

FIGURE 17-19

The RIT must be cognizant of all surrounding features when evaluating the fireground for collapse potential. Features such as overhead power lines, unsupported canopies, window A/C units, and unprotected utilities, as seen on this particular photo, need to be recognized.

FIGURE 17-20

All collapse types will project debris beyond the height of the building when they occur. No set distance away is guaranteed to be safe.

Establishing a Collapse Zone

A wall will collapse in one of three different ways. One is the **90-degree angle collapse,** which involves the entire height of the wall falling outward.

The second type is the **inward/outward collapse,** in which the wall partially falls inward at the top as the bottom falls outward. This usually occurs under excessive force or pressure provided by an explosion such as a backdraft.

The third type is the **curtain wall collapse,** which is the end product of a wall crumbling or collapsing upon itself straight downward.

All of the collapse types will project debris beyond the height of the building when they occur **(Figure 17-20).** There is no absolute safe zone from projected debris. As a minimum, it is recommended that a distance of at least one and one-half times the building's height be designated as a collapse zone **(Figure 17-21).** For further protection, aerial devices being utilized for master streams should be placed outside of the collapse zone at the corners of the building, as these are considered to be the most stable parts of the structure. If defensive operations are taking place, there is no reason that master stream devices need to be attended inside of the collapse zone—set them up and leave them. A firefighter's life is not worth a building that is going to be destroyed.

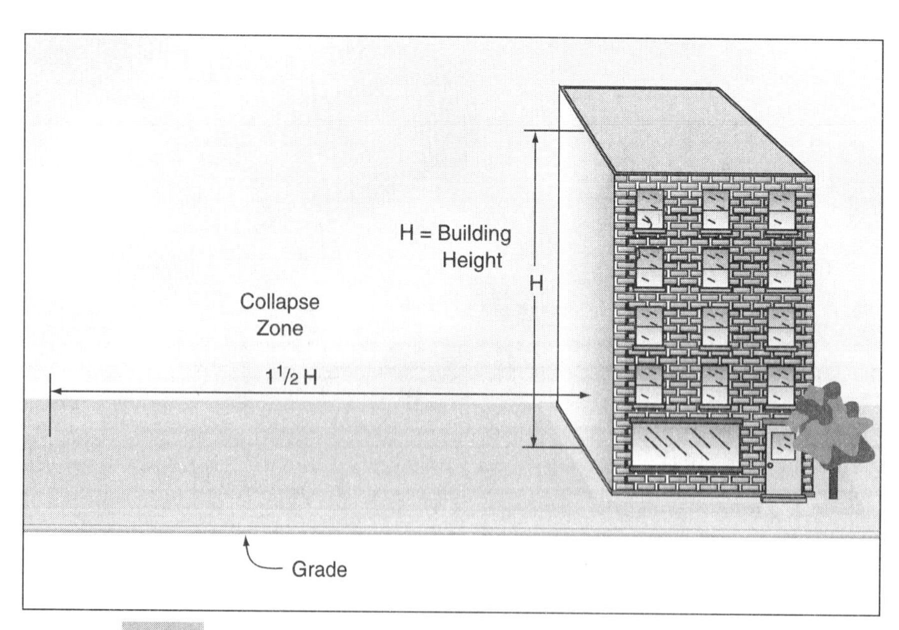

FIGURE 17-21

As a minimum, it is recommended that a distance of at least one and one-half times the building's height be designated as a collapse zone.

The Operational Plan and Rapid Intervention Teams

When a collapse situation is realized, **emergency radio traffic**—which signifies that a priority message is to follow—must be utilized over the fireground along with a signal to evacuate the area. The universal evacuation signal is five short blasts of an air horn on the fire apparatus. Once evacuation takes place, a PAR should take place on the fireground to determine if everyone is accounted for.

The first priority for the RIT after a collapse occurs is to acquire information quickly and assess the extent of the problem. With collapse, a technical rescue team should be called immediately in order to have the best chance possible of extricating trapped firefighters. While help is called and the trapped firefighter is located, efforts need to continue to suppress any fire that is working toward void spaces. Be cognizant that excess water could possibly fill a void space and drown a trapped firefighter, just as a broken pipe that is not controlled. If possible, get the trapped firefighter an additional air source and provide additional lighting (if safe).

The operational plan for RITs should initially be the same as it is for technical rescue companies that are beginning to look for victims or downed firefighters in the immediate collapse environment. Strict operational guidelines should be adhered to in order to set the right operational plan in action. Collapse incidents will take place in five stages:

1. **Reconnaissance and rescue of surface victims**
2. **Void search**
3. **Selected debris removal**
4. **General debris removal**
5. **Debriefing**

The RIT will more than likely be involved, to a degree, through the void search stage and possibly into selected debris removal. The RIT will inevitably be working with the technical rescue team if the downed firefighter cannot be reached within the immediate early stages of the collapse.

The initial sizeup by the RIT will set the tone for the entire operation. The RIT is mainly responsible for an overall safety assessment of the situation and the finite location of the downed firefighter if at all possible. It is the time from the moment of collapse through the minutes that go by until a technical rescue team can be assembled or brought in that is of crucial importance. In some fire departments, especially those in rural areas or those with inadequate mutual aid support, it can be a great deal of time before an adequate technical rescue effort is begun. It is essential for the RIT to have a basic knowledge of collapse and the ability to start searching immediate voids that a downed firefighter may be in. The RIT members must be knowledgeable enough to not cause an additional collapse or injure themselves in their rescue efforts.

Every action taken near a collapse building will have a reaction.

Reconnaissance and Rescue of Surface Victims

The RIT's initial size-up should include the survey of the entire structure, which includes the gathering of information of who the lost firefighters are and their last known locations. The RIT will need to take a six-sided approach to surveying the structure. The six-sided approach consists of the front, rear, sides, top, and bottom. An aerial device set up outside of the collapse zone can provide a valuable location from which to size up the collapse area. Once acknowledged, hazards should be handled in one of four ways:

1. by avoiding the hazard area.
2. by removing the hazard.
3. by shoring or supporting the hazard.
4. by monitoring the hazard for deterioration over time.

An important observation to be made by the RIT is the status of utilities. With collapse, live wires and broken pipes can be distributed amongst the debris pile. Utilities may not be able to be controlled in the immediate vicinity of the building due to the collapse. This may require utilities to be controlled at other points remote from the incident site. Monitoring the collapse area for hazards such as natural gas, flammable vapors, and hazardous materials needs to be continuous.

Fires may be intensified throughout the debris from the collapse. The RIT should obtain hoseline protection for all phases of the rescue effort, which may involve spot fires and other hazards. **Positive pressure ventilation (PPV)** fans may be required to provide fresh air to firefighters that are trapped and to increase visibility.

The most important assessment is recognition of the possibility of secondary collapse and making certain that it is carefully weighed and considered prior to rescue attempts. Sounds, sagging floors, walls out of plumb, and materials shifting are some of the most prominent indicators of secondary collapse.

One of the first places that the RIT can begin to search for firefighters in a collapse is on or near the surface of the debris pile itself. If multiple firefighters are missing, a place where a surface victim is found is a good place to start looking for the others. Portable radio communications are the most obvious aid in helping locate the trapped firefighter. Once communication is established it is important for the RIT to maintain communications with the victim on the specific channel that he calls on. A radio-assisted feedback search and PASS devices, if operable, can also be utilized to help locate trapped firefighters. Using the **"hailing" system** can also be used to help locate a trapped firefighter at this stage. In this system, rescuers are placed around the debris pile and silence is ordered on the incident scene. One rescuer will call for some sort of acknowledgement from a trapped firefighter if he can be heard. After the first rescuer takes a turn, silence should be exercised by the incident personnel for a time period to give the trapped firefighter a chance to reply. After that time period, the next rescuer should repeat the process. If a reply is heard, the RIT should attempt another call from a different location or angle to help pinpoint the location of the trapped firefighter. It is important to note that a reply may not be voice communication but sounds such as tapping objects or other noises.

The efforts of the RIT become a high priority when collecting information in this stage. By pinpointing information, the RIT can save technical rescue teams critical time in initiating advanced rescue efforts.

Rapid Intervention and the Initial Void Search

The physical search in a collapse environment is the most dangerous process of the rescue because rescuers will be involved with moving over and around the collapse environment while additionally committing rescuers into areas that may be unstable. Because of the high danger involved in collapse situations, personnel operating in the collapse area should be kept to a minimum and should have strong supervision. Void space search will not occur unless there is adequate equipment and trained personnel on scene to shore or crib the void from further collapse while searching **(Figure 17-22)**.

Technical rescue support will be equipped with the proper specialized equipment in order

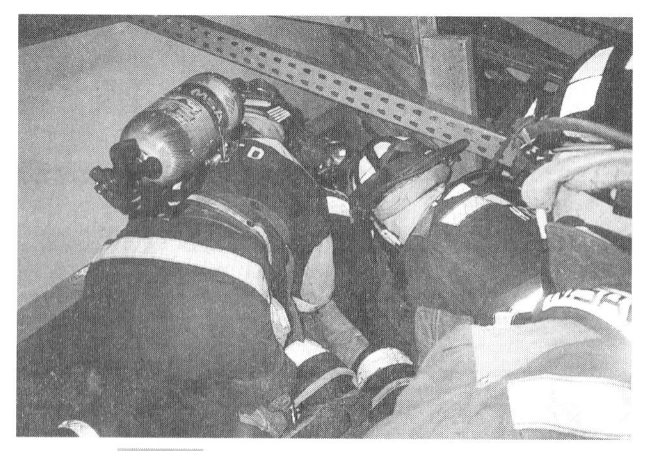

FIGURE 17-22

Void space search for downed firefighters should only take place if there are an adequate number of trained rescuers with the proper equipment on scene.

to stabilize and tunnel through the debris and must be called for as early as possible. The RIT will only be working with minimal tools that are readily available. Some of the basic tools that can assist the RIT in collapse operations are:

- four-gas monitor
- portable radio
- thermal imaging camera
- 4 × 4 short wood shoring and cribbing materials
- screw jacks
- rescue struts
- assortment of large and small wedges
- hand lights
- portable hydraulic tools such as Porta Power
- hydraulic spreaders and cutters
- 100-ft rope bag
- air bags
- identification and marking materials
- power saws
- irons and pike poles
- EMS supplies
- charged hoseline

This equipment is usually readily available and on the scene of the incident so that it can be retrieved quickly and placed into operation with minimal manpower.

In order for the RIT to be able to respond appropriately in these environments, members will need a basic understanding of the five characteristic ways collapse will present itself. These are:

- **pancake**
- **supported lean-to**
- **unsupported lean-to**
- **V shaped**
- **A frame**

These collapse types will present themselves in different ways depending on how the original structural elements were affected and eventually failed. Each collapse type will create voids in and throughout the rubble piles. The voids are created when the structures or objects within the building disrupt the collapse as it is coming down, causing various shapes and sizes of void spaces. It is important for the RIT to recognize the five types of collapses because each creates its own characteristic void spaces depending on the type of collapse.

The pancake collapse is the result of the failure of a bearing wall of the structure. The failure of the bearing wall will cause the roof and floors to collapse upon each other. Large objects found in the building will interrupt the collapse and create void spaces where a downed firefighter may be trapped **(Figure 17-23).** The RIT should use existing openings into the structure such as stairways in order to begin a surface check or void search. The extent of these collapses sometimes makes it impossible to access void spaces and additional alternative routes will have to be found by RITs and technical rescue teams. Technical rescue teams will many times have to begin cutting through building materials on the debris pile to make access to voids contained in a pancake-type collapse.

Lean-to collapses are caused when the supports for the roof or floors of a building fail on one side. The supported lean-to collapse is a failure of one side of the roof or floor that falls until it rests and is supported by substantial objects in the structure. There is a good chance that a trapped firefighter will be found at the bottom of the

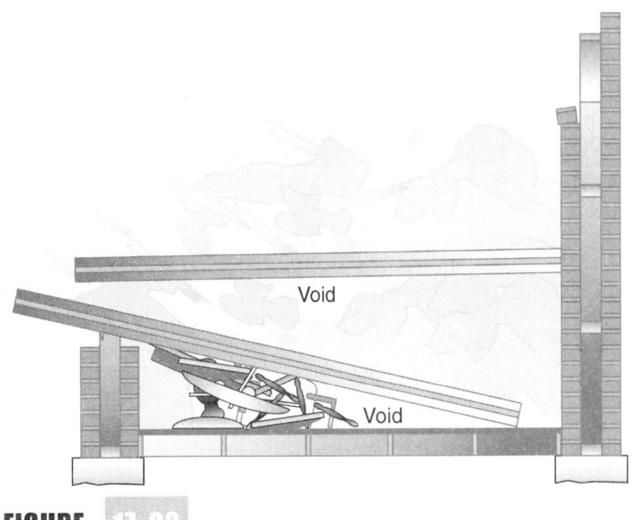

FIGURE 17-23
Pancake collapse.

failed end of the collapse. This type of collapse usually provides a void area that is able to be stabilized for rescuers to get into to retrieve the victim **(Figure 17-24).**

The unsupported lean-to collapse is a very unstable form of collapse for rescuers. The unsupported lean-to collapse is similar to the supported lean-to collapse except the one side that has failed is without support; it floats freely above the objects and floors below. The RIT should understand that any additional weight, movement, or sudden shift of weight from nearby debris can cause a secondary collapse. No attempts at void entry should be made unless the void is able to be stabilized with the proper equipment. The equipment, manpower, and materials that are needed for this type of collapse are usually well beyond the scope of the RIT's involvement. The RIT's efforts should concentrate on establishing the exact location of the trapped firefighter and protecting him in whatever way possible without causing further collapse until the needed resources arrive **(Figure 17-25).**

The V-shaped collapse is often caused by the failure of an interior support such as a load-bearing wall or column. In this type of collapse, the roof or floor will fail toward the center and collapse until the two sides rest on the level or objects below it. In this case, voids are created on both sides of the failure toward each bearing wall. The downed firefighter that was working on the collapsed level will be located toward the center of the "V" and may be covered by debris **(Figure 17-26).**

In the A-frame type of collapse, the floors separate from the exterior walls but as they come down they are still supported by interior load-bearing walls. This creates void spaces toward the load-bearing wall or column that did not fail **(Figure 17-27).**

Understanding the different types of collapses will help the RIT determine where to begin the search for the trapped firefighter.

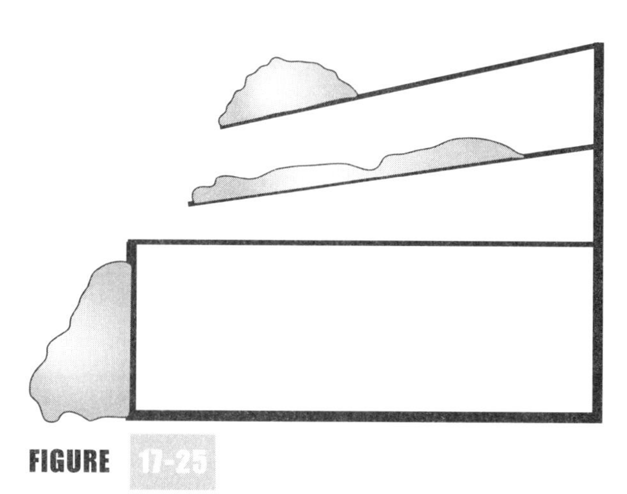

FIGURE 17-25

Unsupported lean-to collapse.

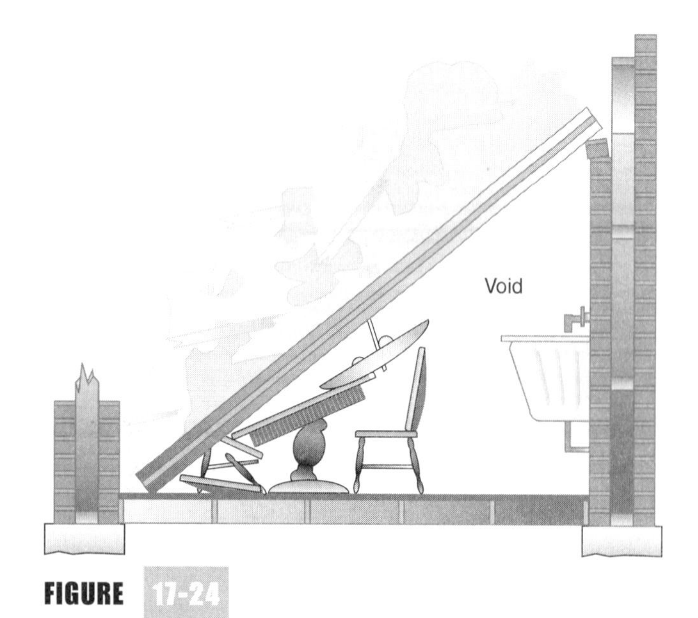

FIGURE 17-24

Supported lean-to collapse.

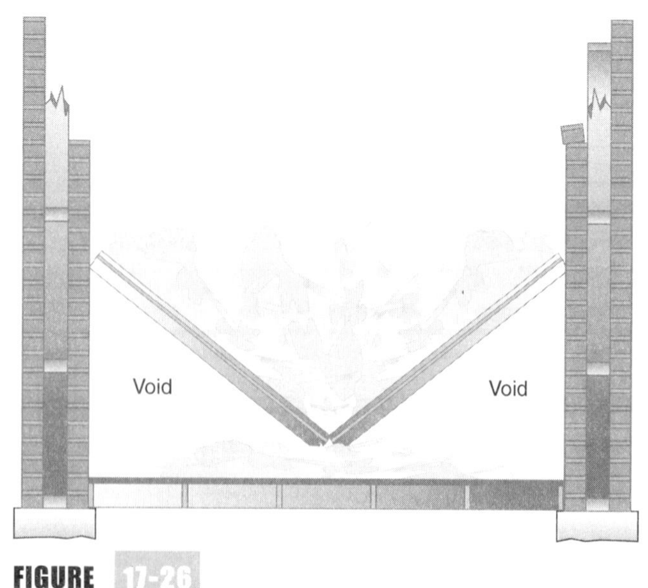

FIGURE 17-26

V-type collapse.

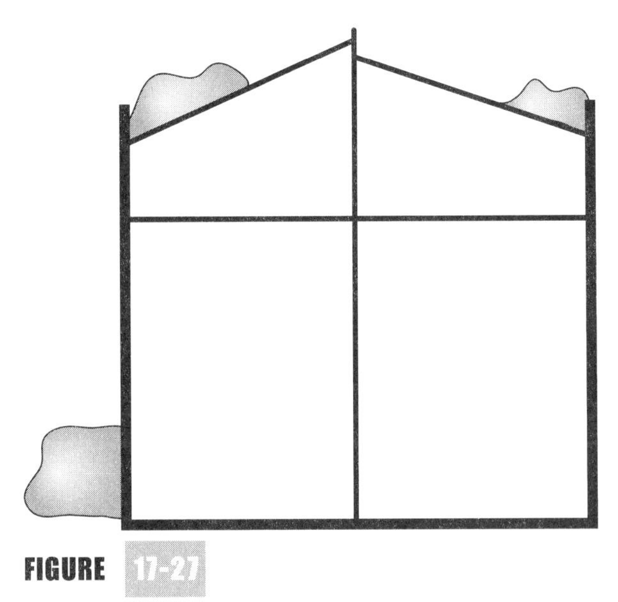

FIGURE 17-27

A-frame collapse.

FIGURE 17-28

Cribbing can be used to stabilize areas that are 3 feet or less in height. Box cribbing is the preferred technique and can be built into the load by using timber wedges and thinner pieces of timber.

Void spaces should not be explored or entered unless the proper equipment and trained people are present to carry out the operation. A technical rescue team that can provide this should be called to the scene as soon as possible. Knowing where these spaces are located and the location of the trapped firefighter is necessary to establish operations. The RIT should be made up of properly trained personnel who will begin these operations. Minimal equipment located on scene may be able to help start the rescue process. One of the greatest hazards when working in a collapse environment is the further failure of building components and objects that are creating void spaces. **Shoring** or **cribbing** must be in place to prevent further collapse. Both of these measures are meant to give rescuers an opportunity to perform a rescue. Wood timbers, ellis clamps, screw jacks, air bags, power spreaders, hydraulic jacks, and rescue struts can be used to shore up a collapse. Cribbing can be used to stabilize areas that are 3 feet or less in height. Box cribbing is the preferred technique and can be built into the load by using timber wedges and thinner pieces of timber (**Figure 17-28**). It is important to stress, however, that any attempt to move building components may cause further failure. Efforts and equipment need to be concentrated on supporting the environment.

Shoring and cribbing are not used to lift but to stabilize or support the collapse.

Access to the basement level of a collapsed structure should be explored as soon as possible. Often, void spaces created in the collapse will be readily accessible from below. Because the collapsed load may be resting on the structural components in the basement, shoring and cribbing should take place on that level in an aggressive manner.

Selected Debris Removal

Selected debris removal may involve the RIT to a limited degree. In selected debris removal, a specific location where a trapped firefighter may be found is approached by removing debris and obstacles. It must be remembered that every action that takes place will have a reaction in the collapse environment. Anything that may be suspected of holding a weight load should not be removed. It is during the selected debris removal stage that wall breaching may take place to access voids spaces. Work in this stage will require rescuers to take their time and be absolutely certain of every action that they will take.

General Debris Removal

General debris removal will take place well after a RIT is relieved of its responsibilities. Heavy equipment will be used to perform general debris removal. It is at this point that the search is no longer for survivors.

Debriefing

Debriefing is the final stage of a collapse situation. A critique of incident operations will take place during this stage. The stress experienced by rescuers during a collapse situation involving trapped firefighters is a traumatic incident outside the scope of what can be considered normal in a firefighter's work environment. This type of stress can lead to **post-traumatic stress disorder (PTSD).** Symptoms of PTSD can include repeatedly reliving the event, alcohol or drug dependency, nightmares, insomnia, and mood swings such as anger and depression. For this reason it is important to have critical incident stress debriefing teams available for rescuers during this stage. A critical incident stress debriefing team should be available anytime that a RIT is put into action in all situations, not only collapse. It is even highly recommended in situations that end in successful rescues.

Summary

Collapse is more times than not preceded by signs indicating that it will occur. However, sometimes these indications will not be evident until it is too late. The RIT has a responsibility to have a thorough knowledge of building construction and collapse indications. RIT members serve as an additional set of eyes on the fireground.

Once a collapse occurs, it is the RIT that will attempt the initial rescue efforts of trapped firefighters. The RIT's involvement in trying to locate a downed firefighter buried in a collapse is very stressful and challenging. As stated before, the RIT is not a technical rescue team and should not be treated as such. Technical rescue teams bring with them additional abilities such as high-technology searching devices, specialized equipment, training, and in some cases, the addition of canine assistance.

Rescuers involved in RIT operations at collapses prior to the technical rescue team's arrival should be able to provide a minimal amount of basic shoring and cribbing if needed when accessing and locating void spaces with downed firefighters. If there is a doubt as to whether the surface can be entered, it should be avoided and dealt with when technical rescue support is present. It should be remembered that the RIT's duties are to locate and not necessarily retrieve a downed firefighter unless members have proper training and proper equipment is available.

Review the considerations for collapse situations:

- Immediately call for a technical rescue team that has the training and equipment needed.
- Control utilities immediately.
- Monitor continuously for the presence of hazardous materials and flammable gases.
- Hoselines should always be provided to address fire conditions.
- Any breaching, shoring, tunneling, or trenching should be conducted by tech-

nical rescue teams if an immediate removal of the victim cannot be obtained.

- Do not attempt to move or lift structural components back to their preexisting conditions; it is better to stabilize the material the way you find it.
- RITs can provide shoring or cribbing to help stabilize an immediate void opening to a downed firefighter.
- Once shoring is in place it should never be removed.
- An appropriate and knowledgeable safety officer should be assigned for the duration of the operation.
- RIT members should gain access to a basement of a collapse structure whenever possible and shore it aggressively.
- RIT members should make certain that they are accounted for.
- Avoid too many rescuers working in and around the collapse.
- If a risk is taken to search an immediate void with the known presence of a trapped firefighter, a search rope should be utilized by the rescuer for the purposes of identifying his path to the victim so that he can be located in case of secondary collapse.
- When working in a collapse environment, it is important to remember that every action will cause a reaction.

The RIT's abilities in the collapse environment are limited by members' training and available equipment. The techniques covered in retrieving firefighters through drags, carries, elevation differences, and up through floors are all maneuvers available in rescuing a downed firefighter immediately in a collapse environment if conditions allow it. The potential for rescues by the RIT is very good in the collapse environment if that environment is respected and understood by all RIT members.

KEY TERMS

90-degree angle collapse
A-frame collapse
backdraft
Balloon-frame construction
Cribbing
Curtain wall collapse
Dead load
Debriefing
Emergency radio traffic
General debris removal
Hailing system
Inward/outward collapse
Lightweight wooden I beam
Pancake collapse
Platform construction
Positive pressure ventilation (PPV)
Post-traumatic stress disorder (PTSD)
Reconnaissance and rescue of surface victims
Selected debris removal
Shoring
Supported lean-to collapse
Unsupported lean-to collapse
V-shaped collapse
Void search
Wood-frame construction

REVIEW QUESTIONS

1. What are three key factors when trying to minimize hazards presented to firefighters in regard to collapse?

    1. ______________________________
    2. ______________________________
    3. ______________________________

2. Houses built prior to _____ are more than likely to be balloon-frame construction.

    a. 1800
    b. 1950
    c. 1960
    d. 1980

3. A fire in the basement or cellar in __________ construction has a high likelihood of extending to the attic.
   a. lightweight truss
   b. platform frame
   c. balloon frame
   d. none of the above

4. What is considered paramount to controlling a fire in balloon-frame construction?
   1. ____________________
   2. ____________________
   3. ____________________

5. ______________ construction gets its strength from tension and compression of materials used in its construction.
   a. Lightweight truss
   b. Platform-frame
   c. Balloon-frame
   d. none of the above

6. Steel building components will expand or twist at
   a. 700°F.
   b. 850°F.
   c. 1,000°F.
   d. 1,200°F.

7. Heavy smoke under pressure is a possible indication of
   a. backdraft.
   b. flashover.
   c. rollover.
   d. positive pressure ventilation.

8. As a minimum, it is recommended that a distance of _____ the building's height be designated as the collapse zone.
   a. three times
   b. two times
   c. one and one-half times
   d. three-quarters of

9. Once identified, what are the four ways that a hazard can be mitigated?
   1. ____________________
   2. ____________________
   3. ____________________
   4. ____________________

10. List four signs or indicators of secondary collapse.
    1. ____________________
    2. ____________________
    3. ____________________
    4. ____________________

11. The pancake collapse is a result of a failure of a(n)
    a. bearing wall or sudden impact load onto a floor.
    b. interior wall or shearing load onto the roof of the building.
    c. torsional load or failure of interior walls.
    d. none of the above

12. Void space search should not occur unless there is what on-scene?
    1. ____________________
    2. ____________________

13. A ______________-type collapse occurs when supports for roofs or floors fail on one side.

14. A ______________-type collapse results when there is a failure of an interior support or column.

15. What are some possible symptoms of post-traumatic stress disorder?

ADDITIONAL RESOURCES

Dodson, D., *Fire Department Incident Safety Officer.* Clifton Park, NY: Thomson Delmar Learning, 1999.

Dunn, V., *Safety and Survival on the Fireground.* New Jersey: Penwell Publishing, 1992.

Dunn, V., *Collapse of Burning Buildings.* New York: Penwell Publishing, 1988.

Federal Emergency Management Agency, *Urban Search and Rescue Response System-Rescue Specialist Training Manual.* Emmitsburg, MD: Federal Emergency Management Agency, 1996.

Firefighter's Handbook, 2nd ed. Clifton Park, NY: Thomson Delmar Learning, 2004.

Norman, J., *Fire Officers Handbook of Tactics,* 2nd ed. Saddlebrook, NJ: PennWell Publishing,1998.

O'Connell, J., Collapse Search and Rescue Operations: Tactics and Procedures, Part 2—Size Up and Safety. *Fire Engineering,* June 1993.

O'Connell, J., Collapse Search and Rescue Operations: Tactics and Procedures, Part 1—Collapse Voids and Initial Void Search. *Fire Engineering,* May 1993.

Smith, J., Masonry Wall Collapse. *Firehouse,* December 1999.

Smith, J., Building Collapse Indicators Continued. *Firehouse,* June 1997.

Smith, J., Building Collapse Indicators. *Firehouse,* April 1997.

CHAPTER

18 Rapid Intervention Training

Learning Objectives

Upon completion of this chapter, you should be able to:

- explain the importance of being creative and open minded when instructing in RIT techniques and self-survival.
- discuss what motivates students to learn.
- list and describe the five types of adult learners.
- list and describe the seven intelligences that compose the multiple intelligence theory.
- talk about safety parameters that are essential for RIT and self-survival training.
- discuss the qualifications necessary for a RIT and self-survival instructor.
- clarify what objectives are important when setting up RIT and self-survival training.
- explain how different props can be constructed and used for training, including wall stud simulators, sono tubes, wall-breaching simulators, entanglement simulators, A-frame simulators, and Denver simulators.
- discuss some basic ideas or concepts that can be used or modified for the purpose of RIT or self-survival training.
- talk about the advantages of scenario-based training evolutions.

CASE STUDY

"On June 16, 1999, a 38-year-old male firefighter/captain (the victim) died after falling approximately 20 feet from the top of a ladder which had been previously raised to the second-story window of a fire building at a fire training center. On the morning of the incident, several fire departments were involved in a multi-jurisdictional, multi-company training exercise. The exercise was conducted by three divisions performing separate evolutions simultaneously. Division A demonstrated proper tactics and procedures during live fire-attack operations and proper search-and-rescue techniques within a simulated single-family residential occupancy. Division B demonstrated proper search-and-rescue techniques and ladder-rescue operations from a second-story elevated platform and/or window, and Division C demonstrated proper tactics and procedures for advancing a fire-attack hoseline to gain access to a third-floor fire by entering a second-floor window via a ladder and extending the hoseline up a stairwell to the fire. The evolutions were to be performed twice a day over a 3-day period (session 1 in the morning and session 2 in the afternoon). The incident occurred near the end of session 1 on the first day of training. The victim, who was acting as a proctor for the training exercise and monitoring Division C, was positioned on the second floor of the training facility with Division C, which had just completed their evolution when the incident occurred.

"The victim and firefighters from Division C were assembled on the second story when the air horn sounded to evacuate the building as previously planned. At that time, and for unknown reasons, the victim announced he was going to attempt a new procedure he had learned previously at a Rescue Intervention Training Course, which was referred to as the "bail out." The new procedure involved a head-first advance over the top of the ladder, hooking an arm through a ladder rung, and grasping a side rail, swinging the legs around to the side of the ladder and sliding down the ladder to the ground. Without hesitation or comment, the victim, who was about 3 feet away from the top of the ladder, took one step and leaped over the top of the ladder. The victim was unable to adequately hook the ladder rungs or grasp a ladder side rail and fell about 20 feet headfirst to the concrete landing. Although the victim received immediate attention from firefighters and medics in the area, the victim was transported to the local hospital where he was pronounced dead about 40 minutes after the incident. NIOSH investigators concluded that to minimize similar occurrences, fire departments should:

- *ensure that all new training programs undergo a comprehensive review prior to the implementation of the program.*
- *collaborate with other fire-related organizations regarding the feasibility of all new training procedures before the programs are implemented.*
- *ensure that all aspects of safety are adhered to per established standards and recommendations while training is being conducted.*
- *designate individual safety officers at all significant training exercises to observe operations and ensure that safety rules and regulations are followed."*

—Case Study taken from NIOSH Firefighter Fatality Report # 1999-25,* "Firefighter (Captain) Dies after Fall from Ladder during a Training Exercise—California," *full report available on-line at http://www.cdc.gov/niosh/face9925.html.

Introduction

After safety, training is the most important aspect of rapid intervention. Without proper training, a RIT will never be able to carry out its task of providing for the safety and rescuing of one of their own. Rescuing a downed firefighter will be one of the most taxing and demanding events that take place on the fireground, both physically and mentally. Some situations may not allow for a second chance. Are you ready to meet this challenge?

Adult Learning and Rapid Intervention

Adults have acquired many experiences in their lives and each experience will dictate how they perceive certain issues. In the fire service, it is often said that "Experience is the best teacher." In some cases this may be true, but it must also be remembered that experience is the roughest type of teaching aid because it presents the test first and the lesson afterward. Training in rapid intervention will need to be a constant, ongoing process that is evaluated and kept on a pace with the ever-changing conditions and challenges that are being faced by today's fire service. This training needs to appeal to and captivate today's firefighters to reach its highest degree of success.

Greater employee involvement to reach the best decisions and ideas inside the workplace is a growing trend in large corporations. The fire service cannot ignore this when it comes to continuous improvement of services rendered by its members. Fire service instructors are required to teach more interactively, challenging both themselves and the students to raise questions and examine assumptions when solving problems. Feeding off the student's experiences and ideas will allow an instructor to have more to offer the next time a particular subject area is approached. This is especially true when speaking about rapid intervention. Again, there is no room for egos when it comes to RIT training. It will be important to be creative and open minded in moving forward to generate new ideas and concepts that will produce solutions to the problems that have been and could be encountered. Safety must never be compromised in any manner when training and should be the focus of all training exercises.

Adults are motivated to learn for different reasons, including gaining knowledge, improving self-esteem, gaining financial reward, or advancing in their career. More often than not, the students that attend an RIT course outside of the department are there because they are aggressive and hungry to gain knowledge and "get dirty." Sometimes they are there because they witnessed an event or even participated at a scene where they or someone they knew had a "close call" or, even worse, got seriously hurt or killed. Whatever the case, it is important for the instructor to learn about and adapt to the students' motivations and learning types so that interaction takes place in full capacity.

There are five types of adult learners (Endorf and McNeff, 1991). To be successful, learning will have to be problem centered for all five types. The five types are type I, type II, type III, type IV, and type V. Descriptions of each of the five types are as follows:

- **Type I,** the **goal-oriented learner.** This type of person is in competition with themself. Type I learners are very confident and self-directed. Identifying their personal needs is one of their strengths. They prefer and excel in an interactive or participative type of classroom.
- **Type II** is the **affective learner.** Type II learners will take a cooperative role in the learning experience and will show excitement toward their learning environment. They regard their instructor as a source of knowledge and expertise. They do not feel comfortable in questioning their instructors. Providing one-on-one instruction outside the classroom may help in raising type II students' comfort level.
- **Type III** or **transitional learners** are looking for the way in which they can connect their prior education and experience with what they are learning. They will prefer a learning environment that encourages interaction and discussion of their ideas. They favor a sense of equality with their instructor and will not readily accept being fed information through means such as a video presentation or one-sided lecture.
- **Type IV** or **integrated learners** are primarily interested in their personal success. They are self-directed learners who prefer a collaborative learning environment in which they can stand out.
- The **risk taker** or **type V learners** are very self-confident and eager to learn new concepts. They will stray away from guidelines if there is an opportunity to gain new knowledge.

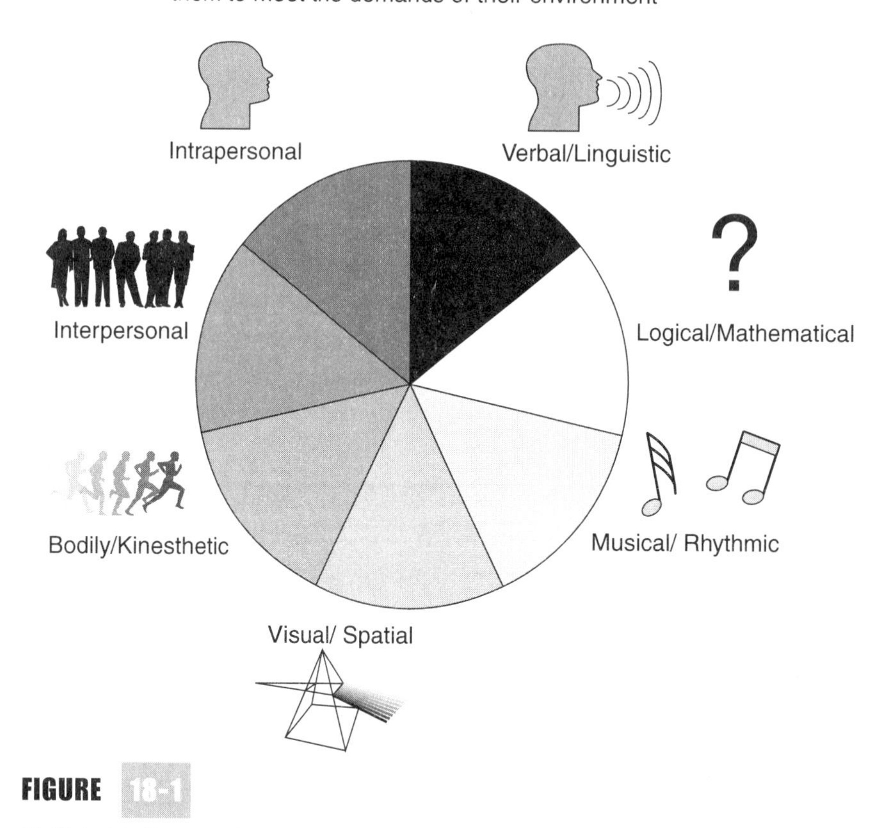

FIGURE 18-1

Multiple intelligence theory states that all humans are born with at least seven intelligences and will demonstrate different levels of ability or weakness in each.

Adult students will also learn best in an atmosphere that is created using the **multiple intelligence theory.** According to this theory, all humans are born with at least seven intelligences and will demonstrate different levels of ability or weakness in each (Brougher, 1997; **Figure 18-1).**

Verbal/linguistic intelligence allows people to process complex ideas through the use of language. Reading and writing are considered the standard norm for this form of intelligence. Appealing to the verbal/linguistic intelligence can be enhanced by making students use puzzles, develop questionnaires, or give oral presentations in regard to the material. Sharing or delegating responsibility in preparing and presenting the training will aid the instructor in reaching out to people who are strong in this area.

The **logical/mathematical intelligence** enables individuals to analyze and engage in higher-order thinking. It centers on problem solving and allows people to perceive meaning from logical patterns, functions, and processes. All classroom experiences will require the student to use logical thought processes to some degree. Tactical simulations that challenge students to employ theories and methods priorly introduced will appeal to and be most beneficial to people that rely heavily on this intelligence.

The **musical/rhythmic intelligence** allows individuals to create, communicate, and form understandings by utilizing the pitch, rhythm, and tones of sounds. Music modifies the mood, sharpens focus of thought, and deepens insight. How often do we see athletes listening to music through earphones just prior to competition? Music can be very effective if it is utilized as a background to introduce a new topic or stimulate discussion. The type of music or sound will be dependent upon the situation that an instructor is trying to convey. As an example, during drills such as SCBA survival or rapid intervention training, recorded fireground transmissions, an activated personal alert safety system (PASS) device, or hard rock music can be successful in creating a heightened chaotic or confusing state in drill participants. Sound or music does not have to be at a high volume to accomplish a desired effect.

Visual/spatial intelligence enables students to learn through graphic images such as pictures, diagrams, or maps. The mental transformation of colors, shapes, space, and the relation between them allows people to understand concepts. To people relying on visual or spatial intelligence, a picture truly is "worth a thousand words." Students with strong visual/spatial intelligence may be confused by a constant flow of words in sentences and paragraphs, but can understand a concept more easily if it is graphically illustrated. Utilizing photos, drawings, and diagrams is the way to appeal to students relying on this intelligence. A technique introduced in training cannot be oversimplified by using too many illustrations or photos.

Bodily/kinesthetic intelligence centers on the fact that some students can think or reason best when their bodies are active. They have a better understanding when they are involved with hands-on applications of theories. Most firefighters tend to learn best when this intelligence is referred to. Firefighters for the most part are action-oriented people. Actually performing the techniques and seeing the results firsthand will engrain the concept into the individual's experience base. Having performed a task or technique in training and seeing that it works will cause the individual to have the necessary confidence to employ it safely and with confidence on the fireground.

Interpersonal intelligence centers on students' awareness and interpretation of feelings, emotions, motivations, and goals of the people around them. Students that favor this intelligence thrive best when working in small groups to complete projects or assignments. Skills and methods incorporated into drill sessions that rely on teamwork and trust will help provide the best learning experience for people relying upon this intelligence and will also help in developing these qualities that are demanded on the fireground.

Intrapersonal intelligence allows individuals to form a mental model of themselves. This model is then used to make decisions concerning that individual. It is a personal understanding of the person's own emotions, motivations, and learning goals. People that favor this intelligence will give much thought to decisions while focusing on capitalizing their strengths. The key point is to tie a personal value to the instructor's objective in the training. Unfortunately, the "what's in it for me" attitude has taken its toll on our society as a whole. People need to be shown the worth of the training that they are going to receive. Explaining the reasoning or origin of why a technique is being introduced will clarify its importance. Case studies are an excellent way for instructors to reinforce concepts that they are presenting and should be utilized whenever it is feasible—this cannot be overemphasized when talking about firefighter self-survival or rapid intervention.

Individuals are endowed with each of these seven intelligences. The way individuals blend and utilize these intelligences are what makes them unique learners. It is through an environment that utilizes all of the intelligences that adult students will get the most out of their educational experience. For this reason an instructor or course developer should consider all seven when designing lesson plans and presenting material.

All students will not exhibit the same level of background skills or understanding when it comes to rapid intervention. The student's skill level should not be taken for granted. Even the most experienced firefighters will be weak in certain areas; we are all human and that is why we have training sessions. We have personally witnessed veteran firefighters complain about basic training but then not be able to carry out a simple task such as donning an SCBA properly. Everything needs to be explained and performed from square one. This includes the reason why a particular skill or maneuver is being introduced. This is where case studies and experience come into play.

Adults vary tremendously in how they acquire knowledge. There is not one single technique or method that can adequately fulfill the diverse needs of learners. A combination and understanding of all the concepts listed will be necessary to create a successful learning environment. Though it sounds overwhelming, the authors take all of these factors into consideration when setting up and arranging training sessions for firefighters. Although the material and end results are the same, the instructional approach taken needs to be different in each situation. This becomes extremely difficult when dealing with large groups of students from different areas, but with some effort it can be done.

Safety and Rapid Intervention Training

Rapid intervention training is some of the roughest and most demanding training that firefighters will have to complete in their careers. It is unfortunate that a significant number of firefighters get seriously injured and even killed while performing training each year. For the most part, training fatalities can be prevented. There is never any legitimate reason for someone to get seriously hurt while training. Training is the one and only time that we have control over the conditions and environment that we work in.

A set of rules that should be adhered to without exception for RIT training are as follows:

1. Instructors are in charge. The authority and responsibility of the drill lies with the instructor.
2. No horseplay or freelancing.
3. Pay attention, understand what is going on, and if you are unsure, ask.
4. Always wear full protective gear.
5. Always work in pairs.
6. Communicate with partner and instructors; a real emergency should be designated by the words "for real." (The word "Mayday" might get confused with the drill scenario.)
7. Stay alert and use all senses. Conditions can change quickly in evolutions or training sessions.
8. Stay low and move cautiously in evolutions, especially when working with limited visibility.
9. Stay within your limits.
10. Learn and have fun while staying focused.

Never perform any rapid intervention training that deals with elevation differences without a rated safety harness being worn by the participant or having a proper safety belay line in place and attended. Safety lines for evolutions dealing with elevation differences should be checked prior to

Safety

Never perform any rapid intervention training that deals with elevation differences without a rated safety harness being worn by the participant or having a proper safety belay line in place and attended.

each and every participant taking their turn. Incorporating safety lines for training into some of the skills introduced can be referenced in **Figures 18-2** through **18-4.**

Rapid intervention training with live victims should never be done under real fire condi-

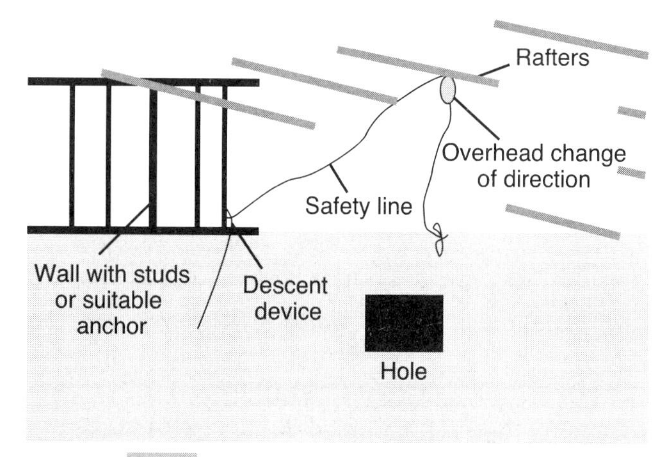

FIGURE 18-2

Safety line setup for through-the-floor evolutions.

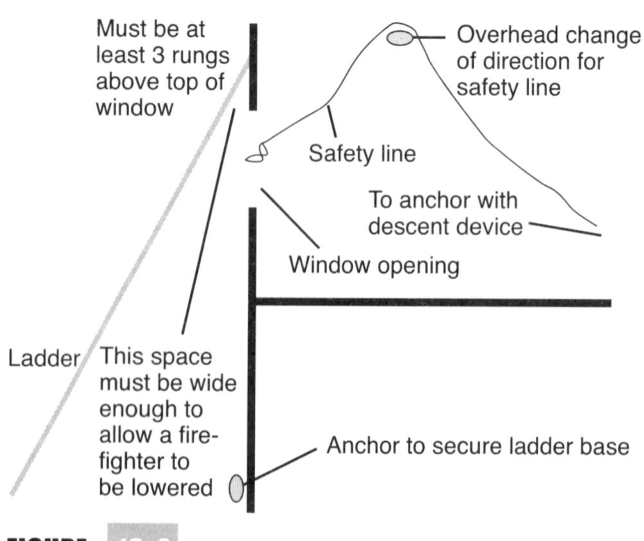

FIGURE 18-3

Safety line setup for modified 2-to-1 mechanical advantage with a ladder.

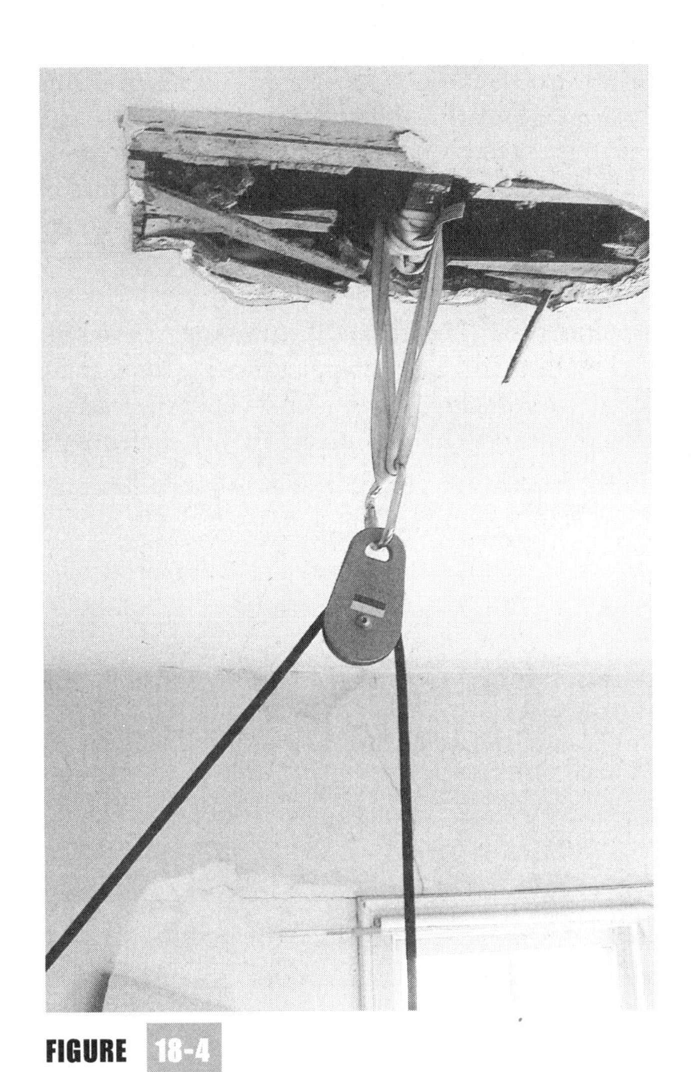

FIGURE 18-4

Overhead change of direction for safety line using a pulley.

tions. There have been numerous documented cases of firefighters losing their lives while posing as victims during live fire training. The variables present during training fires can easily lead to a disaster. **NFPA 1403** (Standard on Live Fire Evolutions in Structures) should be strictly adhered to as a minimum whenever live fire is used for training.

Rapid intervention training with live victims should never be done under real fire conditions.

However, other training aspects must be performed as realistically as possible. All students must be required to wear all of their gear just as if the training were a real incident. This includes breathing air from the SCBA and wearing helmet chin straps, protective firefighting hoods, and firefighting gloves. If you cannot perform the skills required with these items in place, how will you ever be able to perform them on the real fireground? If realism is desired, theatrical smoke machines using water-based solution, wax paper fitted over the outside of SCBA facepieces, strobe lights, activated PASS devices, and recorded fireground radio traffic are just a few of the options available to give a training scenario a shot of adrenaline and realism **(Figure 18-5).** Use common sense when setting up the training session. For example, do not have firefighters using power tools such as saws with their SCBA facepieces obscured with wax paper applied to the outside of the lens or working around live hazards such as holes in the floor with zero visibility. Actions such as these are only asking for trouble.

As with actual firefighting, make certain that the resources and personnel are available to set up an adequate rehab area for training participants. Adhere to the same guidelines that exist for real incidents. EMS should be available to check vital signs and treat any minor injuries that may occur and fluids for hydration should be available. Make certain that students are well rested before they are ready to go back into the training scenario.

Finally, the instructor must be qualified to be able to teach the material being presented. Being a certified instructor, training program manager, member of a training organization, or company

FIGURE 18-5

Theatrical smoke machines can give rapid intervention training a sense of realism when used properly in a safe manner.

officer by itself does not qualify one to teach RIT. Similarly, having observed a technique or having read about it does not mean that an individual knows enough about it to teach it. Instructors must be well versed in the material and be trained on it by actually attending courses and performing the skills in a hands-on manner, or **train the trainer format.** Experience with the subject material is also an important aspect that needs to be considered for a safe training session.

Training Objectives

Just like other firefighter training, rapid intervention training must be progressive in nature. A good foundation of basic firefighting skills is essential prior to a firefighter learning the idiosyncrasies of RIT. A good foundation of firefighting basics will also keep the firefighter from getting into trouble on the fireground. Skills should progress to scenarios. Both skills and scenarios need to progress from easy to relatively difficult. Take a simple skill and build upon it. Just make certain that a scenario or skill station is attainable. There is nothing more discouraging for a student than being faced with a training scenario that is both unrealistic and unattainable. To keep on track in this regard, make certain that the particular exercise is in fact valid. Send instructors that were not involved in the set up of the exercise through it. If they cannot make it through, then it is unreasonable to expect the students to be successful.

There are currently no set standards for the level of rapid intervention training required by firefighters. As a guideline for departmental training, the terminal objectives listed in **Table 18-1** can provide a starting point to get off on the right foot.

Rapid Intervention Training Props

Money should not be a factor in getting fire department members trained in self-survival or rapid intervention techniques. Most skills can be performed with standard equipment that should be present in a fire company and do not require a specialized facility. A dedicated departmental training facility is definitely a plus but is not required to hold good rapid intervention training sessions **(Figure 18-6).** Company officers and training directors should be aggressive in seeking out buildings that are slated for demolition or are vacant. A good majority of property owners are more than happy to allow the fire department to use a structure slated for demolition for training purposes **(Figure 18-7).** In some cases, this may even qualify as a tax savings for the person donating the structure. Again, live fire training is not necessary when using an acquired structure and is highly discouraged for RIT training.

FIGURE 18-6

Dedicated departmental training facilities are a great asset but are not required to hold rapid intervention training.

FIGURE 18-7

Acquired structures can be valuable for conducting RIT and self-survival training.

RIT Training Terminal Objectives

1. Given the course material, the student shall be able to function as a member of a working RIT.
2. Given the course material, the student shall be able to assemble and utilize necessary equipment to carry out rapid intervention operations.
3. Given the source material, the student shall be able to operate specialized tools such as air bags and hydraulics in low-visibility environments to effect rescues of downed firefighters.
4. Given the course material, the student shall be able to troubleshoot problems with SCBA that may be encountered in hostile environments.
5. Given the course material, the student shall be able to assess and change over a downed firefighter's air supply.
6. Given the course material, the student shall be able to navigate through entanglement hazards with proficiency.
7. Given the course material, the student shall be able to identify and set up suitable anchors for performing self-rescue or rapid intervention operations.
8. Given the course material, the student shall be able to perform the window hang and rope slide self-rescue maneuvers.
9. Given the course material, the student shall be able to perform advanced search techniques involved in rescuing downed firefighters.
10. Given the course material, the student shall be able to carry out large-area searches.
11. Given the course material, the student shall be capable of moving downed firefighters through buildings including stairs, elevation differences, large areas, and constricted spaces.
12. Given the course material, the student shall be able to remove a downed firefighter from a constricted space as presented in the Denver drill.
13. Given the course material, the student as a member of a RIT team shall be capable of moving a downed firefighter over and through obstacles.
14. Given the course material, the student shall be capable of removing downed firefighters from buildings by three different methods utilizing ladders.
15. Given the course material, the student shall be able to set up high-anchor point 2-to-1 mechanical advantage hauling and lowering systems utilizing ground ladders.
16. Given the course material, the student shall be capable of removing a downed firefighter from a peaked root.
17. Given the course material, the student shall be capable of performing the removal of a downed firefighter from sublevel elevation differences by utilizing the handcuff and handcuff cradle maneuver.
18. Given the course material, the student shall be capable of performing the removal of a downed firefighter from sublevel elevation differences by utilizing a charged hoseline.
19. Given the course material, the student shall be capable of performing the removal of a downed firefighter from sublevel elevation differences by utilizing the "W" technique.
20. Given the course material, the student shall be capable of performing the removal of a downed firefighter from sublevel elevation differences by utilizing modified 2-to-1 mechanical advantage systems.

TABLE 

RIT Training Terminal Objectives.

The possibilities for high-quality training are only limited by the imagination and safety with acquired structures. When able to use an acquired structure for training purposes, be generous to neighboring fire departments and invite them to partake in the training also. Remember: they may be the RIT coming to get you if you have trouble on the fireground! Maximize the time and training potential afforded to you by the building owner. With the proper setup and precautions taken, a large number of firefighters can be cycled through the training. Make certain that structural stability is maintained after the completion of each drill and make sure the structure is left in a safe condition when the fire department is through. This includes covering any holes in floors or windows and locking doors as well as removing broken glass. Also make certain that the end result of the training will not be an eyesore for the community. Not taking this into consideration can make future training acquisitions difficult.

Property owners are also sometimes willing to let the fire department train in a building if it is vacant. We have found this to be especially true in the cases of commercial or warehouse spaces for large-area search drills. Just remember to be respectful if allowed to use a vacant space. Do not destroy anything; clean it after use, and remove any trash that may have been left behind by the firefighters.

The only way to take advantage of buildings being demolished or vacant is to aggressively seek out owners and ask them for use of the building. A lot of people are not aware of the fire department's need and are more than willing to help out when they discover that need. The worst thing that can happen is that they will say no. If they say yes, be prepared to provide them with a permission to use form and a release of liability in case anyone gets injured, and make certain that the building itself is safe for firefighters to train in. Things to be aware of include deteriorated structural stability, presence of asbestos tile, hazardous insulation, and so on.

If push comes to shove, the engine room or parking lot at the firehouse can also be easily used for rapid intervention training. Many training props or simulators can be constructed for minimal cost. A lot of the materials needed for their construction can be obtained from scrap at construction sites with permission of the builder and only basic carpentry skills are required. Again, imagination and safety are the only limiting factors.

Some props that can be easily built for training are discussed in the following sections.

FIGURE 18-8

Wall stud simulator.

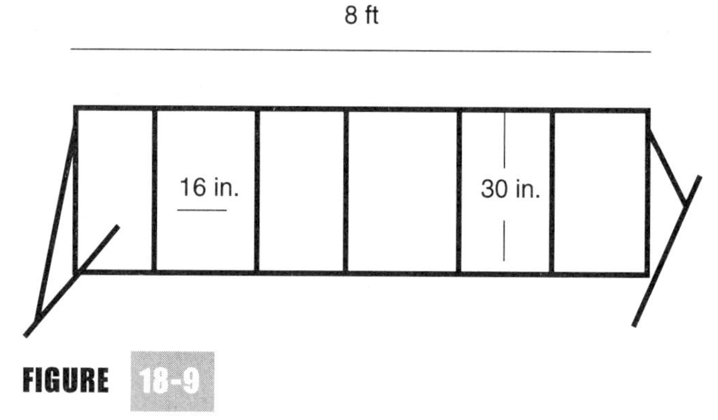

FIGURE 18-9

Low-profile wall stud simulator.

Wall Stud Simulator

A wall stud simulator can be utilized for various self-survival and RIT drills. It is built with the wall studs spaced 16 inches on center to simulate a wall that may be encountered by the firefighter. A low-profile simulation may be built into the wall to make the prop more diverse in its use **(Figure 18-8).** The height and overall width of the wall will be entirely up to the department using it. A variation of this prop can be built to be only 30 inches high, which will force firefighters to remain low when navigating through the prop **(Figure 18-9).**

Sono Tubes

Sono tubes or concrete forms can be easily obtained and readily used for numerous training purposes. They are available in various diameters—one in the range of 28 to 30 inches seems to be the most useful for self-survival and RIT training. They can be used to simulate tight or constricted spaces for SCBA skills or moving a downed firefighter **(Figure 18-10).** Sono tubes have even been used to successfully simulate a collapsed staircase in a RIT scenario on several occasions. Again, imagination and safety are the limiting factors.

Wall Breach Simulator

The wall breach simulator provides a realistic way for firefighters to practice going through gypsum board (drywall) that is commonly found. The prop is 4 feet high and 4 feet wide with wall studs 16 inches on center **(Figure 18-11).** Low-profile spaces

FIGURE 18-10

Sono tubes can be used to simulate tight or constricted spaces for SCBA skills or moving a downed firefighter.

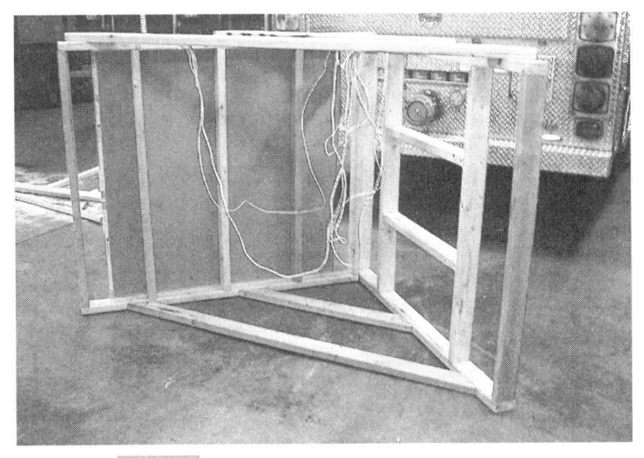

FIGURE 18-11

Wall breach simulator.

FIGURE 18-12

Gypsum board is slid into a channel to be held in place on the wall breach simulator.

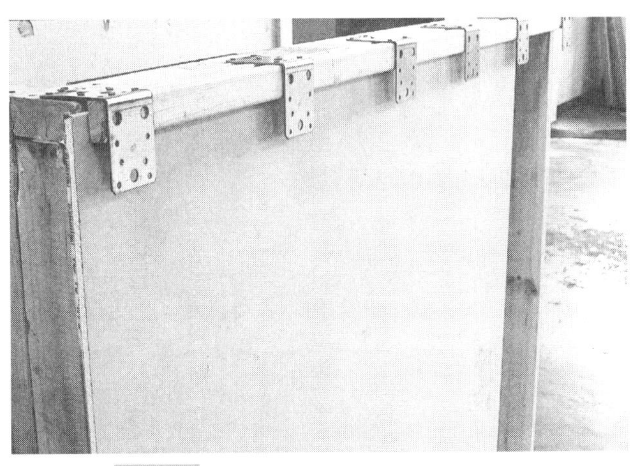

FIGURE 18-13

Close-up view of channels on the wall breach simulator.

can be included and are also spaced 16 inches apart. Pieces of 4-foot by 4-foot drywall are used with the simulator. The drywall is slid into a channel (made to accommodate the size of drywall) on the top and bottom of the outside of the wall **(Figures 18-12** and **18-13)**. The drywall can also be screwed in place for additionally stability if desired. A piece of new drywall is slid into the channels of the simulator for each participant. Wires, romex, or conduit can also be positioned through the walls for additional realism or difficulty.

Entanglement Simulator

The entanglement simulator provides a place for firefighters to practice and perfect their techniques for navigating through hazards that may exist in a drop ceiling collapse **(Figures 18-14, 18-15,** and **18-16)**. The overall length of the prop can be 6 to 12 feet, with a width of 4 feet. The height of the legs are 30 inches with the rafters spaced 24 inches apart, 2 × 6 lumber is best for this prop. Wires and entanglement hazards are attached into the prop by use of eye hooks screwed into the 2 × 6s or the wires can be tied directly to the framing. The types of wires can vary (we highly suggest using a mixture of wires similar to category 5, coaxial cable, romex, and any other wire that you can obtain). Plastic-coated clothesline is an economical alternative to use for wiring in this prop. Wires should be looped low to ground as well as high.

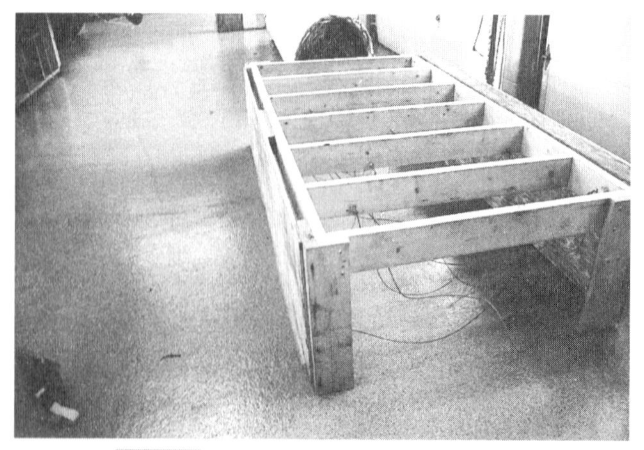

FIGURE 18-14

View of entanglement simulator.

FIGURE 18-16

Multiple entanglement simulators can be placed together to challenge students.

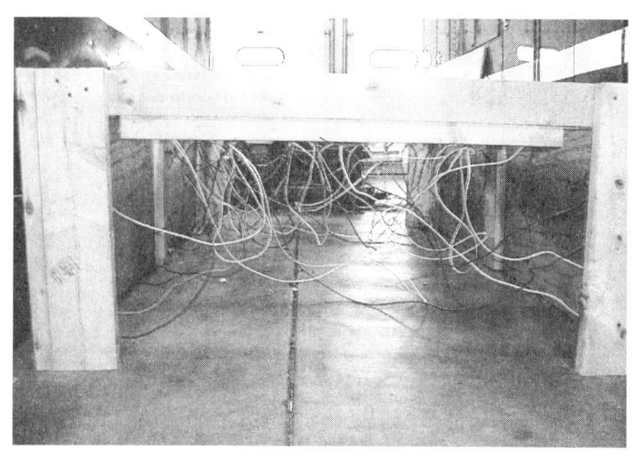

FIGURE 18-15

Inside view of the entanglement simulator.

FIGURE 18-17

A-frame training obstacle.

A-Frame Simulator

The A frame is another versatile training prop. It can be used for SCBA training (going under) or moving a downed firefighter (over). The height of the A frame should be 42 inches to simulate a high windowsill. The prop must be able to support the weight of a firefighter—4 × 4s for the legs and 2 × 4s for the rungs were used in the prop pictured in **Figure 18-17.**

Denver Simulator

The "Denver simulator" is meant to create a constricted space that a firefighter must be moved through and then lifted over a windowsill. The walls are 28 inches apart with the windowsill located 42 inches off the floor level. The window opening should be made 22 inches wide. The side walls are framed with 2 × 4s and sheeted with 4 × 8 plywood. The window portion is 22 in. wide and 42 in. tall and framed with 2 × 4s and sheeted with plywood cut to size. Two 2 × 4s cut at 45 degree angles are used to stabilize each side of the prop and a 28-inch 2 × 4 ties the walls together toward the rear of the prop (**Figure 18-18**).

Training Sessions and Drills

Training sessions must be progressive in nature. Rapid intervention and self-survival training will be new to some students. Any student involved in

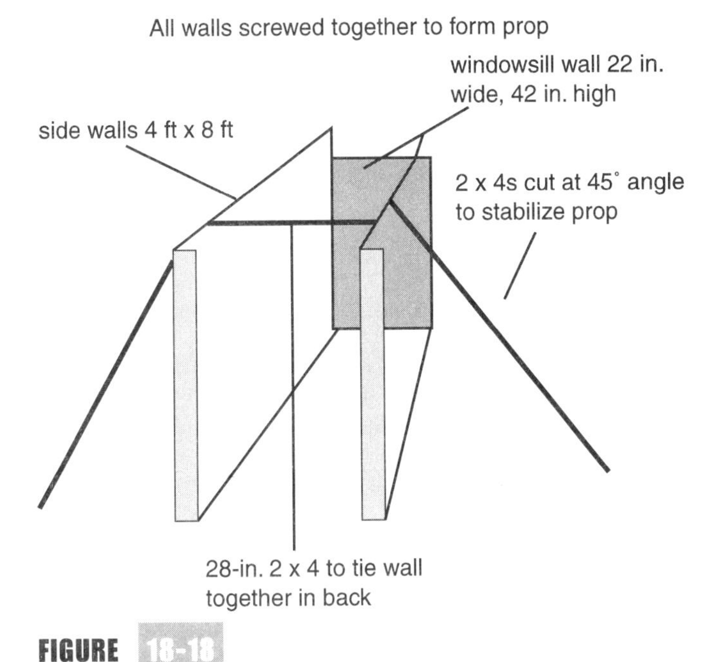

FIGURE 18-18

Denver simulator.

rapid intervention training should have a sound background and skills in firefighting basics.

Firefighters should be provided with an explanation of why a particular technique or idea is important. This can take place in a classroom setting and open discussion should be encouraged among the group. Perhaps one of the students has had an experience related to or similar in nature that can be shared with the whole group. With rapid intervention, it is important to learn from each other. This is the only way that we can prevent our people suffering the same consequences. This is especially true when discussing case studies.

When out on the training ground, all safety facets should be covered thoroughly so that the students have an understanding of their purpose and how they are meant to prevent harm. It should never be taken for granted that any student already knows something, no matter what rank or experience level. Start from square one with everyone; explain the "nuts and bolts" of the procedure and progress from there. Demonstrating the procedure is also a definite must. Break the skill down into steps or parts and have an instructor slowly demonstrate each step. The skill should then be demonstrated at full speed by an instructor. Students should slowly demonstrate the steps to the instructor while providing explanation of the procedure. This will help to maximize skill retention by the student. Only after this procedure should students be expected to perform the skills at full speed.

It is important that the instructor pay close attention to the drill participants at all times. Direction must be provided by instructors to keep the students both on track and safe **(Figure 18-19)**. Injuries can take place very easily during rapid intervention training if the instructor misses something or turns away. An unsafe act or omission will not wait to happen. Any act that compromises safety should be stopped immediately by an instructor. When stopping an evolution or skill station, provide an explanation to the student as to why it was stopped and how to make it safe. Self-survival and rapid intervention techniques may be new or difficult to students. Be prepared to spend as much time as needed to get students to an acceptable level of understanding.

Rapid intervention training needs to be broken down into sections; it cannot all be covered in one training session. Each consecutive session needs to progress in the level of skills to be performed. Each drill session must be broken down into training stations where a particular skill or area is introduced and practiced. A basic, introductory type of training session can be found in **Table 18-2.**

Some possible basic training sessions that are simple to set up and can be done inside with limited room include the following:

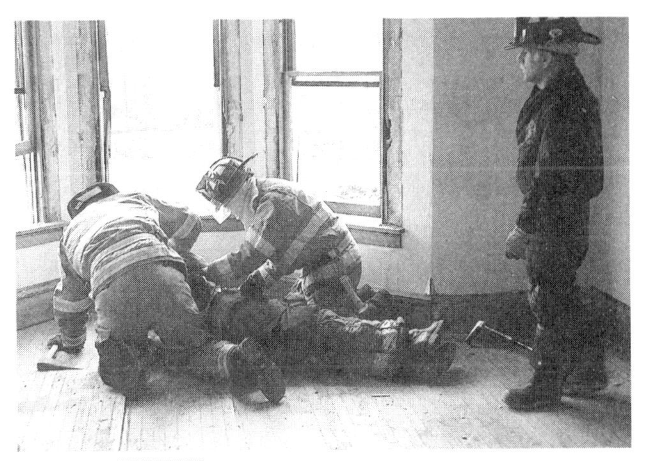

FIGURE 18-19

Instructors must carefully watch students during rapid intervention drills to ensure their safety at all times.

RICO (Rapid Intervention Company Operations) Basic Training

Abstract

The need for the eyes and ears of RITs is now and should be a priority for every fireground situation. The numerous risks faced by today's fire service dictate that well-trained and skilled firefighters be available to rescue one of their own. The information provided within this training session establishes the basic principles for the students to function as part of the RIT.

The program begins with a short presentation explaining the basic concepts of rapid intervention including responsibilities of the RIT, Mayday management, and policy development.

Following the presentation, students are broken up into groups to complete the following training stations.

Training Station 1—Moving the Downed Firefighter The biggest challenge for any rapid intervention operation is actually moving the downed firefighter. Skills and concepts covering basic drags and carries are practiced, including moving up and down stairs.

Equipment Needed

Safety cones (1), Halligan bar (1), Close hook (2), Pieces of webbing (1), Rope bag (2), Large carabiners (1), Stokes basket (1)

Training Station 2—Entanglement/SCBA Review

Equipment Needed

Wax paper, SCBA facepiece covers or hoods turned backwards to diminish visibility (1), Short (6–8 ft) pike pole

Entanglement props with additional wire will be necessary for this station. Wire cutters may also be utilized if firefighter is unable to navigate prop with conventional methods shown. SCBA techniques such as familiarization and low-profile techniques will be reviewed. A wall or simulator with studs 16 in. on center will be required. If using an existing wall, have a stud space cleared out for demonstration and practice.

Option

If more realism is desired, sheets of drywall cut to size can be screwed to the studs to allow the firefighter to breach the wall prior to going through the stud space.

Additional Equipment Needed for This Option

Screw gun (1), Drywall screws, Drywall cut to size, (1) cut piece for each firefighter, Halligan bar (1)

Numerous case studies have been presented involving firefighters falling though floors or staircases and not being able to be rescued. The handcuff knot has been the standard method instructed across the country for firefighter removal in these situations. Stations 3–5 will introduce numerous variations and techniques that have proven to be successful and are very easy to perform with minimal equipment.

Training Station 3—Handcuff, Handcuff Cradle, and "W" Method of Removing a Firefighter from an Elevation Difference

Equipment Needed

Rope bags (different colors preferred for demonstration purposes, (2) Rope bag for safety line (1), Pulley for safety line (1), Extra-large carabiners (1), Harnesses or belts rated for safety use (3)

A hole in the floor (4 ft × 4 ft) will be necessary for this training station.

Training Station 4—Hoseline Retrievals for Removing a Firefighter from Elevation Difference

Equipment Needed

Charged Hoseline (1 3/4 preferred), Rope bag for safety line (1), Pulley for safety line (1), Pieces of webbing (2), Large carabiners (3), Harnesses or belts rated for safety use (3)

A hole in the floor (4 ft × 4 ft) will be necessary for this training station. Joists or rafters above the hole will need to be exposed for setting up a pulley that will serve as a change of direction for the safety line. The safety line will run to a suitable anchor such as the frame of the doorway. A Munter hitch will be utilized for this purpose. Another option to use is tandem soft Prussik cords.

TABLE 18-2

Eight Hour Training Program.

Training Station 5—Denver Drill

Equipment Needed

Bag of rope (1), Large carabiners (4), Roof or straight ladder (must be able to fit it inside of hole in floor,) Harnesses or belts rated for safety use (3)

Different variations of techniques that can be utilized for rescuing downed firefighters from constricted areas will be discussed and practiced.

Training Station 6—Removing a Firefighter from an Upper-Floor Window Utilizing a Modified 2-to-1 System

Equipment Needed

Rope bags (2), Extra-large carabiners (big enough to fit onto a ladder rung (3), Pulleys (4), Pieces of webbing (2), Large carabiners (3), Extension ladders (2)

Firefighters will be shown how to utilize this technique by using a prerigged system and by making one "on the fly." Joists or rafters above the window and set back a little will need to be exposed for setting up a pulley that will serve as a change of direction for the safety line. The safety line will run to a suitable anchor such as the frame of the doorway. A Munter hitch will be utilized for this purpose. Another option to use is tandem soft Prussik cords. A window on an upper floor where a ladder can be placed approximately three rungs above the top will be necessary. All glass will need to be removed from this window for safety.

Learning Objectives

1. Given the training material, the student shall be able to function as a member of a working rapid intervention team.
2. Given the training material, the student shall be able to assemble and utilize necessary equipment to carry out rapid intervention operations.
3. Given the training material, the student shall be capable of moving downed firefighters through buildings including stairs, elevation differences, large areas, and constricted spaces.
4. Given the training material, the student shall be capable of removing downed firefighters from buildings by three different methods utilizing ladders.
5. Given the training material, the student shall be able to set up high-anchor point 2-to-1 mechanical advantage hauling and lowering systems utilizing ground ladders.
6. Given the training material, the student shall be capable of performing the removal of a downed firefighter from sublevel elevation differences by utilizing the handcuff and handcuff cradle maneuver.
7. Given the course material, the student shall be capable of performing the removal of a downed firefighter from sublevel elevation differences by utilizing a charged hoseline.
8. Given the training material, the student shall be capable of performing the removal of a downed firefighter from sublevel elevation differences by utilizing the "W" technique.
9. Given the training material, the student shall be able to remove a downed firefighter from a constricted space as presented in the Denver drill.

Target Audience

All firefighters are considered the target audience. The eight-hour training session will serve as a comprehensive introduction to the basic skills needed to operate as part of the RIT or as a skills refresher for the more experienced student. It is preferred that firefighters attending this session be familiar with firefighting basics and self-survival practices.

TABLE 

Eight Hour Training Program. *(continued)*

- Reference a NIOSH case study on-line. Discuss the case study openly among the crew. What could be done by your department or crew to prevent the same outcome?
- Pull out a preplan. Discuss the preplan with the group and possible scenarios that may be encountered.
- Take a photo of all sides of the crew's apparatus with the compartment doors closed. Photocopy it and have the crew write down on the photocopies what exact tools and equipment are located in each compartment. When finished, go out to the apparatus floor and check.

Have students pull off each tool and give a short talk about each.

- Take photos of a building in your town and use them to perform a table-top tactical simulation. Assign members to different functions. Have them communicate as if they were talking on the radio.
- Run a simulated Mayday radio drill with the members.
- Review tying knots and setting up simple mechanical advantage systems. Once the group is confident, set time limits for the skills. Just make certain that students' procedures are indeed correct.
- Have firefighters switch out each other's SCBA masks with their facepiece blacked out.
- Have firefighters connect into each other's EBSS with their facepiece blacked out and while wearing firefighting gloves.
- Perform an air consumption test (see Chapter 5).
- Have students don their turnout gear with SCBA. With facepieces blacked out, have them search a room in a team of two. When they come out, provide them with a pencil and paper to draw what they encountered. When all groups are done, allow them back inside to compare their drawing with what is really present.
- Borrow an SCBA from a neighboring department (if different) and familiarize the crew with it.
- Practice a toxic or "hot" bottle change.
- Don SCBA for times.
- Lay out several hoselines on the apparatus floor. While in full turnout gear, with SCBA and blacked out facepiece, have the students locate a line and determine which direction leads to the exit.
- Have neighboring department visit with their aerial device and explain how it operates and show what equipment and tools are carried and their location.

Scenario-Based Training

Scenario-based training is an excellent way to allow students to test their skills. Scenarios must be set up with goals and objectives that are attainable. The following are offered as suggestions that may be modified to conduct scenario-based training once self-survival and RIT skills have been trained. **Post-incident analysis (PIA)** of the training scenario should be conducted with the group when the scenario is complete to find areas that need improvement as well as what worked and what did not.

Simulated Debris Collapse with Entanglement

In this scenario students will be asked to conduct a large-area search in either an office space or manufacturing space that involves a prestabilized loose debris collapse simulating a partial ceiling collapse. This will also involve electrical and cable wiring causing an entanglement problem involving the downed firefighter or the rescuers. The placement of the downed firefighter can be within the collapse itself or the downed firefighter may be trapped in a small room or area on the other side of the collapse. (It is recommended to use a training mannequin outfitted in firefighter turnouts and SCBA for the downed firefighter.) Whichever placement is used, it will be the RIT's responsibility to enter and leave the structure with the victim through the same entry and exit point. This will require the RITs to be cognizant of manpower, materials, and emergency air supply that may be needed to accomplish this evolution. This particular evolution may require having to remove light debris in order to gain access to the downed firefighter as well as bring him back through the same area. This exercise should be conducted in near-zero visibility. This can be attained through either the use of mechanical smoke machines or blacking out the students' facepieces. All exercises should maintain as a minimum the 2-in-2-out rule. If a student's low-air alarm is activated, he should determine the amount of working time he has left and then exit the structure with a partner. All students involved with the scenario and working inside the structure shall always be on-air.

Large-Area Search—Lost Four-Man Company

In this particular scenario students will be asked to conduct a large-area search involving a large office or manufacturing space that may involve multiple floors for the purpose of rescuing a lost fire company made up of four members. This will require the students to establish good communications along with sound search and accountability techniques. Locating and coordinating the rescue of these four members will require several RITs. It will also require team leaders to account for multiple searches going on at one time.

Four firefighters outfitted in turnout gear and SCBA equipment will be needed. Their placement should be such that two of the victims are together with one conscious and the other with severe injuries. The other two victims should be separated from each other by approximately 30 feet. One of these victims will be unconscious while the other is disoriented.

This exercise should be conducted in near-zero visibility. This can be attained through either the use of mechanical smoke machines or blacking out the students' facepieces. All exercises should maintain as a minimum the 2-in-2-out rule. If a student's low-air alarm is activated, he should determine the amount of working time he has left and then exit the structure with a partner. All students involved with the scenario and working inside the structure shall always be on-air.

Downed Firefighters with Wall Breach

The RIT is given an assignment of locating a team of two downed firefighters inside of a structure. The RIT is to employ a conventional search to find the missing firefighters. Once found, it will be determined that the downed firefighters will need air. Because of fire extension the RIT will have to breach an interior wall to reach the firefighters. Due to collapse or further fire extension, the RIT will be unable to bring the downed members back out the way that they entered and must radio another RIT crew to breach an exterior wall to facilitate exit.

Firefighter Removal from Basement—Collapsed Staircase

The RIT is given assignment of locating two disoriented firefighters in the basement. Upon searching, the RIT discovers that the staircase to the basement has collapsed. The RIT team is to access basement and remove the downed firefighters.

Firefighters Trapped beneath Collapse

The RIT will be assigned to locate and retrieve two downed firefighters who were operating on a hoseline when a partial collapse of the floor above occurred, leaving them trapped beneath the collapse. The RIT will first have to search and locate the downed firefighters. Once located, the victims will need to be accessed by wall breaching and lifting of debris with the use of hydraulics and air bags.

Firefighters through the Floor with Restricted Space Removal

Two firefighters operating on a hoseline go through the floor into the lower level. The RIT is assigned to locate and remove the downed firefighters. The RIT will enter through a window with constricted space (28-in. opening, 46 in. off the floor) and will search to find the hoseline that the crew was operating. Once found, they will have to determine the proper direction to go to reach the downed firefighters. While searching they will have to navigate through obstacles such as wire entanglements and narrow openings to reach the victims. After locating the victims, the RIT members will be required to remove the downed firefighters through the same path that they entered.

Summary

The only way to prepare for a Mayday on the fireground is to train. RIT training is the most demanding training that a firefighter will have to complete. When people are in trouble or stressed, they will rely on their instincts. Constant training will provide that instinct for firefighters. Instructors that teach RITs must be knowledgeable, experienced, enthusiastic, patient, and safety oriented.

Safety should be the number one priority and never be compromised for any reason when training on rapid intervention and self-survival skills.

Training programs should take into account the fact that there are different types of adult learners—programs should appeal to each type of learner and each type of the seven intelligences.

Training is secondary only to safety in the role of the successful rapid intervention.

KEY TERMS

Bodily/kinesthetic intelligence
Interpersonal intelligence
Intrapersonal intelligence
Logical/mathematical intelligence
Multiple intelligence theory
Musical/rhythmic intelligence
NFPA 1403
Post-incident analysis (PIA)
Train the trainer format
Type I (goal-oriented) learner
Type II (affective) learner
Type III (transitional) learner
Type IV (integrated) learner
Type V (risk taker) learner
Verbal/linguistic intelligence
Visual/spatial intelligence

REVIEW QUESTIONS

1. Adults are motivated to learn for four different reasons. These are:
   1.
   2.
   3.
   4.

2. Match the following:

   ____ goal-oriented learner
   ____ affective learner
   ____ transitional learner
   ____ integrated learner
   ____ integrated learner
   ____ risk taker

   a. self-directed
   b. excels in participative environment
   c. strays away from guidelines to learn something new
   d. may need one-on-one instruction
   e. needs to connect prior learning

3. According to the multiple intelligence theory, all humans are born with seven intelligences and have a different level or ability in each.

   True   False

4. Match the following:

   ____ verbal/linguistic
   ____ logical/mathematical
   ____ musical/rhythmic
   ____ visual/spatial
   ____ bodily/kinesthetic
   ____ interpersonal
   ____ intrapersonal

   a. finds teamwork appealing
   b. learns through graphics and images
   c. prefers hands-on applications
   d. strong reading/writing skills
   e. allows individual to form understanding through sounds
   f. centers on problem solving
   g. case studies are considered helpful for reinforcement

5. Safety lines for training should be checked
   a. after each training session.
   b. before each participant's use.
   c. when set up prior to drill.
   d. none of the above

6. Live victims are acceptable when using live fire conditions as long as certain safety parameters are in place.

   True   False

7. NFPA _____ should be followed when using live fire conditions for training.
   a. 1403
   b. 1910.120
   c. 472
   d. 13

8. Being certified as a training program manager qualifies an individual to teach rapid intervention training.

   True   False

9. How can an instructor validate a training exercise?

10. Explain the process for teaching a rapid intervention skills to a student.

■ ADDITIONAL RESOURCES

Angle, J., *Occupational Safety and Health in the Emergency Services,* 2nd ed. Clifton Park, NY: Thomson Delmar Learning, 2005.

Brougher, J., Creating a Learning Environment for Adults Using Multiple Intelligence Theory. *Adult Learning,* April, 1997.

Endorf, M., and McNeff, M., The Adult Learner: Five Types. *Adult Learning,* 1991.

NFPA 1403 Standard on Live Fire Evolutions in Structures. Quincy, MA: National Fire Protection Association, 2002.

Glossary

0 degrees Kelvin Known as absolute zero (−273°C, −460°F). Heat energy produces molecular motion except at absolute zero.

2-in-2-out rule Provision in OSHA 1910.134 that outlines deployment of first arriving fire personnel and the provision of personnel available for their rescue if needed.

10-10-10 rule A baseline parameter to determine working time for a RIT involved in large-area search operations. The rule implies ten minutes to get in, ten minutes to get out, and ten extra minutes for safety.

90-degree angle collapse A type of wall collapse where a large portion or even the entire wall falls outward at a 90-degree angle.

a-frame collapse A type of collapse that results when the floors separate from the exterior walls. The floors are held up by interior walls within the structure, creating void spaces toward the center of the structure.

accountability system A system or process used by fire departments to track personnel while operating at emergency incidents.

active emitter Objects that generate their own thermal energy, such as animals and humans. Active emitters can vary in relation to mass and density. They can also be suppressed by barriers such as clothing and construction features inside structures.

aerial ladder A power-driven ladder that is affixed to a truck.

anchor The point used to secure or attach rope for use in performing a maneuver.

backdraft Sometimes referred to as a smoke explosion. Occurs when oxygen is introduced into an area holding superheated products of incomplete combustion.

balloon-frame construction Wood-stud framing system that runs continuously from the ground level or foundation to the roof eaves. Balloon-frame construction provides open continuous channels that allow rapid fire extension.

blanket carry A type of carry that utilizes a blanket or tarp for rescuers to lift and move a downed firefighter.

blanket drag A type of drag that utilizes a blanket or tarp for rescuers to pull a downed firefighter to safety.

bodily/kinesthetic intelligence Learning through hands-on applications of theories. Students acquire a better understanding when their bodies are active. Most firefighters tend to learn best when this intelligence is referred to.

body anchor Consists of a team member acting as a stable point to secure a rope for the use of moving, lowering, or raising. The firefighter acting as the anchor should brace himself behind a wall or at a door frame.

body ramp A technique used to enable a single rescuer to raise a downed firefighter up and over a windowsill.

box cut A technique consisting of three cuts to open a sectional or tilt-up overhead door.

brick veneer A single layer of brick used as an aesthetic facing on the outside of a frame wall. The brick in a veneer wall is tied to the framing by metal ties and supports only its own weight.

buddy breathing An emergency procedure in which two firefighters share a single air source due to one of them experiencing a problem with their own source.

butterfly knot Used when a three-directional pull or attachment point is needed in a rope.

butterfly ladder bailout An emergency self-survival maneuver used by firefighters to exit a widow onto a ladder.

captive carabiner A carabiner that contains a rod across the inside at the tapered end away from the gate. This creates a separate enclosed space inside the carabiner.

carabiner Devices used as attachment points for rope or webbing.

consensus standards Expected rules or performance levels that are developed by members of a specific trade, industry, or profession.

contrast The images that are provided because of heat differential on the screen of the thermal imaging camera.

control/entry supervisor RIT member stationed at entry point that is responsible for tracking the RIT's progress and working air time.

convected heat currents Heat transfer by the movement of air currents.

corner window anchor Anchor that is established by a tool positioned across one of the lower corners of a window.

cradle carry A type of victim carry. When performed properly, the victim is moved in a seated position above any obstacles or debris.

cribbing The use of different-sized lumber assembled in a box-type manner to support weakened or unstable loads.

curtain wall collapse A type of wall collapse that results when a wall crumbles or collapses upon itself straight downward.

dead load The weight of a building and any permanent features. This includes all structural members and building materials such as floors, columns, roofs, protected surfaces, interior walls, exterior walls, and fixed equipment such as HVAC systems.

debriefing The final stage of an emergency situation. A critique of incident operations should take place in this step as well as an opportunity for rescue workers to speak with qualified individuals to help them deal with the stress factors presented.

Denver Drill Rapid intervention drill for firefighters consisting of removing a downed firefighter from a constricted space that is 28 inches wide out of a 22-inch wide window that is 42 inches above the floor.

direct source emitter Provides the most thermal energy and can be easily visualized through thermal imaging. Examples are flame, moving fire gases, or preflashover thermal layering within a room or structure.

disorientation Mental state that a firefighter may experience if he becomes lost or confused while working in a hazardous atmosphere that may involve fire, smoke, heat, and darkness.

doggy door cut A form of forcible entry for outward-swinging doors in which a cut is made from one side of a door jamb to the other below the locking mechanism. A Halligan bar is then used to pry the door to swing free.

drywall ladder technique A self-survival maneuver that allows a firefighter to raise himself up a wall constructed of gypsum board to a window or point of egress.

EFIS (External Insulating Finishing Systems) An acronym for a building finishing process called External Insulating Finish Systems. It is meant to resemble traditional stucco and does not support any structural weight load of a building.

electrical conduit An outer flexible or fixed metal casing for electrical wire to run through. The conduit usually runs through the interior of a wall connecting to electrical outlets, fixtures, and hardwired appliances. These flexible or fixed metal casings complicate wall breaching and can also cause entanglement problems for the firefighter.

emergency air supply unit A modified air supply including a facepiece that is used for rapid intervention maneuvers that require a transfer or exchange of air for the downed firefighter as well as any member of the RIT.

emergency breathing support system (EBSS) A feature on some SCBA's that allows one user to share their air supply with another.

emergency hose slide A rapid egress technique used by firefighters to exit an untenable or rapidly deteriorating environment from an upper floor. Firefighters descend the hoseline, controlling the rate of descent with their legs. This is a high-risk maneuver and should only be performed in an extreme circumstance.

emergency procedure check A standardized method of checking an SCBA that will find, locate, and remedy a malfunction or failure while in a hostile environment.

emergency radio traffic Term used to signify that a priority message is to follow on the radio.

emergency rope slide A rapid egress technique used by firefighters to exit an untenable or rapidly deteriorating environment from an upper floor. The firefighter utilizes his body as a friction control device to lower himself to the ground level. This is a high-risk maneuver and should only be performed in an extreme circumstance.

exhalation valve A component of the SCBA facepiece that acts as a one-way valve, allowing air to be exhaled and released outside the facepiece.

expansion joints Cuts or separations placed in large concrete slabs to control cracking and breaking of the slab when it expands and contracts due to temperature differences. The lines could indicate a direction of travel to a disoriented firefighter in order to assist in finding a wall or column.

extremity carry A type of victim carry requiring two rescuers. One rescuer is located behind the head of the downed firefighter, lifting from beneath the arms, while the other rescuer lifts from the legs.

false or drop ceiling A suspended type of ceiling involving tiles and metal strips that hold the tiles suspended from an original interior roof or ceiling. The ceilings provide an air space for unrecognized fire as well as an easily accessed place to run wires and cables.

fascia board The trim board that runs along the base of a roof and parallel to the eaves.

figure eight on a bight Knot that is used for connecting objects or people to a rope.

firefighter assist and support team (FAST) Name designating a team of specially trained firefighters who are solely responsible for the safety, search, and rescue of trapped or lost firefighters at an emergency incident.

four-way haul Four lines provided to distribute the weight of a load, improving the ease and maneuverability of the raising of a downed firefighter.

friction device Any device or object that causes resistance between a rope and the object it is rubbing against.

gang nails Name for connecting points of structural members used in lightweight wood truss construction. They are usually made of 18-gauge stamped steel and penetrate the wood only 1/2 to 3/8 of an inch.

general debris removal Stage of collapse situation that involves the removal of general debris with heavy equipment. At this point, the search is no longer for survivors.

girth hitch A loop of rope or webbing brought back through itself while it is secured around an object.

glass block A block used in construction that is made of glass to provide light and security. It is set into place with mortar and metal ties. Requires special forcible entry and breaching considerations.

gusset plates Name for connecting points of structural members used in lightweight wood truss construction. They are usually made of 18-gauge stamped steel and penetrate the wood only 1/2 to 3/8 of an inch.

hailing system A system used to help locate a downed firefighter or civilian in a collapse situation. Rescuers are placed at strategic areas of a debris pile while ordering complete silence of people and machinery. One rescuer will then call out for a response from the victim, hoping to receive an acknowledgment. If none is given the next positioned rescuer will call out. If a reply is heard, another rescuer at a different position should call out in order to pinpoint the position of the response.

Halligan anchor lifting system A rapid intervention maneuver that is used to raise a downed firefighter who has fallen into an elevation difference back up to the level at which he fell from. It provides a quick anchoring system that can supply a simple 2-to-1 mechanical advantage. It can also be adapted for lowering.

Halligan floor/roof anchor A type of anchor in which a Halligan bar is driven into the floor or roof. A rope is secured to the Halligan bar for use after it is driven.

handcuff knot A simple knot formed by passing two loops of rope to the inside of each other, which are then placed on a victim's extremities and cinched down.

hang and drop A rapid egress technique used by firefighters to exit an untenable or rapidly deteriorating environment from an upper floor. This is a high-risk maneuver and should only be performed in an extreme circumstance. By hanging in a fully extended position, a firefighter can decrease his fall distance by his height. This is an absolute last-resort technique that can result in serious injury even if performed correctly from a lower-level window.

harness conversion The process of taking an SCBA harness and configuring it into a modified body harness for the purposes of assisting rescuers in moving a downed firefighter.

haul line A rope that force is applied to in moving a downed firefighter.

header course A line of brick in masonry construction that is laid perpendicular to other brick for the length of the wall. The header course will provide a structural bond for two vertical sections of masonry. Commonly a header course is seen every six to seven courses in true masonry construction.

high-point anchor An anchor that is placed above the load being raised.

hook-and-go ladder bailout A rapid egress technique used by firefighters to exit an untenable or rapidly deteriorating environment from an upper floor with the use of a ladder. This is a high-risk maneuver and should only be performed in an extreme circumstance.

image inversion Objects and firefighters that may not be discernable on a viewing screen of a thermal imaging device. This is due to higher temperatures and infrared energy that are being created by a fire environment than those given off by active emitters, which may appear darker on a viewing screen.

immediately dangerous to life of health (IDLH) An atmosphere that is immediately dangerous to life and health.

Incident management system (IMS) An organizational system used to manage resources at an emergency incident.

infrared energy Heat is a form of infrared energy and is unable to be seen by the unaided eye. It is the heat that the technological function of thermal imaging devices capitalizes on.

initial rapid intervention crew (IRIC) As outlined in NFPA 1710, two members of the initial attack crew must be assigned as a rapid deployment rescue team for the purposes of rescuing lost or trapped firefighters.

interpersonal intelligence Learning through the awareness and interpretation of feelings, emotions, motivations, and goals of other people. Working in small groups to complete projects is an aspect of this type of learning. Incorporating skills and methods into drill sessions that rely on teamwork is a reflection of this type of learning.

intrapersonal intelligence Allows individuals to form a mental model of themselves. It is a personal understanding of one's own emotions, motivations, and learning goals. This type of learning intelligence allows individuals to give much thought into making decisions while also capitalizing on their strengths.

inward/outward collapse A type of wall collapse in which the top of the wall partially falls inward while the bottom part of the wall falls outward. Most often caused by excessive force or pressure as produced by an explosion.

kicking out Refers to the inappropriately positioned ladder that will suddenly move outward when weight is loaded onto it. The prevention of this reaction is to have the ladder heeled at all times.

knee-method window removal A technique for raising a victim up and out a window in which two rescuers utilize their knees to position and lift the victim headfirst, face down to awaiting rescuers on the exterior.

ladder fulcrum method A method of lowering an injured or downed firefighter from a elevated area. The maneuver allows rescuers the ability to lower the victim in a horizontal position to the ground.

ladder slide A rapid egress technique used by firefighters to exit an untenable or rapidly deteriorating environment from an upper floor with the use of a ladder when multiple firefighters are exiting rapidly and consecutively.

landmark search Type of search utilizing the thermal imaging camera where the operator identifies a specific landmark and brings the search team to that point as the rescuers search along the way.

lathe and plaster walls A form of wall construction involving layers of plaster troweled over thin wood strips secured horizontally to stud framing. Most commonly seen in older buildings.

lean-to collapse Type of collapse caused when the supports for the roof or floors of a building fall to one side.

lift-and-lead-drag A basic drag that utilizes one fire- fighter to drag the downed firefighter from the upper body while another team member leads the way, providing for the safety of their egress.

lightweight truss construction Type of construction that obtains its strength from compression and tension of materials used in its construction as opposed to mass. The pieces of the truss are connected together by the use of gusset plates or glue.

lightweight wooden I beam Structural member utilized in some modern construction where the center portion is fabricated from chips and wood particles held in place with glue. I beams allow for large, unobstructed areas in buildings but fail very rapidly when exposed to heat and flame.

logical/mathematical intelligence Learning that enables individuals to analyze and engage in higher-order thinking. It centers on problem solving and allows people to perceive meaning from logical patterns, functions, and processes.

low-pressure/25% alarm An integrated alarm system in an SCBA indicating to the wearer that a 25 percent residual air supply remains in the cylinder. A firefighter's mental and physical condition will determine the actual amount of time provided by that amount of air.

main line rope-assisted team search Search concept in which a rope is anchored in a nonhazardous area while team members with specific assignments hold and follow the rope in as it is paid out of a rope bag controlled by the team leader.

masked Blocking of the thermal emission from the sensor of the thermal imaging device. Objects can be easily masked when wet or covered by dense objects.

Mayday A term used *only* to signify that a person is in a life-threatening situation and needs immediate assistance. A Mayday can be declared by anyone having knowledge of a person in distress.

mechanical advantage system Rope system that is designed to decrease a load for the purposes of dragging, lifting, or raising.

modified diaper harness A rescue harness made from webbing that can be used for dragging, raising, or lowering firefighters.

mule kicking A technique performed by a distressed firefighter for the purpose of penetrating a wall for emergency egress from one area to another when the use of a tool is not available.

multiple application service tool (MAST) A sling device that is utilized in moving a downed firefighter. Made out of five loops that are interconnected.

multiple intelligence theory The theory that all humans are born with at least seven intelligences and will demonstrate different levels of ability or weakness in each area.

multiple rescuer staircase lift Technique for carrying a downed firefighter up a flight of stairs using multiple firefighters. Firefighters at the head lift the downed member by the SCBA shoulder straps while another rescuer lifts the lower torso by being positioned high into the groin area of the victim. The main objective is to raise the SCBA cylinder valve over the stair treads.

musical/rhythmic intelligence Allows individuals to create, communicate, and form understandings by utilizing pitch, rhythm, and tones of sounds. Can modify mood, sharpens focus of thought, and may deepen insight.

National Fire Protection Association (NFPA) The NFPA uses a consensus process to establish codes and standards for the fire service.

NFPA 1403 National Fire Protection Standard on Live Fire Evolutions in Structures. This standard addresses all safety aspects of preliminary setups and considerations in regard to recommendations on water supply, hoseline applications, emergency ingress and egress, ventilation, materials burned, and many other considerations for practical learning involving the live fire training environment.

NFPA 1500 National Fire Protection Association Standard on Fire Department Occupational Safety and Health. This standard and program specifically outlines requirements and procedures for the safety of fire personnel in order to assist fire departments in addressing numerous safety concerns and issues within the occupation of firefighting.

NFPA 1710 National Fire Protection Association Standard on Organization and Deployment of Fire Suppression Operations, Emergency Medical Operations, and Special Operations to the Public for Career Fire Departments.

NFPA 1720 National Fire Protection Association Standard on Organization and Deployment of Fire Suppression Operations, Emergency Medical Operations, and Special Operations to the Public for Volunteer Fire Departments.

NFPA 1981 National Fire Protection Standard on Open Circuit Self-Contained Breathing Apparatus for Firefighters.

NFPA 1983 National Fire Protection Association Standard on Fire Service Life Safety Rope and System Components.

NFPA 704 system Identification system for signifying the presence of hazardous materials at fixed facilities.

Occupational Safety and Health Administration (OSHA) A federal agency that provides for the occupational safety of employees.

OSHA 29 CFR 1910.134 OSHA standard that covers the parameters required for a respiratory protection program. It contains the 2-in-2-out rule, which outlines deployment of first-arriving fire personnel and the provision of personnel available for their rescue if needed.

pancake collapse A type of structural collapse resulting from the failure of the bearing walls of a structure, causing the roof and floors to collapse upon each other.

PASS device Personal alert safety system. It is a small device worn by firefighters that is motion sensitive. The motion device will go into an audible alarm, signaling that the firefighter may be in trouble if there is no movement for a specified time period. It can also be manually activated by the firefighter if desired.

passive emitter A inanimate object whose temperature will vary depending upon the environment and time frame that it is exposed.

penciling A technique utilized during fire attack to determine heat levels within a room. Short bursts of water are directed at the ceiling overhead to determine heat conditions above. If no water returns down, flashover conditions are approaching. Effective penciling techniques can help prevent flashovers from occurring.

personal accountability report (PAR) Roll call of companies operating at an emergency incident. Commonly performed at specified time intervals as well as when mode of operations on the fireground change (i.e., offensive to defensive) or a signifigant event takes place (Mayday, collapse, etc.).

personal alert safety system (PASS) Also known as a personal accountability safety system. A small device worn by firefighters that is motion sensitive. The motion device will go into an audible alarm, signaling that the firefighter may be in trouble, if there is no movement for a specified time period. It can also be manually activated by the firefighter if desired.

platform construction Type of wood-frame construction that is fabricated so that the floor acts as a platform for the floor above. The platform provides a horizontal break in the vertical channels of the wall, which aids in controlling extension of fire.

point of no return Refers to a firefighter's individual management of his air supply. It is determined by the amount of air consumed going into a structure versus the amount of residual air needed to exit the structure.

positive pressure ventilation (PPV) Utilization of fans to displace air in an enclosure or void by creating pressure differentials.

post-incident analysis (PIA) A recap and in-depth review of an incident or training session. Its purpose is to seek information to address areas that need improvement as well as areas that provided positive results.

post-traumatic stress disorder Condition that can occur when a firefighter is exposed to a critical incident outside of what can be considered normal in a firefighter's work environment.

pre-fire planning A walk-through, inspection, or survey that provides knowledge of a building prior to an emergency taking place.

Prussik or rescue loop A loop of cord used for pulling, hauling, or raising a downed firefighter. It is comprised of a double fisherman's knot applied in a cord to form a loop.

pulley A device that decreases friction while having the ability to transfer force from one direction to another when using ropes.

push-and-pull drag A type of victim drag performed by one rescuer pulling the downed firefighter by the SCBA straps at the head while a second rescuer pushes from a position high in the groin area.

rabbit tool A handheld forcible entry tool that operates through hydraulics.

radio-assisted feedback A procedure using two portable radios that are placed closely together and keyed to the talk position, producing a high-pitched feedback sound. This sound can be heard over the victim's portable radio and can be used to assist a search team in finding the location of the downed firefighter.

rapid intervention Actions taken during fire suppression and other incidents that provide continual evaluation of incident scene safety concerns or rescue of firefighters.

rapid intervention company operations (RICO) A function and/or training that involves techniques and maneuvers for the individual, single, and multiple team operations of firefighters dedicated to the safety, search, and rescue of trapped or lost firefighters at an emergency incident.

rapid intervention crew (or company) (RIC) Name designating a team of specially trained firefighters who are solely responsible for the safety, search, and rescue of trapped or lost firefighters at an emergency incident.

Rapid Intervention Crew Universal Air Coupling (RIC UAC) A special connection that will allow an SCBA cylinder that is low on air to be transfilled from another SCBA cylinder regardless of the manufacturer's make or model.

rapid intervention team (RIT) A team of specially trained firefighters who are solely designated to provide for the safety, search, and rescue of trapped or lost firefighters at an emergency incident.

rebar Metal cylindrical rods used to add strength and reinforcement to concrete.

reconnaissance and rescue of surface victims An initial sizeup of a collapse situation through a six-sided approach to the collapse area while beginning to search for victims that are on or near the surface.

rehab Branch within IMS that provides for the rest, medical concerns, and water and food needs of firefighters working on the fireground.

rescue assistant team (RAT) Name designating a team of specially trained firefighters who are solely responsible for the safety, search, and rescue of trapped or lost firefighters at an emergency incident.

rescue branch officer Sometimes referred to as the RIT Operations Chief. It is a position within the IMS that provides for the safety, accountability and support of the RIT. It allows the RIT Leader/Officer to focus on the rescue efforts. This position reports to the Safety Section Officer.

rescue litter ladder slide method A technique used to remove a downed firefighter who may have serious injuries from an elevated area through the use of a Stokes basket and a ground ladder. A rope is attached to the head of the lifter and the basket is then slid along the rails of the ladder, being guided and controlled by rescuers on the rope as well as on the ladder.

rescue loops Loops of cord used for pulling, hauling, or raising a downed firefighter. They are comprised of a double fisherman knot applied in a cord to form a loop.

ridge board Wood structural member located at the highest point of a roof where rafters are attached.

RIT officer/leader (RITLO) Responsible for direct supervision of the RIT at the task level this position reports to the RIT operations chief.

RIT operations chief Sometimes referred to as the rescue branch officer. It is a position within the IMS that provides for the safety, accountability, and support of the RIT. It allows the RIT leader/officer to focus on the rescue efforts. This position reports to the safety section officer.

roll-up overhead door A commercial door constructed of metal strips that interlock with each other that when raised roll up and around a drum located at the top of the door opening.

roof decking Refers materials making up the construction of a roof before covering materials such as shingles or tar paper are applied.

roof ladder A type of single ladder that has folding hooks at its tip. The hooks are turned out to provide a means of placing the ladder over the ridge board of a roof. This stabilizes the ladder in order to allow a firefighter to work on a pitched roof.

safety officer Person assigned by Command that is responsible for recognizing and controlling immediate dangers to all incident personnel.

safety section officer (SSO) Command team position that is directly in command of RIT operations in the case of a mayday. Incident safety officer, rehab division, accountability division, and the RIT-chief will report to the SSO.

SCBA remote gauge Gauge that is readily visible by a firefighter that tells how much air is remaining in the cylinder when the SCBA is on. Commonly found on the shoulder strap of the SCBA but may be found mounted on the waist belt of the SCBA harness with some older models.

scrub area Refers to the area that can be covered by an aerial device at a multistory building or structure.

seated window removal A technique performed by two rescuers to lift a downed firefighter over and out of a windowsill in a face-forward, seated position to rescuers on the exterior.

sectional overhead door Overhead doors that are constructed in sections that are hinged to one another. When the door is raised it rolls along a stationary track located at each side of the opening.

selected debris removal Stage of collapse incident that involves the RIT on a limited basis. When a location of a victim or downed firefighter is known, he may be approached or freed from his entrapment by removing selected debris from the collapse.

sheathing A material applied to wood-frame structures over the structural framing to which siding or other exterior material finish is applied. It provides rigidity to frame walls.

shoring A process in which lumber or other equipment such as ellis clamps, screw jacks, air bags, and hydraulics are used to support a weakened area for the purpose of safer rescue operations.

simple oriented search A systematic search technique that provides good coverage of a medium-size or small open-area room in place of a complete right- or left-handed search.

skip breathing A technique used in an emergency by a firefighter to maximize the air supply contained inside the SCBA.

slice tool Also known as a metal cutting torch. Uses rods and pure oxygen to generate heat in thousands of degrees to cut and penetrate metal.

snorkel A power-driven articulating boom that is affixed to a truck. This type of apparatus has a basket attached to the end of the boom. Snorkels can provide elevated master streams and maneuverability around overhead objects for rescue.

sole plate The horizontal structural member on the bottom of a framed wall.

stair raise with handcuff knot A technique where multiple rescuers can raise a downed firefighter up a set of stairs through utilizing a handcuff knot cinched onto the downed firefighter's forearms. This technique is useful when faced with a narrow stairway.

stair raise with tool A technique where multiple rescuers can lift a downed firefighter up a set of stairs through utilizing a tool such as a Halligan bar or axe placed through the upper shoulder straps of the victim.

standard operating procedures (SOPs) Written directives that enhance firefighter safety and facilitates the decisions that are made by all personnel on the fireground. Having good standard operating procedures provides both control and accountability.

stokes basket Another name for a rescue litter. A wire or plastic basket used for carrying injured people to safety.

strip ventilation A technique in which firefighters working on a roof cut a hole that is long and narrow between roof rafters for the purpose of ventilation and controlling the advancement of fire. The long, narrow furrow usually runs the entire width of a flat roof or runs from ridge to a soffit on a pitched roof. Also called trench cutting.

stucco A portland-based form of concrete containing sand and water that is applied over a wood frame and metal lathe wall. Stucco can also be applied directly over a masonry wall.

supported lean-to collapse Type of collapse that results from the failure of one side of a roof or floor that falls until it rests and is supported by substantial objects inside the structure.

tactical worksheet A checksheet used by the IC to help keep track of the tactics employed to accomplish the strategic goals of the incident. The tactical worksheet can be utilized by the RIT officer to collect information without taking the IC's attention away from the incident.

teepee cut A technique where an inverted "V" cut is made into an overhead door. It allows the door to remain in the lowered position while the cut section is peeled downward to allow entry or egress.

tensionless hitch Used to establish an anchor point. It consist of a figure eight on a bight wrapped around the object to be used as an anchor. It is recommended that it be wrapped around the object a minimum of four times. It is then left to hang freely, where a carabiner can be attached.

thermal imaging camera (TIC) Device that is based on infrared energy technology that allows firefighters the ability to ascertain objects by shape in conditions that do not allow normal vision.

TIC lead search The thermal imaging camera operator leads the search team through the area to be searched while directing and relating information to the areas needed to be searched as well as monitoring fire conditions around the team.

tilt-up door A type of overhead door consisting of one section that is a slab of wood. The door is hinged at the sides and center allowing it to pivot open from the bottom.

tool drag A technique where rescuers can drag a downed firefighter by utilizing a tool such as a Halligan bar or axe placed through the upper shoulder straps of the victim as a handle.

tower ladder A power-driven ladder with a basket on the end that is affixed to a truck.

train the trainer format A training format where firefighters are trained to teach the material presented to other firefighters.

trench cut A technique in which firefighters working on a roof cut a hole that is long and narrow between roof rafters for the purpose of ventilation and controlling the advancement of fire. The long, narrow furrow usually runs the entire width of a flat roof or runs from ridge to a soffit on a pitched roof. Also called strip ventilation.

true mechanical advantage The actual coefficient of work being performed by a rope system meant to decrease the weight load when moving or lifting a downed firefighter. Friction added into such a system will decrease the mechanical advantage potential.

type I (goal-oriented) learner A confident and self-directed learner. The type of person who is in competition with himself. Excels in an interactive or participative type of classroom.

type II (affective) learner The type of learner who takes a cooperative role in the learning experience. Regards the instructor as a source of knowledge. Do not feel comfortable questioning the instructor.

type III (transitional) learner A learner that looks for a way to connect prior education and experience with what is being learned now. Prefer interaction and discussion of ideas. Feel a sense of equality with the instructor.

type IV (integrated) learner Primarily interested in personal success. Self-directed learners who prefer a collaborative learning environment.

type V (risk taker) learner A student that is self-confident and eager to learn new ideas. May stray away from guidelines to gain new knowledge.

unsupported lean-to collapse A type of collapse where the one side that has failed is without any support. The failed side floats freely and is very unstable; any slight movement may cause a secondary collapse.

verbal/linguistic intelligence Allows people to process complex ideas through the facility of language. Reading and writing apply to this form of intelligence. Puzzles, questionnaires, and oral presentations regarding the material to be learned can enhance the learning environment.

visual/spatial intelligence Learning that takes place through graphic images such as pictures, diagrams, or maps. Allows people to understand concepts through mental transformations of colors, shapes, and space. "A picture is worth a thousand words" relates to this type of learning.

void search A physical search that is conducted in a collapse environment by rescuers moving over and around debris piles, locating spaces created by the materials involved in the collapse.

v-shaped collapse A type of collapse that is caused by the failure of an interior support, resulting in void spaces on both sides of the collapse toward the bearing walls.

v-type collapse A type of collapse that is caused by the failure of an interior support, resulting in void spaces on both sides of the collapse toward the bearing walls.

"W" technique A technique that provides the RIT the ability to raise a downed firefighter from a below-grade area back up to the rescuers. It is given its name because of the appearance of the system as it looks laid out on the ground, which also reminds rescuers of its proper assembly.

walk-out basement Refers to an architectural feature in the construction of a building in which an elevation difference between the grade at the front of the structure differs with the elevation or grade at the sides or the rear of the structure. It usually includes an alternate entry/exit.

wall breaching The penetration of a wall through various methods for the purposes of emergency egress or rescue.

water or tape knot Knot that is used to join two ends of webbing together.

wood-frame construction Type of construction in which the structural members that provide framework and support are fabricated out of wood. Most common type of construction encountered.

Bibliography

Angle, J., *Occupational Safety and Health in the Emergency Services,* 2nd ed. Clifton Park, NY: Thomson Delmar Learning, 2005.

Angulo, R., I Say "to Bail." *Fire Engineering,* 32–40, August 2000.

Avillo, A., and Nasta, M., Lessons Learned from Mask Confidence Training. *Fire Engineering,* 73–80, August 2000.

Baker, M., and Ross, J., Firefighter Personal Escape Training: Ladder Bailout Safety. *Fire Engineering,* 20–26, January 2002.

Baker, M., and Ross, J., The 3 "Ws" of Saving Our Own. *Firehouse,* 82–85, July 2000.

Bastian, J., The 5 Don'ts of Thermal Imaging. *Firehouse,* 70–72, June 2003.

Beirne, M., and Simpson, W., Practice Like You Play: Rapid Intervention Teams. *Firehouse,* 76–78, March 2001.

Brannigan, F., *Building Construction for the Fire Service,* 3rd ed. Quincy, MA: National Fire Protection Association, 1992.

Brougher, J., Creating a Learning Environment for Adults Using Multiple Intelligence Theory. *Adult Learning,* 28–29, April 1997.

Brown, M., *Engineering Practical Rope Rescue Systems.* Clifton Park, NY: Thomson Delmar Learning, 2004.

Brunacini, A., *Fire Command,* 2nd ed. Quincy, MA: National Fire Protection Association, 2002.

Casey, J., *The Fire Chief's Handbook,* 4th ed. Saddlebrook, NJ: PennWell Publishing, 1987.

Cobb, R., Rapid Interventions Teams: A Fireground Safety Factor, *Firehouse,* 52–56, May 1998.

Cobb, R., Rapid Intervention Teams: They May Be Your Only Chance. *Firehouse,* 54–57, July 1996.

Coleman, J., Searching Smarter, Part 3: Advanced Oriented Search. *Fire Engineering,* 113–120, April 2001.

Coleman, J., Searching Smarter, Part 2: The Oriented Search. *Fire Engineering,* 93–98, March 2001.

Coleman, J., Searching Smarter, Part 1: The Basics. *Fire Engineering,* 99–116, February 2001.

Coleman, J., and Lasky , R., Managing the Mayday. *Fire Engineering,* 51–62, January 2002.

Crawford, J., Rapid Intervention Teams: Are You Prepared for the Search? *Firehouse,* 50–54, April 1999.

Dodson, D., *Fire Department Incident Safety Officer.* Clifton Park, NY: Thomson Delmar Learning, 1999.

Dorgan, S., Search Rope Basics. *Fire Engineering,* 99–105, January 2000.

Downey, R., *The Rescue Company.* Saddlebrook, PA: PennWell Publishing, 1992.

Dube, R., Rescue Rope for Rapid Intervention Teams. *Fire Engineering,* 14, October 2000.

Dunn, V., 50 Causes for All Firefighter Death and Injury. *Firehouse,* 26–36, March 2001.

Dunn, V., Smoke (Backdraft) Explosions. *Firehouse,* 18–24, September 2000.

Dunn, V., Disorientation: A Firefighter Killer. *Firehouse,* 18–21, November 1999.

Dunn, V., *Safety and Survival on the Fireground.* New Jersey: PennWell Publishing, 1992.

Dunn, V., *Collapse of Burning Buildings.* New York: PennWell Publishing, 1988.

Eisner, H., The ABC's of Thermal Imaging. *Firehouse,* 54–56, November 2000.

Elliott, E., Mayday! Mayday! Chief Officer Down. *Firehouse,* 34–36, April 2001.

Endorf, M., and McNeff, M., The Adult Learner: Five Types. *Adult Learning,* 20–25, 1991.

Fire Department, New York City, *Ladder Company Operations, Use of Aerial Ladders.* Firefighting Procedures Volume 3, Book 2, 1997.

Firefighter's Handbook, 2nd ed. Clifton Park, NY: Thomson Delmar Learning, 2004.

Fredericks, A., Engine Company Support of RIT/FAST Operations. *Fire Engineering,* 79–96, April 1999.

Hoff, R., and Lasky, R., Saving Our Own: The Firefighter Who Has Fallen through the Floor. *Fire Engineering,* 12–18, March 1998.

Hoff, R., and Kolomay, R., *Firefighter Rescue and Survival.* Tulsa, OK: PennWell Publishing, 2003.

IFSTA, *Fire Service Rescue,* 6th ed. Stillwater: Oklahoma State University, 1996.

Jakubowski, G., and Morton, M., *Rapid Intervention Teams.* Stillwater, OK: Fire Protection Publications, 2001.

Kortlang, C., Five Companies Trapped: A Lesson in Accountability. *Fire Engineering,* 69–76, March 2001.

Kreis, S., Rapid Intervention Isn't Rapid. *Fire Engineering,* 56–66, December 2003.

Lamb, P., The Rapid Intervention Time Line and Crew Survivability. *Fire Engineering,* 93–95, August 2002.

Lasky, R., May 1998. Saving Our Own: Designing a Firefighter Survival Training Aid. *Fire Engineering,* 10–13, May 1998.

Lasky, R., Saving Our Own: The Rapid Intervention Team Officer. *Fire Engineering,* 17–20, July 1997.

Lasky, R., Pressler, R., and Salka, J., The Head-First Ladder Slide: Three Methods. *Fire Engineering,* 59–61, October 2000.

Lasky, R., and Shervino, T., Saving Our Own: Moving the Downed Firefighter up a Stairwell. *Fire Engineering,* 14–18, December 1997.

Lund, P., Tactical Considerations for the Rapid Intervention Team. *Firehouse,* 86, September 1999.

McCormack, J., Mayday: Are You Prepared if It Happens to You? *Fire Engineering,* 16–18, August 2000.

McCormack, J., Rapid Intervention Emergency Air Supply. *Fire Engineering,* 14–18, July 2000.

McLees, M., The Rapid Intervention Rope Bag, Part 3. *Firehouse,* 68–69, October 2001.

McLees, M., The Rapid Intervention Rope Bag, Part 2. *Firehouse,* 50–53, February 2001.

McLees, M., The Rapid Intervention Rope Bag, Part 1. *Firehouse,* 40–42, April 2000.

Mittendorf, J., *Truck Company Operations.* Saddlebrook, NJ: PennWell Publishing, 1998.

Mittendorf, J., *Ventilation Methods and Techniques.* El Toro, CA: Fire Technology Service, 1991.

Morton, M., Fast Teams: Prepare an Action Plan Using a Two-Team Method. *Fire Rescue,* 75–76, January 1999.

NFPA 1403 Standard on Live Fire Evolutions in Structures. Quincy, MA: National Fire Protection Association, 2002.

NFPA 1500 Fire Department Occupational Safety and Health Program. Quincy, MA: National Fire Protection Association, 1997.

NFPA 1710 Organization and Deployment of Fire Suppression Operations, Emergency Medical Operations, and Special Operations to the Public by Career Fire Departments. Quincy, MA: National Fire Protection Association, 2001.

NFPA 1720 Organization and Deployment of Fire Suppression Operations, Emergency Medical Operations, and Special Operations to the Public by Volunteer Fire Departments. Quincy, MA: National Fire Protection Association, 2001.

NFPA, Fire Investigations, *Printing Office Fire—Denver, CO, September 28, 1992.* Quincy, MA: National Fire Protection Association.

Norman, J., *Fire Officers Handbook of Tactics,* 2nd ed. Saddlebrook, NJ: PennWell Publishing, 1998.

Norman, J., Rapid Intervention Techniques. *Fire House,* 48–53, November 1997.

O'Connell, J., Collapse Search and Rescue Operations: Tactics and Procedures, Part 2—Size Up and Safety. *Fire Engineering,* 113–124, June 1993.

O'Connell, J., Collapse Search and Rescue Operations: Tactics and Procedures, Part 1—Collapse Voids and Initial Void Search, *Fire Engineering,* 76–84, May 1993.

Pressler, R., Five-Point Size Up. *Fire Engineering,* 75–86, February 2001.

Rapid Intervention Teams and How to Avoid Needing Them. Technical Report Series, Report #123, U.S. Fire Administration, Washington D.C., 2003.

Richardson, M., Thermal Imaging Training: Covering the Basics. *Firehouse,* 86–88, April 2001.

Salka, J., S.C.B.A. Competence and Confidence. *Fire Engineering,* 25–31, September 1993.

Sitz, T., Rapid Removal of an Unresponsive Firefighter from a Peaked Roof. *Fire Engineering,* 97–102, March 2003.

Smith, J., Firefighter Safety: Our Constant Goal. *Firehouse,* 18–21, December 2000.

Smith, J., Masonry Wall Collapse. *Firehouse,* 15–18, December 1999.

Smith, J., Building Collapse Indicators Continued. *Firehouse,* 20–24, June 1997.

Smith, J., Building Collapse Indicators. *Firehouse,* 28–30, April 1997.

Smith, M., Rapid Intervention: What Is It? *Firehouse,* 18–19, August 2001.

Thiel, A., *Prevention of Self-Contained Breathing Apparatus Failures.* Special Report, U.S. Fire Administration, March 1998.

U.S. Department of Labor, *All about OSHA.* Government Publication Number OSHA 2056-07R, 2003.

U.S. Fire Administration, *Rapid Intervention Teams and How to Avoid Needing Them.* U.S. Fire Administration Technical Report Series, Federal Emergency Management Agency. *Available on-line at http://www.usfa.*

Webb, W., Below-Grade Rescue Kit. *Fire Engineering,* 109–111, May 2001.

Woodworth, S., Thermal Imaging for the Fire Service, Part 6: The Search. *Fire Engineering,* 24–27, August 1997.

Woodworth, S., Thermal Imaging for the Fire Service, Part 4: Thermal Imaging Devices. *Fire Engineering,* 16–18, February 1997.

Woodworth, S., Thermal Imaging for the Fire Service, Part 3: Thermal Characteristics. *Fire Engineering,* 22–26, November, 1996.

Woodworth, S., Thermal Imaging for the Fire Service, Part 1: The Basics of Thermal Imaging. *Fire Engineering,* 22–26, July, 1996.

Index

Note: Italicized page numbers indicate illustrations.